SCHAUM'S OUTLINE OF

THEORY AND PROBLEMS

of

FLUID MECHANICS
and HYDRAULICS

SECOND EDITION

•

BY

RANALD V. GILES, B.S., M.S. in C.E.

Professor of Civil Engineering
Drexel Institute of Technology

•

SCHAUM'S OUTLINE SERIES

McGRAW-HILL BOOK COMPANY

New York, St. Louis, San Francisco, Toronto, Sydney

ISBN 07-023234-2

14 15 SH SH 7 5 4 3 2 1 0 6 9 8

Preface

This book is designed primarily to supplement standard textbooks in fluid mechanics and hydraulics. It is based on the author's conviction that clarification and understanding of the basic principles of any branch of mechanics can be accomplished best by means of numerous illustrative problems.

The previous edition of this book has been very favorably received. In this second edition many chapters have been revised and enlarged to keep pace with the most recent concepts, methods and terminology. Attention is focused earlier on Dimensional Analysis by placing this expanded material in Chapter 5. The most extensive revisions are in the chapters on Fundamentals of Fluid Flow, Fluid Flow in Pipes, and Flow in Open Channels.

The subject matter is divided into chapters covering duly-recognized areas of theory and study. Each chapter begins with statements of pertinent definitions, principles and theorems together with illustrative and descriptive material. This material is followed by graded sets of solved and supplementary problems. The solved problems illustrate and amplify the theory, present methods of analysis, provide practical examples, and bring into sharp focus those fine points which enable the student to apply the basic principles correctly and confidently. Free body analysis, vector diagrams, the principles of work and energy and of impulse-momentum, and Newton's laws of motion are utilized throughout the book. Effort has been made to present original problems developed by the author during many years of teaching the subject. Numerous proofs of theorems and derivations of formulas are included among the solved problems. The large number of supplementary problems serve as a complete review of the material of each chapter.

In addition to the use of this book by engineering students of fluid mechanics, it should be of considerable value as a reference book to the practicing engineer. He will find well-detailed solutions to many practical problems and he can refer to the summary of the theory when necessity arises. Also, the book should serve the professional engineer who must review the subject for licensing examinations or other reasons.

I wish to thank my colleague, Robert C. Stiefel, who carefully checked the solutions to the many new problems. I also wish to express my gratitude to the staff of Schaum Publishing Company, particularly to Henry Hayden and Nicola Miracapillo, for their valuable suggestions and helpful cooperation.

RANALD V. GILES

Philadelphia, Pa.
June, 1962

CONTENTS

CONTENTS

CONTENTS

APPENDIX

DIAGRAMS

SYMBOLS and ABBREVIATIONS

The following tabulation lists the letter symbols used in this book. Because the alphabet is limited, it is impossible to avoid using the same letter to represent more than one concept. Since each symbol is defined when it is first used, no confusion should result.

a acceleration in ft/sec², area in ft²

A area in ft²

b weir length in ft, width of water surface in ft, bed width of open channel in ft

c coefficient of discharge, celerity of pressure wave in ft/sec (acoustic velocity)

c_c coefficient of contraction

c_v coefficient of velocity

C coefficient (Chezy), constant of integration

CG center of gravity

C_p center of pressure, power coefficient for propellers

C_D coefficient of drag

C_F thrust coefficient for propellers

C_L coefficient of lift

C_T torque coefficient for propellers

C_1 Hazen-Williams coefficient

cfs   cubic feet per second

d, D diameter in feet

D_1 unit diameter in in.

e efficiency

E bulk modulus of elasticity in lb/ft² or lb/in², specific energy in ft lb/lb

f friction factor (Darcy) for pipe flow

F force in lb, thrust in lb

g gravitational acceleration in ft/sec² = 32.2 ft/sec²

gpm   gallons per minute

h head in ft, height or depth in ft, pressure head in ft

H total head (energy) in ft or ft lb/lb

H_L, h_L lost head in ft (sometimes LH)

hp   horsepower = $wQH/550$ = 0.746 kw

I moment of inertia in ft⁴ or in⁴

I_{xy} product of inertia in ft⁴ or in⁴

k ratio of specific heats, isentropic (adiabatic) exponent, von Karman constant

K discharge factors for trapezoidal channels, lost head factor for enlargements, any constant

K_c lost head factor for contractions

l mixing length in ft

L length in ft

L_E equivalent length in ft

m roughness factor in Bazin formula, weir factor for dams

M mass in slugs or lb sec²/ft, molecular weight

n roughness coefficient, exponent, roughness factor in Kutter's and Manning's formulas

N rotational speed in rpm

N_s specific speed in rpm

N_u unit speed in rpm

N_F Froude number

N_M Mach number

N_W Weber number

p pressure in lb/ft², wetted perimeter in ft

p' pressure in lb/in²

P force in lb, power in ft lb/sec

P_u unit power in ft lb/sec

psf   lb/ft²

psia   lb/in², absolute

psig   lb/in², gage

q unit flow in cfs/unit width

Q volume rate of flow in cfs

Q_u unit discharge in cfs

r any radius in ft

r_o radius of pipe in ft

R gas constant, hydraulic radius in ft

R_E Reynolds Number

S slope of hydraulic grade line, slope of energy line

S_o slope of channel bed

sp gr   specific gravity

t time in sec, thickness in in., viscosity in Saybolt sec

T temperature, torque in ft lb, time in sec

u peripheral velocity of rotating element in ft/sec

u, v, w components of velocity in X, Y and Z directions

v volume in ft³, local velocity in ft/sec, relative velocity in hydraulic machines in ft/sec

v_s specific volume $= 1/w = $ ft³/lb

v_* shear velocity in ft/sec $= \sqrt{\tau/\rho}$

V average velocity in ft/sec (or as defined)

V_c critical velocity in ft/sec

w specific (unit) weight in lb/ft³

W weight in lb, weight flow in lb/sec $= wQ$

x distance in ft

y depth in ft, distance in ft

y_c critical depth in ft

y_N normal depth in ft

Y expansion factors for compressible flow

z elevation (head) in ft

Z height of weir crest above channel bottom, in ft

α (alpha)   angle, kinetic-energy correction factor

β (beta)   angle, momentum correction factor

δ (delta)   boundary layer thickness in ft

Δ (delta)   flow correction term

ϵ (epsilon)   surface roughness in ft

η (eta)   eddy viscosity

θ (theta)   any angle

μ (mu)   absolute viscosity in lb sec/ft² (or poises)

ν (nu)   kinematic viscosity in ft²/sec (or stokes) $= \mu/\rho$

π (pi)   dimensionless parameter

ρ (rho)   density in lb sec²/ft⁴ or slugs/ft³ $= w/g$

σ (sigma)   surface tension in lb/ft, intensity of tensile stress in psi

τ (tau)   shear stress in lb/ft²

ϕ (phi)   speed factor, velocity potential, ratio

ψ (psi)   stream function

ω (omega)   angular velocity in rad/sec

Conversion Factors

1 cubic foot $=$ 7.48 U.S. gallons $=$ 28.32 liters

1 U.S. gallon $=$ 8.338 pounds of water at 60°F

1 cubic foot per second $=$ 0.646 million gallons per day
$=$ 443.8 gallons per minute

1 pound-second per square foot (μ) $=$ 478.7 poises

1 square foot per second (ν) $=$ 929 square centimeters per second

1 horsepower $=$ 550 foot-pounds per second $=$ 0.746 kilowatts

30 inches of mercury $=$ 34 feet of water $=$ 14.7 pounds per square inch

Chapter 1

Properties of Fluids

FLUID MECHANICS and HYDRAULICS

Fluid mechanics and hydraulics represent that branch of applied mechanics dealing with the behavior of fluids at rest and in motion. In the development of the principles of fluid mechanics, some fluid properties play principal roles, others only minor roles or no roles at all. In fluid statics, specific weight is the important property, whereas in fluid flow, density and viscosity are predominant properties. Where appreciable compressibility occurs, principles of thermodynamics must be considered. Vapor pressure becomes important when negative pressures (gage) are involved, and surface tension affects static and flow conditions in small passages.

DEFINITION of a FLUID

Fluids are substances which are capable of flowing and which conform to the shape of containing vessels. When in equilibrium, fluids cannot sustain tangential or shear forces. All fluids have some degree of compressibility and offer little resistance to change of form.

Fluids may be divided into liquids and gases. The chief differences between liquids and gases are (a) liquids are practically incompressible whereas gases are compressible and often must be so treated and (b) liquids occupy definite volumes and have free surfaces whereas a given mass of gas expands until it occupies all portions of any containing vessel.

AMERICAN ENGINEERING SYSTEM of UNITS

Three selected reference dimensions (fundamental dimensions) are length, force and time. In this book the corresponding three fundamental units used will be the foot of length, the pound of force (or pound weight), and the second of time. All other units may be derived from these. Thus unit volume is the ft³, unit acceleration is the ft/sec², unit work is the ft lb, and unit pressure is the lb/ft². Should data be given in other units, they must be converted to the foot-pound-second system before applying them to the solution of problems.

The unit for mass in this system, the *slug*, is derived from the units of force and acceleration. For a freely falling body in vacuum the acceleration is that of gravity ($g = 32.2$ ft/sec² at sea level) and the only force acting is its weight. From Newton's second law,

$$\text{force in pounds} = \text{mass in slugs} \times \text{acceleration in ft/sec}^2$$

Then

$$\text{weight in pounds} = \text{mass in slugs} \times g(32.2 \text{ ft/sec}^2)$$

or

$$\text{mass } M \text{ in slugs} = \frac{\text{weight } W \text{ in pounds}}{g(32.2 \text{ ft/sec}^2)} \tag{1}$$

SPECIFIC WEIGHT

The specific weight w of a substance is the weight of a unit volume of the substance. For liquids, w may be taken as constant for practical changes of pressure. The specific (unit) weight of water for ordinary temperature variations is $62.4 \, \text{lb/ft}^3$. See Appendix, Tables 1C and 2, for additional values.

The specific weights of gases may be calculated by using the *equation of state* of a gas

or
$$\frac{pv_s}{T} = R \quad \text{(Boyle's and Charles' laws)} \tag{2}$$

where pressure p is absolute pressure in lb/ft^2, specific volume v_s is the volume per unit weight in ft^3/lb, temperature T is the absolute temperature in degrees Rankine $(460° +$ degrees Fahrenheit), and R is the *gas constant* in feet/degree Rankine. Since $w = 1/v_s$, the above equation may be written

$$w = \frac{p}{RT} \tag{3}$$

MASS DENSITY of a BODY ρ (rho) = mass per unit volume = w/g.

In the engineering system of units, the mass density of water is $62.4/32.2 = 1.94$ slugs/ft^3 or lb sec^2/ft^4. In the metric system the density of water is 1 g/cm^3 at 4°C. See Appendix, Table 1C.

SPECIFIC GRAVITY of a BODY

The specific gravity of a body is that pure number which denotes the ratio of the weight of a body to the weight of an equal volume of a substance taken as a standard. Solids and liquids are referred to water (at $39.2°\text{F} = 4°\text{C}$) as standard, while gases are often referred to air free of CO_2 or hydrogen (at $32°\text{F} = 0°\text{C}$ and 1 atmosphere $= 14.7 \, \text{lb/in}^2$ pressure) as standard. For example,

$$\text{specific gravity of a substance} = \frac{\text{weight of the substance}}{\text{weight of equal volume water}} \tag{4}$$

$$= \frac{\text{specific weight of substance}}{\text{specific weight of water}}$$

Thus if the specific gravity of a given oil is 0.750, its specific weight is $0.750(62.4 \, \text{lb/ft}^3)$ $= 46.8 \, \text{lb/ft}^3$.

The specific gravity of water is 1.00 and of mercury is 13.57. The specific gravity of a substance is the same in any system of measures. See Appendix, Table 2.

VISCOSITY of a FLUID

The viscosity of a fluid is that property which determines the amount of its resistance to a shearing force. Viscosity is due primarily to interaction between fluid molecules.

Referring to Fig. 1-1, consider two large, parallel plates at a small distance y apart, the space between the plates being filled with a fluid. Consider the upper plate acted on by a constant force F and hence moving at a con-

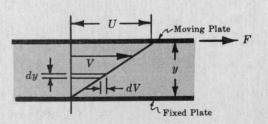

Fig. 1-1

stant velocity U. The fluid in contact with the upper plate will adhere to it and will move at velocity U, and the fluid in contact with the fixed plate will have velocity zero. If distance y and velocity U are not too great, the velocity variation (gradient) will be a straight line. Experiments have shown that force F varies with the area of the plate, with velocity U, and inversely with distance y. Since by similar triangles, $U/y = dV/dy$, we have

$$ F \propto \frac{AU}{y} = A\frac{dV}{dy} \qquad \text{or} \qquad \frac{F}{A} = \tau \propto \frac{dV}{dy} $$

where $\tau = F/A$ = shear stress. If a proportionality constant μ (mu), called the *absolute (dynamic) viscosity*, is introduced,

$$ \tau = \mu\frac{dV}{dy} \qquad \text{or} \qquad \mu = \frac{\tau}{dV/dy} \tag{5} $$

The units of μ are $\dfrac{\text{lb sec}}{\text{ft}^2}$, since $\dfrac{\text{lb/ft}^2}{(\text{ft/sec})/\text{ft}} = \dfrac{\text{lb sec}}{\text{ft}^2}$. Fluids which follow the relation of equation (5) are called *Newtonian fluids* (see Problem 9).

Another viscosity coefficient, the *kinematic coefficient of viscosity,* is defined as

$$ \text{Kinematic coefficient } \nu \text{ (nu)} = \frac{\text{absolute viscosity } \mu}{\text{mass density } \rho} $$

or

$$ \nu = \frac{\mu}{\rho} = \frac{\mu}{w/g} = \frac{\mu g}{w} \tag{6} $$

The units of ν are $\dfrac{\text{ft}^2}{\text{sec}}$, since $\dfrac{(\text{lb sec/ft}^2)(\text{ft/sec}^2)}{\text{lb/ft}^3} = \dfrac{\text{ft}^2}{\text{sec}}$.

Viscosities are reported in handbooks as poises and stokes (cgs units) and on occasion as Saybolt seconds, from viscosimeter measurements. Conversions to the ft-lb-sec system are illustrated in Problems 6-8. A few values of viscosities are given in Tables 1 and 2 of the Appendix.

Viscosities of liquids decrease with temperature increases but are not affected appreciably by pressure changes. The absolute viscosity of gases increases with increase in temperature but is not appreciably changed due to pressure. Since the specific weight of gases changes with pressure changes (temperature constant), the kinematic viscosity varies inversely as the pressure. However, from the equation above, $\mu g = w\nu$.

VAPOR PRESSURE

When evaporation takes place within an enclosed space, the partial pressure created by the vapor molecules is called vapor pressure. Vapor pressures depend upon temperature and increase with it. See Table 1C for values for water.

SURFACE TENSION

A molecule in the interior of a liquid is under attractive forces in all directions, and the vector sum of these forces is zero. But a molecule at the surface of a liquid is acted on by a net inward cohesive force which is perpendicular to the surface. Hence it requires work to move molecules to the surface against this opposing force, and surface molecules have more energy than interior ones.

The surface tension of a liquid is the work that must be done to bring enough molecules from inside the liquid to the surface to form one new unit area of that surface (ft lb/ft²).

This work is numerically equal to the tangential contractile force acting across a hypothetical line of unit length on the surface (lb/ft).

In most problems of introductory fluid mechanics, surface tension is not of particular importance. Table 1C gives values of surface tension σ (sigma) for water in contact with air.

CAPILLARITY

The rise or fall of a liquid in a capillary tube (or in some equivalent circumstance, such as in porous media) is caused by surface tension and depends on the relative magnitudes of the cohesion of the liquid and the adhesion of the liquid to the walls of the containing vessel. Liquids rise in tubes they wet (adhesion > cohesion) and fall in tubes they do not wet (cohesion > adhesion). Capillarity is important when using tubes smaller than about $\frac{3}{8}$ inch in diameter.

FLUID PRESSURE

Fluid pressure is transmitted with equal intensity in all directions and acts normal to any plane. In the same horizontal plane the pressure intensities in a liquid are equal. Measurements of unit pressures are accomplished by using various forms of gages. Unless otherwise stated, gage or relative pressures will be used throughout this book. Gage pressures represent values above or below atmospheric pressure.

UNIT PRESSURE or PRESSURE is expressed as force divided by area. In general,

$$p \text{ (lb/ft}^2 \text{ or psf)} \;=\; \frac{dP \text{ (lb)}}{dA \text{ (ft}^2)}$$

For conditions where force P is uniformly distributed over an area, we have

$$p \text{ (psf)} \;=\; \frac{P \text{ (lb)}}{A \text{ (ft}^2)} \quad \text{and} \quad p' \text{ (psi)} \;=\; \frac{P \text{ (lb)}}{A \text{ (in}^2)}$$

DIFFERENCE in PRESSURE

Difference in pressure between any two points at different levels in a liquid is given by

$$p_2 - p_1 = w(h_2 - h_1) \quad \text{in psf} \tag{7}$$

where w = unit weight of the liquid (lb/ft^3) and $h_2 - h_1$ = difference in elevation (ft).

If point 1 is in the free surface of the liquid and h is positive downward, the above equation becomes

$$p = wh \quad \text{(in psf gage)} \tag{8}$$

To obtain the lb/in^2 pressure unit, we use

$$p' \;=\; \frac{p}{144} \;=\; \frac{wh}{144} \quad \text{(in psi gage)} \tag{9}$$

These equations are applicable as long as w is constant (or varies so slightly with h as to cause no significant error in the result).

PRESSURE VARIATIONS in a COMPRESSIBLE FLUID

Pressure variations in a compressible fluid are usually very small because of the small unit weights and the small differences of elevation being considered in hydraulic calculations. Where such differences must be recognized for small changes in elevation dh, the law of pressure variation may be written

$$dp = -w\,dh \qquad (10)$$

The negative sign indicates that the pressure decreases as the altitude increases, with h positive upward. For applications, see Problems 29-31.

PRESSURE HEAD h

Pressure head h represents the height of a column of homogeneous fluid that will produce a given intensity of pressure. Then

$$h \text{ (ft of fluid)} = \frac{p \text{ (lb/ft}^2)}{w \text{ (lb/ft}^3)} \qquad (11)$$

BULK MODULUS of ELASTICITY (E)

The bulk modulus of elasticity (E) expresses the compressibility of a fluid. It is the ratio of the change in unit pressure to the corresponding volume change per unit of volume.

$$E = \frac{dp'}{-dv/v} = \frac{\text{lb/in}^2}{\text{ft}^3/\text{ft}^3} = \text{lb/in}^2 \qquad (12)$$

COMPRESSION of GASES

Compression of gases may occur according to various laws of thermodynamics. For the same mass of gas subjected to two different conditions,

$$\frac{p_1 v_1}{T_1} = \frac{p_2 v_2}{T_2} = WR \quad \text{and} \quad \frac{p_1}{w_1 T_1} = \frac{p_2}{w_2 T_2} = R \qquad (13)$$

where p = absolute pressure in lb/ft^2, v = volume in ft^3, W = weight in lb,
        w = specific weight in lb/ft^3, R = gas constant in ft/degree Rankine,
        T = absolute temperature in degrees Rankine ($460 + °$F).

FOR ISOTHERMAL CONDITIONS (constant temperature) the above expression (13) becomes

$$p_1 v_1 = p_2 v_2 \quad \text{and} \quad \frac{w_1}{w_2} = \frac{p_1}{p_2} = \text{constant} \qquad (14)$$

Also,   Bulk Modulus $E = p$ (in psf) (15)

FOR ADIABATIC or ISENTROPIC CONDITIONS (no heat exchanged) the above expressions become

$$p_1 v_1^k = p_2 v_2^k \quad \text{and} \quad \left(\frac{w_1}{w_2}\right)^k = \frac{p_1}{p_2} = \text{constant} \qquad (16)$$

Also
$$\frac{T_2}{T_1} \;=\; \left(\frac{p_2}{p_1}\right)^{(k-1)/k} \tag{17}$$

and   Bulk Modulus $E = kp$ (in psf) (18)

where k is the ratio of the specific heat at constant pressure to the specific heat at constant volume.   It is known as the adiabatic exponent.

Table 1A in the Appendix lists some typical values of R and k.   For many gases, R times molecular weight is about 1544.

PRESSURE DISTURBANCES

Pressure disturbances imposed on a fluid move in waves.   These pressure waves move at a velocity equal to that of sound through the fluid.   The velocity, or celerity, in ft/sec is expressed as

$$c \;=\; \sqrt{E/\rho} \tag{19}$$

where E must be in lb/ft². For gases, this acoustic velocity is

$$c \;=\; \sqrt{kp/\rho} \;=\; \sqrt{kgRT} \tag{20}$$

Solved Problems

1. Calculate the specific weight w, specific volume v_s and density ρ of methane at 100°F and 120 psi absolute.

    Solution:

      From Table 1A in the Appendix, $R = 96.3$.

$$\text{Specific weight } w \;=\; \frac{p}{RT} \;=\; \frac{120 \times 144}{96.3(460 + 100)} \;=\; 0.321 \text{ lb/ft}^3$$

$$\text{Specific volume } v_s \;=\; \frac{1}{w} \;=\; \frac{1}{0.321} \;=\; 3.11 \text{ ft}^3/\text{lb}$$

$$\text{Density } \rho \;=\; \frac{w}{g} \;=\; \frac{0.321}{32.2} \;=\; .00997 \text{ slugs/ft}^3$$

2. If 200 ft³ of oil weighs 10,520 lb, calculate its specific weight w, density ρ and specific gravity.

    Solution:

$$\text{Specific weight } w \;=\; \frac{10{,}520 \text{ lb}}{200 \text{ ft}^3} \;=\; 52.6 \text{ lb/ft}^3$$

$$\text{Density } \rho \;=\; \frac{w}{g} \;=\; \frac{52.6 \text{ lb/ft}^3}{32.2 \text{ ft/sec}^2} \;=\; 1.63 \text{ slugs/ft}^3$$

$$\text{Specific gravity} \;=\; \frac{w_{\text{oil}}}{w_{\text{water}}} \;=\; \frac{52.6 \text{ lb/ft}^3}{62.4 \text{ lb/ft}^3} \;=\; 0.843$$

3. At 90°F and 30.0 psi absolute the specific volume v_s of a certain gas was 11.4 ft³/lb. Determine the gas constant R and the density ρ.

Solution:

$$\text{Since } w = \frac{p}{RT}, \text{ then } R = \frac{p}{wT} = \frac{pv_s}{T} = \frac{(30.0 \times 144)(11.4)}{(460 + 90)} = 89.5.$$

$$\text{Density } \rho = \frac{w}{g} = \frac{1/v_s}{g} = \frac{1}{v_s g} = \frac{1}{11.4 \times 32.2} = .00272 \text{ slugs/ft}^3.$$

4. (a) Find the change in volume of 1.00 ft³ of water at 80°F when subjected to a pressure increase of 300 psi. (b) From the following test data determine the bulk modulus of elasticity of water: at 500 psi the volume was 1.000 ft³ and at 3500 psi the volume was 0.990 ft³.

Solution:

(a) From Table 1C in the Appendix, E at 80°F is 325,000 psi. Using formula (12),

$$dv = -\frac{v\, dp'}{E} = -\frac{1.00 \times 300}{325,000} = -.00092 \text{ ft}^3$$

(b) The definition associated with formula (12) indicates that *corresponding* changes in pressure and volume must be considered. Here an increase in pressure corresponds to a decrease in volume.

$$E = -\frac{dp'}{dv/v} = -\frac{(3500 - 500)}{(0.990 - 1.000)/1.000} = 3 \times 10^5 \text{ psi}$$

5. A cylinder contains 12.5 ft³ of air at 120°F and 40 psi absolute. The air is compressed to 2.50 ft³. (a) Assuming isothermal conditions, what is the pressure at the new volume and what is the bulk modulus of elasticity? (b) Assuming adiabatic conditions, what is the final pressure and temperature and what is the bulk modulus of elasticity?

Solution:

(a) For isothermal conditions, $p_1 v_1 = p_2 v_2$

Then $(40 \times 144)12.5 = (p_2' \times 144)2.50$ and $p_2' = 200$ psi absolute

The bulk modulus $E = p' = 200$ psi.

(b) For adiabatic conditions, $p_1 v_1^k = p_2 v_2^k$ and Table 1A in the Appendix gives $k = 1.40$.

Then $(40 \times 144)(12.5)^{1.40} = (p_2' \times 144)(2.50)^{1.40}$ and $p_2' = 381$ psi absolute

The final temperature is obtained by using equation (17):

$$\frac{T_2}{T_1} = \left(\frac{p_2}{p_1}\right)^{(k-1)/k}, \qquad \frac{T_2}{(460 + 120)} = \left(\frac{381}{40}\right)^{0.40/1.40}, \qquad T_2 = 1105° \text{ Rankine} = 645°F$$

The bulk modulus $E = kp' = 1.40 \times 381 = 533$ psi.

6. From the International Critical Tables, the viscosity of water at 20°C (68°F) is .01008 poises. Compute (a) the absolute viscosity in lb sec/ft² units. (b) If the specific gravity at 20°C is 0.998, compute the value of the kinematic viscosity in ft²/sec units.

Solution:

The poise is measured in dyne sec/cm². Since 1 lb = 444,800 dynes and 1 ft = 30.48 cm, we obtain

$$1 \frac{\text{lb sec}}{\text{ft}^2} = \frac{444,800 \text{ dyne sec}}{(30.48 \text{ cm})^2} = 478.7 \text{ poises}$$

(a) μ in lb sec/ft² $= .01008/478.7 = 2.11 \times 10^{-5}$

(b) ν in ft²/sec $= \dfrac{\mu}{\rho} = \dfrac{\mu}{w/g} = \dfrac{\mu g}{w} = \dfrac{2.11 \times 10^{-5} \times 32.2}{0.998 \times 62.4} = 1.091 \times 10^{-5}$

7. Convert 15.14 poises to kinematic viscosity in ft²/sec units if the liquid has specific gravity 0.964.

Solution:

The steps illustrated in Problem 6 may be taken or an additional factor may be established for water from $\dfrac{1}{478.7} \times \dfrac{32.2}{62.4} = .001078$.   Hence ν in ft²/sec $= \dfrac{15.14 \times .001078}{\text{sp gr} = 0.964} = .0169$.

8. Convert a viscosity of 510 Saybolt seconds at 60°F to kinematic viscosity ν in ft²/sec units.

Solution:

Two sets of formulas are given to establish this conversion when the Saybolt Universal Viscosimeter is used:

(a)   for $t \leq 100$, μ in poises $= (.00226t - 1.95/t) \times$ sp gr
        for $t > 100$, μ in poises $= (.00220t - 1.35/t) \times$ sp gr

(b)   for $t \leq 100$, ν in stokes $= (.00226t - 1.95/t)$
        for $t > 100$, ν in stokes $= (.00220t - 1.35/t)$

where $t =$ Saybolt second units. To convert stokes (cm²/sec) to ft²/sec units, divide by $(30.48)^2$ or 929.

Using group (b), and since $t > 100$, $\nu = \left(.00220 \times 510 - \dfrac{1.35}{510}\right)\dfrac{1}{929} = .001205$ ft²/sec.

9. Discuss the shear characteristics of the fluids for which the curves have been drawn in Fig. 1-2.

Solution:

(a)   The Newtonian fluids behave according to the law $\tau = \mu(dV/dy)$, or the shear stress is proportional to the velocity gradient or rate of shearing strain. Thus for these fluids the plotting of shear stress against velocity gradient is a straight line passing through the origin. The slope of the line determines the viscosity.

(b)   For the "ideal" fluid, the resistance to shearing deformation is zero, and hence the plotting coincides with the x-axis. While no ideal fluids exist, in certain analyses the assumption of an ideal fluid is useful and justified.

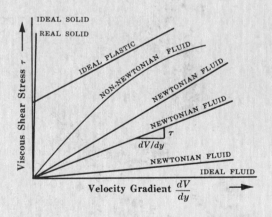

Fig. 1-2

(c)   For the "ideal" or elastic solid, no deformation will occur under any loading condition, and the plotting coincides with the y-axis. Real solids have some deformation and, within the proportional limit (Hooke's law), the plotting is a straight line which is almost vertical.

(d)   Non-Newtonian fluids deform in such a way that shear stress is not proportional to rate of shearing deformation, except perhaps at very low shear stresses. The deformation of these fluids might be classified as plastic.

(e)   The "ideal" plastic material could sustain a certain amount of shearing stress without deformation, and thereafter it would deform in proportion to the shearing stress.

10. Refer to Fig. 1-3. A fluid has absolute viscosity 0.0010 lb sec/ft² and specific gravity 0.913. Calculate the velocity gradient and the intensity of shear stress at the boundary and at points 1 in., 2 in., and 3 in. from the boundary, assuming (a) a straight line velocity distribution and (b) a parabolic velocity distribution. The parabola in the sketch has its vertex at A. Origin is at B.

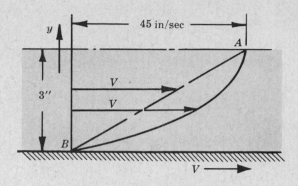

Fig. 1-3

Solution:

(a) For the straight line assumption, the relation between velocity and distance y is $V = 15y$. Then $dV = 15\,dy$, or the velocity gradient is $dV/dy = 15$.

For $y = 0$, $V = 0$, $dV/dy = 15\ \text{sec}^{-1}$ and

$$\tau = \mu(dV/dy) = .0010 \times 15 = .015\ \text{lb/ft}^2$$

Similarly, for other values of y we also obtain $\tau = .015\ \text{lb/ft}^2$.

(b) The equation of the parabola must satisfy the condition that the velocity is zero at the boundary B. The equation of the parabola is $V = 45 - 5(3 - y)^2$. Then $dV/dy = 10(3 - y)$ and tabulation of results yields the following:

y	V	dV/dy	$\tau = .0010(dV/dy)$
0	0	30	0.030 lb/ft²
1	25	20	0.020 lb/ft²
2	40	10	0.010 lb/ft²
3	45	0	0

It will be observed that where the velocity gradient is zero (which occurs at the center line of a pipe flowing under pressure, as will be seen later) the shear stress is also zero.

Note that the units of velocity gradient are sec^{-1}, and so the product $\mu(dV/dy) = (\text{lb sec/ft}^2)(\text{sec}^{-1}) = \text{lb/ft}^2$, the correct dimensions of shear stress τ.

11. A cylinder of 0.40 ft radius rotates concentrically inside of a fixed cylinder of 0.42 ft radius. Both cylinders are 1.00 ft long. Determine the viscosity of the liquid which fills the space between the cylinders if a torque of 0.650 lb ft is required to maintain an angular velocity of 60 revolutions per minute.

Solution:

(a) The torque is transmitted through the fluid layers to the outer cylinder. Since the gap between the cylinders is small, the calculation may be made without integration.

Tangential velocity of the inner cylinder $= r\omega = (0.40\ \text{ft})(2\pi\ \text{rad/sec}) = 2.51\ \text{ft/sec}$.

For the small space between cylinders, the velocity gradient may be assumed to be a straight line and the mean radius can be used. Then $dV/dy = 2.51/(0.42 - 0.40) = 125.5\ (\text{ft/sec})/\text{ft}$ or sec^{-1}.

The torque applied $=$ the torque resisting

$$0.650 = \tau(\text{area})(\text{arm}) = \tau(2\pi \times 0.41 \times 1.00)(0.41) \quad \text{and} \quad \tau = 0.616\ \text{lb/ft}^2.$$

Then $\mu = \tau/(dV/dy) = 0.616/125.5 = .00491\ \text{lb sec/ft}^2$.

(b) The more exact mathematical approach utilizes the calculus, as follows.

As before, $0.650 = \tau(2\pi r \times 1.00)r$ from which $\tau = 0.1034/r^2$.

Now $\dfrac{dV}{dy} = \dfrac{\tau}{\mu} = \dfrac{0.1034}{\mu r^2}$, where the variables are velocity V and radius r. The velocity is 2.51 ft/sec at the inner radius and is zero at the outer radius.

Rearranging the above expression and substituting $-dr$ for dy (the minus sign indicates that r decreases as V increases), we obtain

$$\int_{V_\text{outer}}^{V_\text{inner}} dV = \frac{0.1034}{\mu} \int_{0.42}^{0.40} \frac{-dr}{r^2} \quad \text{and} \quad V_\text{inner} - V_\text{outer} = \frac{0.1034}{\mu}\left[\frac{1}{r}\right]_{0.42}^{0.40}$$

Then $(2.51 - 0) = \dfrac{0.1034}{\mu}\left(\dfrac{1}{0.40} - \dfrac{1}{0.42}\right)$ from which $\mu = .00494$ lb sec/ft^2.

12. Show that the pressure intensity at a point is of equal magnitude in all directions.

Solution:

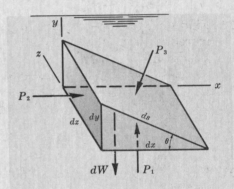

Consider the small triangular prism of liquid at rest, acted upon by the fluid around it. The values of average unit pressures on the three surfaces are p_1, p_2, and p_3. In the z direction the forces are equal and opposite and cancel each other.

Summing forces in the x and y directions, we obtain

$\Sigma X = 0, \quad P_2 - P_3 \sin\theta = 0$

or $\qquad p_2(dy\,dz) - p_3(ds\,dz)\sin\theta = 0$

$\Sigma Y = 0, \quad P_1 - P_3\cos\theta - dW = 0$

or $\qquad p_1(dx\,dz) - p_3(ds\,dz)\cos\theta - w(\tfrac{1}{2}\,dx\,dy\,dz) = 0$

Fig. 1-4

Since $dy = ds\sin\theta$ and $dx = ds\cos\theta$, the equations reduce to the following:

$$p_2\,dy\,dz - p_3\,dy\,dz = 0 \quad \text{or} \quad p_2 = p_3 \tag{1}$$

and $\qquad p_1\,dx\,dz - p_3\,dx\,dz - w(\tfrac{1}{2}dx\,dy\,dz) = 0 \quad \text{or} \quad p_1 - p_3 - w(\tfrac{1}{2}\,dy) = 0 \tag{2}$

As the triangular prism approaches a point, dy approaches zero as a limit, and the average pressures become uniform or even "point" pressures. Then putting $dy = 0$ in equation (2), we obtain $p_1 = p_3$ and hence $p_1 = p_2 = p_3$.

13. Derive the expression $(p_2 - p_1) = w(h_2 - h_1)$.

Solution:

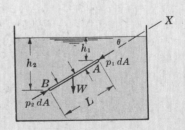

Consider a portion AB of the liquid in Fig. 1-5 as a free body of cross sectional area dA, held in equilibrium by its own weight and the effects of other particles of liquid on the body AB.

At A the force acting is $p_1\,dA$ (the pressure in lb/ft^2 times the area in ft^2); at B it is $p_2\,dA$. The weight of the free body AB is $W = wv = wL\,dA$. The other forces acting on the free body AB are normal to its sides, and only a few of these are shown in the diagram. In taking $\Sigma X = 0$, such normal forces will not enter the equation. Therefore

Fig. 1-5

$$p_2\,dA - p_1\,dA - wL\,dA\sin\theta = 0$$

Since $L\sin\theta = h_2 - h_1$, the above equation reduces to $(p_2 - p_1) = w(h_2 - h_1)$.

14. Determine the pressure in psi at a depth of 20.0 ft below the free surface of a body of water.

Solution:

Using an average value of 62.4 lb/ft³ for w,

$$p' = \frac{wh}{144} = \frac{62.4 \text{ lb/ft}^3 \times 20.0 \text{ ft}}{144 \text{ in}^2/\text{ft}^2} = 8.66 \text{ psi gage}$$

Note: The ratio 62.4/144 will occur often and the reader may save time by using 0.433 as the quotient. The reciprocal is 144/62.4 = 2.31.

15. Determine the pressure in psi at a depth of 30.0 ft in oil of specific gravity 0.750.

Solution:

$$p' = \frac{wh}{144} = \frac{(0.750 \times 62.4)30.0}{144} = 9.76 \text{ psi gage}$$

16. Find the absolute pressure in psi in Problem 14 when the barometer reads 29.90 in. of mercury, sp gr 13.57.

Solution:

Absolute pressure = atmospheric pressure + pressure due to 20.0 ft water

$$= \frac{(13.57 \times 62.4)(29.90/12)}{144} + \frac{62.4 \times 20.0}{144} = 23.32 \text{ psi absolute}$$

17. What depth of oil, sp gr 0.750, will produce a pressure of 40.0 psi? What depth of water?

Solution:

$$h_{\text{oil}} = \frac{p}{w_{\text{oil}}} = \frac{40.0 \times 144}{0.750 \times 62.4} = 123 \text{ ft}, \qquad h_{\text{water}} = \frac{p}{w_{\text{water}}} = \frac{40.0 \times 144}{62.4} = 92.4 \text{ ft}$$

18. (*a*) Convert a pressure head of 15 ft of water to feet of oil, sp gr 0.750.
 (*b*) Convert a pressure head of 24 in. of mercury to feet of oil, sp gr 0.750.

Solution:

(*a*) $h_{\text{oil}} = \dfrac{h_{\text{water}}}{\text{sp gr oil}} = \dfrac{15.0}{0.750} = 20.0 \text{ ft}$ (*b*) $h_{\text{oil}} = \dfrac{h_{\text{water}}}{\text{sp gr oil}} = \dfrac{13.57 \times 24/12}{0.750} = 36.2 \text{ ft}$

19. Prepare a chart so that gage and absolute pressures can be compared readily, with limitations noted.

Solution:

Let A in Fig. 1-6 represent an absolute pressure of 54.7 psi. The gage pressure will depend upon the atmospheric pressure at the moment. If such pressure is standard for sea level (14.7 psi), then the gage pressure for A is 54.7 − 14.7 = 40.0 psi. Should the current barometer read a pressure equivalent to 14.4 psi, then the gage pressure would be 54.7 − 14.4 = 40.3 psi gage.

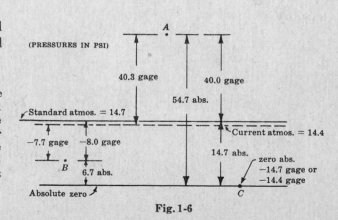

Fig. 1-6

Let B represent a pressure of 6.7 psi absolute. This value plots graphically as less than the standard 14.7 psi and the gage pressure for B is $6.7 - 14.7 = -8.0$ psi gage. If the current atmospheric pressure is 14.4 psi, then the gage pressure for B is $6.7 - 14.4 = -7.7$ psi gage.

Let C represent a pressure of absolute zero. This condition is equivalent to a negative "standard" gage pressure of -14.7 psi and a negative current gage pressure of -14.4 psi.

Conclusions to be drawn are important. Negative gage pressures cannot exceed a theoretical limit of the current atmospheric pressure or a -14.7 psi standard value. Absolute pressures cannot have negative values assigned to them.

20. In Fig. 1-7 the areas of the plunger A and cylinder B are 6.00 and 600 in² respectively and the weight of B is 9000 lb. The vessel and the connecting passages are filled with oil of specific gravity 0.750. What force P is required for equilibrium, neglecting the weight of A?

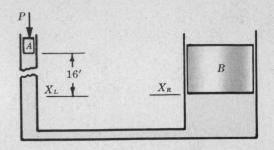

Fig. 1-7

Solution:

First determine the unit pressure acting on the plunger A. Since X_L and X_R are at the same level in the same liquid, then

$$\text{pressure at } X_L \text{ in psi} = \text{pressure at } X_R \text{ in psi}$$

or $\quad$ pressure under A + pressure due to 16 ft oil $\quad = \quad \dfrac{\text{weight of } B}{\text{area of } B}$

Substituting, $\quad\quad\quad\quad\quad\quad\quad\quad p_A' + \dfrac{wh}{144} = \dfrac{9000 \text{ lb}}{600 \text{ in}^2}$

$$p_A' + \frac{(0.750 \times 62.4)16}{144} \text{ psi} = 15.0 \text{ psi} \quad \text{and} \quad p_A' = 9.80 \text{ psi}$$

Force P = uniform pressure $\times$ area = $9.80 \text{ lb/in}^2 \times 6.00 \text{ in}^2 = 58.8 \text{ lb}$.

21. Determine the pressure at A in psi gage due to the deflection of the mercury, sp gr 13.57, in the U-tube gage shown in Fig. 1-8.

Solution:

B and C are at the same level in the same liquid, mercury; hence we equate the pressures at B and C in *psf gage*.

$$\text{Pressure at } B = \text{pressure at } C$$
$$p_A + wh \text{ (for water)} = p_D + wh \text{ (for mercury)}$$
$$p_A + 62.4(12.00 - 10.00) = 0 + (13.57 \times 62.4)(12.65 - 10.00)$$

Solving, $p_A = 2115$ psf and $p_A' = 2115/144 = 14.7$ psi gage.

An alternate solution using pressure heads in feet of water usually requires less arithmetic, as follows:

$$\text{Pressure head at } B = \text{pressure head at } C$$
$$p_A/w + 2.00 \text{ ft water} = 2.65 \times 13.57 \text{ ft water}$$

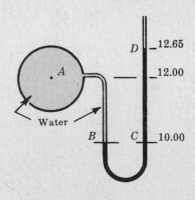

Fig. 1-8

Solving, $p_A/w = 34.0$ ft water and $p_A' = (62.4 \times 34.0)/144 = 14.7$ psi gage, as before.

22. Oil of specific gravity 0.750 flows through the nozzle shown in Fig. 1-9 below and deflects the mercury in the U-tube gage.   Determine the value of h if the pressure at A is 20.0 psi.

Solution:

$$\text{Pressure at } B \;=\; \text{pressure at } C$$

or, using psi units,

$$p'_A + \frac{wh}{144} \text{ (oil)} \;=\; p'_D + \frac{wh}{144} \text{ (mercury)}$$

$$20.0 + \frac{(0.750 \times 62.4)(2.75 + h)}{144} = \frac{(13.57 \times 62.4)h}{144} \quad \text{and} \quad h = 3.77 \text{ ft}$$

Another method:

Using convenient feet of water units,

$$\text{Pressure head at } B \;=\; \text{pressure head at } C$$

$$\frac{20.0 \times 144}{62.4} + (2.75 + h)0.750 \;=\; 13.57h \quad \text{and} \quad h = 3.77 \text{ ft, as before}$$

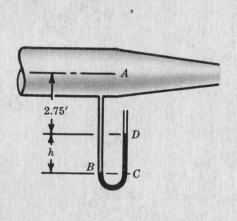

Fig. 1-9

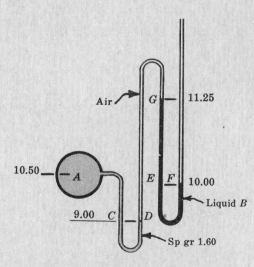

Fig. 1-10

23. For a gage pressure at A of -1.58 psi, find the specific gravity of the gage liquid B in Fig. 1-10 above.

Solution:

$$\text{Pressure at } C \;=\; \text{pressure at } D$$

or, using psf units,
$$p_A + wh = p_D$$
$$-1.58 \times 144 + (1.60 \times 62.4)1.50 \;=\; p_D = -77.8 \text{ psf}$$

Now $p_G = p_D = -77.8$ psf, since the weight of 2.25 ft of air can be neglected without introducing significant error. Also $p_E = p_F = 0$ in psf gage.

Thus,   pressure at G = pressure at E − pressure of $(11.25 - 10.00)$ft of gage liquid

or $$p_G = p_E - (\text{sp gr} \times 62.4)(11.25 - 10.00)$$
$$-77.8 = 0 - (\text{sp gr} \times 62.4)1.25 \quad \text{and} \quad \text{sp gr} = 1.00$$

24. For a gage reading at A of -2.50 psi, determine (a) the elevation of the liquids in the open piezometer columns E, F, and G and (b) the deflection of the mercury in the U-tube gage in Fig. 1-11.

Solution:

(a) Since the unit weight of the air (about $.08\ \text{lb/ft}^3$) is very small compared with that of the liquids, the pressure at Elevation 49.00 may be considered to be -2.50 psi without introducing significant error in the calculations.

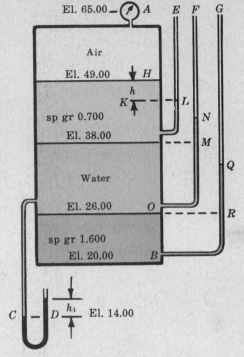

Fig. 1-11

For Column E:

The elevation at L being assumed as shown, we have

in psf gage $\qquad\qquad p_K = p_L$

Then $\qquad\qquad p_H + wh = 0$

or $\quad -2.50 \times 144 + (0.700 \times 62.4)h = 0$

and $h = 8.25$ ft.

Hence the elevation at L is $49.00 - 8.25$ $= 40.75$ ft.

For Column F:

Pressure at El. 38.00 $=$ pressure at El. 49.00 $+$ pressure of liquid of sp gr 0.700

$$= -2.50 + \frac{(0.700 \times 62.4)(49.00 - 38.00)}{144} = 0.830\ \text{psi}$$

which must equal the pressure at M. Thus the pressure head at M is $\dfrac{0.830 \times 144}{62.4} = 1.92$ ft of water, and column F will rise 1.92 ft above M or to Elevation 39.92 at N.

For Column G:

Pressure at El. 26.00 $=$ pressure at El. 38.00 $+$ pressure of 12 ft of water

or $\qquad\qquad p_O = 0.830 + \dfrac{62.4 \times 12}{144} = 6.03\ \text{psi}$

which must be the pressure at R. Then the pressure head at R is $\dfrac{6.03 \times 144}{1.600 \times 62.4} = 8.72$ ft of the liquid, and column G will rise 8.72 ft above R or to Elevation 34.72 at Q.

(b) For the U-tube gage, using feet of water units,

pressure head at D $=$ pressure head at C.

$13.57h_1 =$ pressure head at El. 38.00 $+$ pressure head of 24.00 ft water

$13.57h_1 = 1.92 + 24.00$

from which $h_1 = 1.91$ ft

25. A differential gage is attached to two cross sections A and B in a horizontal pipe in which water is flowing. The deflection of the mercury in the gage is 1.92 ft, the level nearer A being the lower one. Calculate the difference in pressure in psi between Sections A and B. Refer to Fig. 1-12 below.

Solution:

Note: A sketch will be of assistance in clarifying the analysis of all problems as well as in reducing mistakes. Even a single-line diagram will help.

$$\text{Pressure head at } C = \text{pressure head at } D$$

or, using ft of water units,
$$p_A/w - z = [p_B/w - (z + 1.92)] + 13.57(1.92)$$

Then $p_A/w - p_B/w = \text{difference in pressure heads} = 1.92(13.57 - 1) = 24.1 \text{ ft water}$

and $p_A' - p_B' = 24.1(0.433) = 10.46 \text{ psi.}$

If $(p_A' - p_B')$ had been negative, the proper interpretation of the sign would be that the pressure at B was larger than the pressure at A by 10.46 psi.

Differential gages should have the air extracted from all tubing before readings are taken.

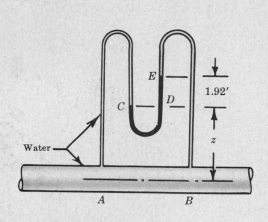

Fig. 1-12

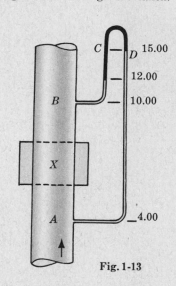

Fig. 1-13

26. The loss through a device X is to be measured by a differential gage using oil of specific gravity 0.750 as the gage fluid. The liquid flowing has specific gravity 1.50. Find the change in pressure head between A and B for the deflection of the oil shown in Fig. 1-13 above.

Solution:

$$\text{Pressure at } C \text{ in psf} = \text{pressure at } D \text{ in psf}$$
$$p_B - (1.50 \times 62.4)2.00 - (0.750 \times 62.4)3.00 = p_A - (1.50 \times 62.4)11.00$$

Then $p_A - p_B = 702 \text{ psf,}$ and difference in pressure heads $= \dfrac{702}{w} = \dfrac{702}{1.50 \times 62.4} = 7.50 \text{ ft of liquid.}$

Another method:

Using feet of liquid (sp gr 1.50) units,

$$\text{pressure head at } C = \text{pressure head at } D$$

$$\frac{p_B}{w} - 2.00 - \frac{0.750 \times 3.00}{1.50} = \frac{p_A}{w} - 11.00$$

Then $p_A/w - p_B/w = \text{difference in pressure heads} = 7.50 \text{ ft of the liquid, as before.}$

27. Vessels A and B contain water under pressures of 40.0 psi and 20.0 psi respectively. What is the deflection of the mercury in the differential gage in Fig. 1-14 below?

Solution:

$$\text{Pressure head at } C = \text{pressure head at } D$$

$$\frac{40.0 \times 144}{62.4} + x + h = \frac{20.0 \times 144}{62.4} - y + 13.57h \quad \text{(ft of water units)}$$

Rearranging, $(144/62.4)(40.0 - 20.0) + x + y = (13.57 - 1)h$. Substituting $x + y = 6.00$ ft and solving, we obtain $h = 4.15$ ft.

The reader should note that the choice of psf or psi units will involve more arithmetic, but the probability of making fewer mistakes may recommend the use of such units rather than pressure head units.

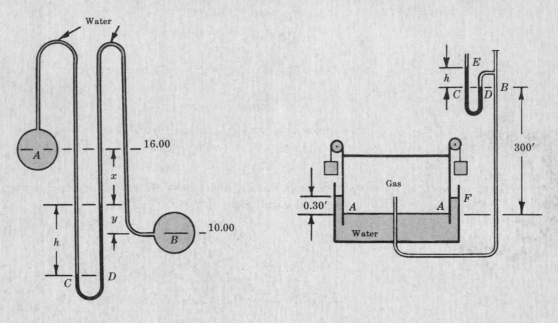

Fig. 1-14   Fig. 1-15

28. The pressure head at level A-A is 0.300 ft of water and the unit weights of gas and air are 0.0350 and 0.0787 lb/ft³ respectively. Determine the reading of the water in the U-tube gage which measures the gas pressure at level B in Fig. 1-15 above.

Solution:

Assume the values of w for air and gas remain constant for the 300 ft difference in elevation. Because the unit weights of gas and air are of the same order of magnitude, the change in atmospheric pressure with altitude must be taken into account. The use of absolute pressure units is recommended.

$$\text{Absolute } p_C = \text{absolute } p_D \quad \text{(psf)}$$

$$\text{atmospheric } p_E + 62.4h = \text{absolute } p_A - 0.0350 \times 300 \quad (A)$$

The absolute pressure at A will next be evaluated in terms of the atmospheric pressure at E, obtaining first the atmospheric pressure at F and then p_A.

$$\text{Absolute } p_A = [\text{atmos. } p_E + 0.0787(h + 300 - 0.30)] + 0.30 \times 62.4 \quad \text{(psf)}$$

Substituting this value in (A), cancelling atmospheric p_E and neglecting very small terms gives

$$62.4h = 300(0.0787 - 0.0350) + 0.30(62.4) \quad \text{and} \quad h = 0.511 \text{ ft} \quad \text{or} \quad 6.13 \text{ in. of water}$$

29. What is the intensity of pressure in the ocean at a depth of 5000 ft, assuming (a) salt water is incompressible and (b) salt water is compressible and weighs 64.0 lb/ft³ at the surface? $E = 300,000$ psi (constant).

Solution:

(a) Intensity of pressure $p = wh = 64.0(5000) = 320,000$ psf gage.

(b) Since a given mass does not change its weight as it is compressed, $dW = 0$; then

$$dW = d(wv) = w \, dv + v \, dw = 0 \quad \text{or} \quad dv/v = -dw/w \tag{A}$$

From equations (10) and (12), $dp = -w \, dh$ and $dv/v = -dp/E$. Substituting in (A),

$$dp/E = dw/w \tag{B}$$

Integrating, $p = E \log_e w + C$. At surface, $p = p_o$, $w = w_o$; then $C = p_o - E \log_e w_o$ and

$$p = E \log_e w + p_o - E \log_e w_o \quad \text{or} \quad (p - p_o) = E \log_e (w/w_o) \tag{C}$$

Putting $dp = -w \, dh$ in (B), $\dfrac{-w \, dh}{E} = \dfrac{dw}{w}$ or $dh = -\dfrac{E \, dw}{w^2}$. Integrating,

$$h = E/w + C_1 \tag{D}$$

At the surface, $h = 0$, $w = w_o$; then $C_1 = -E/w_o$, $h = (E/w - E/w_o)$, and hence

$$w = \frac{w_o E}{w_o h + E} = \frac{(64.0)(300,000 \times 144)}{(64.0)(-5000) + (300,000 \times 144)} = 64.5 \text{ lb/ft}^3$$

remembering that h is positive upwards and using psf units for E. From (C),

$$p = (300,000 \times 144) \log_e (64.5/64.0) = 323,000 \text{ psf gage}$$

30. Compute the barometric pressure in psi at an altitude of 4000 ft if the pressure at sea level is 14.7 psi. Assume isothermal conditions at 70°F.

Solution:

The unit weight of air at 70°F is $w = \dfrac{p}{53.3(460 + 70)}$. Then from equation (10),

$$dp = -w \, dh = -\frac{p}{53.3(530)} dh \quad \text{or} \quad \frac{dp}{p} = -.0000354 \, dh \tag{A}$$

Integrating (A), $\log_e p = -.0000354h + C$, where C is the constant of integration.

To evaluate C: when $h = 0$, $p = 14.7 \times 144 = 2116$ psf absolute. Hence $C = \log_e 2116$ and

$$\log_e p = -.0000354h + \log_e 2116 \quad \text{or} \quad .0000354h = \log_e 2116/p \tag{B}$$

Transforming (B) to $\log_{10}$

$$2.3026 \log_{10} 2116/p = .0000354(4000), \quad \log_{10} 2116/p = .0616, \quad 2116/p = \text{antilog } 0.0616 = 1.152$$

from which $p = \dfrac{2116}{1.152}$ psf and $p' = \dfrac{2116}{1.152 \times 144} = 12.7$ psi.

31. Derive a general expression for the relation between pressure and elevation for isothermal conditions, using $dp = -w \, dh$.

Solution:

For isothermal conditions, the equation $\dfrac{p}{wT} = \dfrac{p_o}{w_o T_o}$ becomes $\dfrac{p}{w} = \dfrac{p_o}{w_o}$ or $w = w_o \dfrac{p}{p_o}$.

Then $dh = -\dfrac{dp}{w} = -\dfrac{p_o}{w_o} \times \dfrac{dp}{p}$. Integrating, $\displaystyle\int_{h_o}^{h} dh = -\frac{p_o}{w_o} \int_{p_o}^{p} \frac{dp}{p}$ and

$$h - h_o = -\frac{p_o}{w_o}(\log_e p - \log_e p_o) = +\frac{p_o}{w_o}(\log_e p_o - \log_e p) = \frac{p_o}{w_o} \log_e \frac{p_o}{p}$$

Actually, the temperature of the atmosphere decreases with altitude. Hence an exact solution requires knowledge of the temperature variations with altitude and the use of the gas law $p/wT = a$ constant.

32. Develop the expression for the relation between the gage pressure p inside a droplet of liquid and the surface tension σ.

Solution:

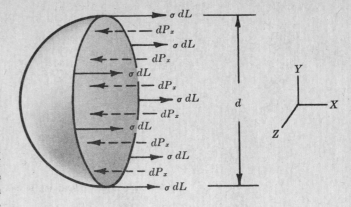

Fig. 1-16

The surface tension in the surface of a small drop of liquid causes the pressure inside the drop to be greater than the pressure outside.

Fig. 1-16 shows the forces which cause equilibrium in the X direction of half of a small drop of diameter d. The forces $\sigma\,dL$ are due to surface tension around the perimeter and the forces dP_x are the X components of the $p\,dA$ forces (see Chapter 2). Then, from $\Sigma X = 0$,

$$\text{sum of forces to the right} = \text{sum of forces to the left}$$
$$\sigma \int dL = \int dP_x$$
$$\text{surface tension} \times \text{perimeter} = \text{pressure} \times \text{projected area}$$
$$\sigma(\pi d) = p(\pi d^2/4)$$

or $p = 4\sigma/d$ in psf gage. The units of surface tension are lb/ft.

It should be observed that the smaller the droplet, the larger the pressure.

33. A small drop of water at 80°F is in contact with the air and has diameter .0200 in. If the pressure within the droplet is .082 psi greater than the atmosphere, what is the value of the surface tension?

Solution:
$$\sigma = \tfrac{1}{4}pd = \tfrac{1}{4}(.082 \times 144)\,\text{lb/ft}^2 \times (.0200/12)\,\text{ft} = .00492\,\text{lb/ft}$$

34. Calculate the approximate height to which a liquid which wets glass will rise in a capillary tube exposed to the atmosphere.

Solution:

The rise in a tube of small diameter may be approximated by considering the mass of liquid $ABCD$ in Fig. 1-17 as a free body.

Since ΣY must equal 0, we obtain

$$\text{components of force due to surface tensions up} - \text{weight of volume } ABCD \text{ down}$$
$$+ \text{ pressure force on } AB \text{ up} - \text{pressure force on } CD \text{ down} = 0.$$

or
$$+ (\sigma \int dL)\sin\alpha - w(\pi d^2/4 \times h) + p(\text{area } AB) - p(\text{area } CD) = 0$$

It can be seen that the pressures at level AB and level CD are both atmospheric. Thus the last two terms on the left side of the equation cancel and, since $\sigma \int dL = \sigma(\pi d)$, we obtain

$$h = \frac{4\sigma \sin\alpha}{wd} \quad \text{in feet}$$

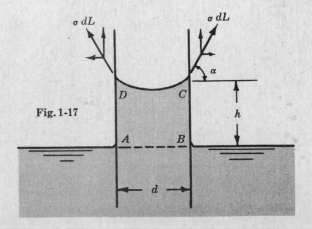

Fig. 1-17

For complete wetting, as with water on clean glass, the angle α is essentially 90°. Further refinement is not warranted herein.

In experimental work, to avoid serious errors due to capillarity, tubes about $\tfrac{3}{8}$ in. in diameter or larger should be used.

35. Estimate the height to which water at 70°F will rise in a capillary tube of diameter 0.120 in.

Solution: From Table 1C, $\sigma = .00497$ lb/ft. Assuming angle $\alpha = 90°$ for a clean tube,

$$ h = \frac{4\sigma}{wd} = \frac{4 \times .00497 \text{ lb/ft}}{62.4 \text{ lb/ft}^3 \times 0.120/12 \text{ ft}} = .0319 \text{ ft} = 0.383 \text{ in.} $$

Supplementary Problems

36. If the density of a liquid is 1.62 slug/ft³, find its specific weight and specific gravity.
    Ans. 52.2 lb/ft³, 0.837

37. Check the values of the density and specific weight of air at 80°F shown in Table 1B.

38. Check the values of the specific weights of carbon dioxide and nitrogen in Table 1A.

39. At what pressure will air at 120°F weigh 0.119 lb/ft³?   Ans. 25.5 psia

40. Two cubic feet of air at atmospheric pressure are compressed to 0.50 ft³. For isothermal conditions, what is the final pressure?   Ans. 58.8 psia

41. In the preceding problem, what would be the final pressure if no heat is lost during compression?
    Ans. 102 psia

42. Determine the absolute viscosity of mercury in lb sec/ft² if the viscosity in poises is .0158.
    Ans. 33.0×10^{-6} lb sec/ft²

43. If an oil has an absolute viscosity of 510 poises, what is its viscosity in the fps system?
    Ans. 1.07 lb sec/ft²

44. What are the absolute and kinematic viscosities in fps units of an oil having a Saybolt viscosity of 155 seconds, if the specific gravity of the oil is 0.932?   Ans. 646×10^{-6} and 358×10^{-6}

45. Two large plane surfaces are 1 in. apart and the space between them is filled with a liquid of absolute viscosity .0200 lb sec/ft². Assuming the velocity gradient to be a straight line, what force is required to pull a very thin plate of 4.00 ft² area at a constant speed of 1.00 ft/sec if the plate is $\frac{1}{3}$ in. from one of the surfaces?   Ans. 4.32 lb

46. The tank in Fig. 1-18 contains oil of specific gravity 0.750. Determine the reading of gage A in psi.
    Ans. −1.16 psi

47. A closed tank contains 2 ft of mercury, 5 ft of water, 8 ft of oil of specific gravity 0.750 and an air space above the oil. If the pressure at the bottom of the tank is 40.0 psi gage, what should be the reading of the gage in the top of the tank?   Ans. 23.4 psi

48. Refer to Fig. 1-19. Point A is 1.75 ft below the surface of the liquid, sp gr 1.25, in the vessel. What is the pressure at A in psi gage if the mercury rises 13.5 in. in the tube?   Ans. −5.66 psi

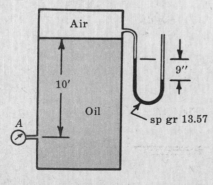

Fig. 1-18

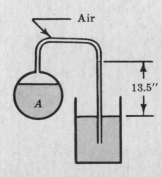

Fig. 1-19

49. Refer to Fig. 1-20. Neglecting friction between the piston A and the gas tank, find the gage reading at B in ft of water. Assume gas and air to be of constant specific weight and equal to .0351 and .0750 lb/ft³ respectively.
Ans. 17.9 in. of water

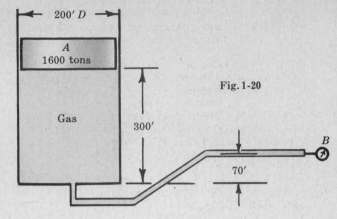

Fig. 1-20

50. Tanks A and B, containing oil and glycerin of specific gravities 0.780 and 1.25 respectively, are connected by a differential gage. The mercury in the gage is at Elevation 1.60 on the A-side and at Elevation 1.10 on the B-side. If the elevation of the surface of the glycerin in tank B is 21.10, at what elevation is the surface of the oil in tank A?
Ans. Elevation 24.90

51. Vessel A, at Elevation 8.00, contains water under 15.0 psi pressure. Vessel B, at Elevation 12.00, contains a liquid under 10.0 psi pressure. If the deflection of the differential gage is 12″ of mercury, lower level on the A-side at Elevation 1.00, determine the specific gravity of the liquid in vessel B.
Ans. 0.500

52. In the left-hand tank in Fig. 1-21, the air pressure is −9 in. of mercury. Determine the elevation of the gage liquid in the right hand column at A.
Ans. Elevation 86.7

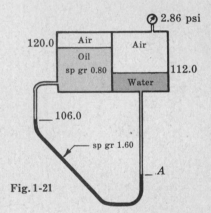

Fig. 1-21

53. The compartments B and C in Fig. 1-22 are closed and filled with air. The barometer reads 14.5 psi. When the gages A and D read as indicated, what should be the value of x for gage E (mercury in each U-tube gage)?
Ans. 5.96 ft

54. The cylinder and tubing shown in Fig. 1-23 contain oil, sp gr 0.902. For a gage reading of 31.2 psi, what is the total weight of piston and weight W?   *Ans.* 136,800 lb

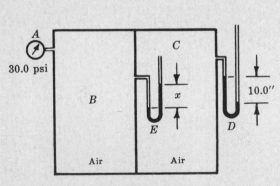

Fig. 1-22

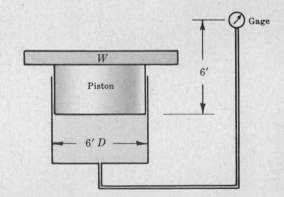

Fig. 1-23

55. Referring to Fig. 1-24, what reading of gage A will cause the glycerin to rise to level B? The specific weights of oil and glycerin are 52.0 and 78.0 lb/ft³ respectively.   *Ans.* 5.06 psi

56. A hydraulic device is used to raise a 10-ton truck. If oil, sp gr 0.810, acts on the piston under a pressure of 177 psi, what diameter is required?   *Ans.* 12 in.

57. If the specific weight of glycerin is 79.2 lb/ft³, what suction pressure is required to raise the glycerin 9 in. vertically in a $\frac{1}{2}$ in. diameter tube?
    Ans. −0.412 psi

58. What is the pressure inside a raindrop which is .06 in. in diameter when the temperature is 70°F?
    Ans. .0276 psi gage

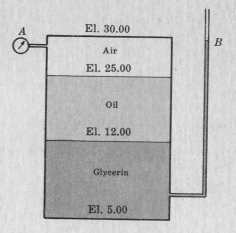

Fig. 1-24

Chapter 2

Hydrostatic Force on Surfaces

INTRODUCTION

Engineers must calculate the forces exerted by fluids in order to design the constraining structures satisfactorily. In this chapter all three characteristics of the hydrostatic force will be evaluated: magnitude, direction and sense. In addition, the location of the force will be found.

FORCE EXERTED by a LIQUID on a PLANE AREA

The force P exerted by a liquid on a plane area A is equal to the product of the specific weight w of the liquid, the depth h_{cg} of the center of gravity of the area, and the area. The equation is

$$P = w\, h_{cg}\, A \tag{1}$$

the units being

$$\text{lb} = \text{lb/ft}^3 \times \text{ft} \times \text{ft}^2$$

Note that the product of the specific weight w and the depth of the center of gravity of the area yields the intensity of pressure at the center of gravity of the area.

The line of action of the force passes through the center of pressure which can be located by applying the formula

$$y_{cp} = \frac{I_{cg}}{y_{cg}\, A} + y_{cg} \tag{2}$$

where I_{cg} is the moment of inertia of the area about its center of gravity axis. The distances y are measured along the plane from an axis located at the intersection of the plane and the liquid surface, both extended if necessary.

The horizontal component of the hydrostatic force on any surface (plane or irregular) is equal to the normal force on the vertical projection of the surface. The component acts through the center of pressure for the vertical projection.

The vertical component of the hydrostatic force on any surface (plane or irregular) is equal to the weight of the volume of liquid above the area, real or imaginary. The force passes through the center of gravity of the volume.

HOOP or CIRCUMFERENTIAL TENSION

Hoop tension or circumferential tension (psi) is created in the walls of a cylinder subjected to internal pressure. For thin-walled cylinders ($t < 0.1d$),

$$\text{Intensity of stress } \sigma \text{ (psi)} = \frac{\text{pressure } p' \text{ (psi)} \times \text{radius } r \text{ (in.)}}{\text{thickness } t \text{ (in.)}} \tag{3}$$

22

LONGITUDINAL STRESS in THIN-WALLED CYLINDERS

Longitudinal stress (psi) in thin-walled cylinders closed at the ends is equal to half the hoop tension.

Solved Problems

1. Develop (a) the equation for the hydrostatic force acting on a plane area, and (b) locate the force.

Solution:

(a) Let trace AB represent any plane area acted upon by a fluid and making an angle θ with the horizontal, as shown in Fig. 2-1. Consider an element of area such that every particle is the same distance h below the surface of the liquid. The horizontal strip shown cross-hatched is such an area, and the pressure is *uniform* over this area. Then the force acting on the area dA is equal to the uniform intensity of pressure p times the area dA, or

$$dP \ = \ p\,dA \ = \ wh\,dA$$

Summing all the forces acting on the area and considering that $h = y \sin \theta$,

$$P \ = \ \int wh\,dA \ = \ \int w(y \sin \theta)\,dA$$
$$= \ (w \sin \theta)\int y\,dA \ = \ (w \sin \theta)y_{cg}A$$

where w and θ are constants and, from statics, $\int y\,dA = y_{cg}A$. Since $h_{cg} = y_{cg} \sin \theta$,

$$P \ = \ w\,h_{cg}\,A \tag{1}$$

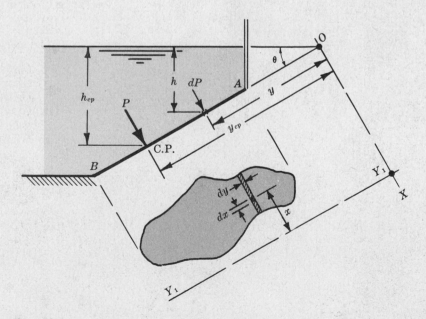

Fig. 2-1

(b)   To locate this force P, proceed as in static mechanics by taking moments. Axis O is chosen as the intersection of the plane area and the water surface, both extended if necessary.   All distances y are measured from this axis and the distance to the resultant force is called y_{cp}, which is the distance to the center of pressure. Since the sum of the moments of all the forces about axis O = the moment of the resultant force, we obtain

$$\int (dP \times y) = P \times y_{cp}$$

But $dP = wh\, dA = w(y \sin \theta)dA$ and $P = (w \sin \theta)y_{cg}A$.   Then

$$(w \sin \theta) \int y^2\, dA = (w \sin \theta)(y_{cg}A)y_{cp}$$

Since $\int y^2\, dA$ is the moment of inertia of the plane area about axis O,

$$\frac{I_o}{y_{cg}A} = y_{cp}$$

In more convenient form, from the parallel axis theorem,

$$y_{cp} = \frac{I_{cg} + Ay_{cg}^2}{y_{cg}A} = \frac{I_{cg}}{y_{cg}A} + y_{cg} \tag{2}$$

Note that the position of the center of pressure is always *below* the center of gravity of the area, or $(y_{cp} - y_{cg})$ is always positive since I_{cg} is always positive.

2.   **Locate the lateral position of the center of pressure.   Refer to Fig. 2-1.**

Solution:

While in general the lateral position of the center of pressure is not required to solve most engineering problems concerning hydrostatic forces, occasionally this information may be needed. Using the sketch in the preceding problem, the area dA is chosen as $(dx\, dy)$ so that the moment arm x is properly used. Taking moments about any non-intersecting axis Y_1Y_1,

$$P\, x_{cp} = \int (dP\, x)$$

Using values derived in Problem 1 above,

$$(wh_{cg}A)x_{cp} = \int p(dx\, dy)x = \int wh(dx\, dy)x$$

or
$$(w \sin \theta)(y_{cg}A)x_{cp} = (w \sin \theta) \int xy(dx\, dy) \tag{3}$$

since $h = y \sin \theta$. The integral represents the product of inertia of the plane area about the X and Y axes chosen, designated by I_{xy}. Then

$$x_{cp} = \frac{I_{xy}}{y_{cg}A} = \frac{(I_{xy})_{cg}}{y_{cg}A} + x_{cg} \tag{4}$$

Should *either* of the centroidal axes be an axis of symmetry of the plane area, I_{xy} becomes zero and the lateral position of the center of pressure lies on the Y-axis which passes through the center of gravity (not shown in the figure). Note that the product of inertia about the center of gravity axes, $(I_{xy})_{cg}$, may be positive or negative, so that the lateral position of the center of pressure may lie on *either* side of the centroidal y-axis.

3. Determine the resultant force P due to the water acting on the 3 ft by 6 ft rectangular area AB shown in Fig. 2-2 below.

Solution:

$$P = wh_{cg}A = (62.4 \text{ lb/ft}^3) \times (4+3) \text{ ft} \times (6 \times 3) \text{ ft}^2 = 7850 \text{ lb}$$

This resultant force acts at the center of pressure which is at a distance y_{cp} from axis O_1 and

$$y_{cp} = \frac{I_{cg}}{y_{cg}A} + y_{cg} = \frac{3(6^3)/12}{7(3 \times 6)} + 7 = 7.43 \text{ ft from } O_1$$

4. Determine the resultant force due to water acting on the 4 ft by 6 ft triangular area CD shown in Fig. 2-2 below. The apex of the triangle is at C.

Solution:

$$P_{CD} = 62.4(3 + \tfrac{2}{3} \times 0.707 \times 6)(\tfrac{1}{2} \times 4 \times 6) = 4370 \text{ lb}$$

This force acts at a distance y_{cp} from axis O_2 and is measured along the plane of the area CD.

$$y_{cp} = \frac{4(6^3)/36}{(5.83/0.707)(\tfrac{1}{2} \times 4 \times 6)} + \frac{5.83}{0.707} = 0.24 + 8.24 = 8.48 \text{ ft from axis } O_2$$

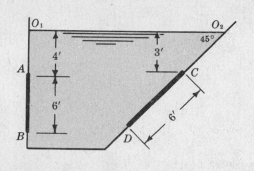

Fig. 2-2

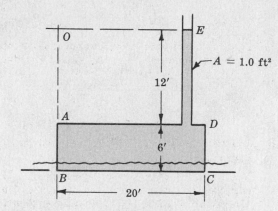

Fig. 2-3

5. Water rises to level E in the pipe attached to tank $ABCD$ in Fig. 2-3 above. Neglecting the weight of the tank and riser pipe, (a) determine and locate the resultant force acting on area AB which is 8 ft wide, (b) determine the total force on the bottom of the tank, and (c) compare the total weight of the water with the result in (b) and explain the difference.

Solution:

(a) The depth of the center of gravity of area AB is 15 ft below the free surface of the water at E. Then

$$P = whA = 62.4(12 + 3)(6 \times 8) = 45,000 \text{ lb} \quad \text{acting at distance}$$

$$y_{cp} = \frac{8(6^3)/12}{15(6 \times 8)} + 15 = 15.20 \text{ ft from } O$$

(b) The pressure on the bottom BC is uniform; hence the force

$$P = pA = (wh)A = 62.4(18)(20 \times 8) = 179,700 \text{ lb}$$

(c) The total weight of the water is $W = 62.4(20 \times 6 \times 8 + 12 \times 1) = 60,700 \text{ lb}.$

A free body of the lower part of the tank (cut by a horizontal plane just above level BC) will indicate a downward force on area BC of 179,700 lb, vertical tension in the walls of the tank, and the reaction of the supporting plane. The reaction must equal the total weight of water or 60,700 lb. The tension in the walls of the tank is caused by the upward force on the top AD of the tank which is

$$P_{AD} = (wh)A = 62.4(12)(160-1) = 119,000 \text{ lb upward}$$

An apparent paradox is thus clarified since, for the free body considered, the sum of the vertical forces is zero, i.e.,
$$179,700 - 60,700 - 119,000 = 0$$

and hence the condition for equilibrium is satisfied.

6. Gate AB in Fig. 2-4(a) below, is 4 ft wide and is hinged at A. Gage G reads -2.17 psi and oil of specific gravity 0.750 is in the right-hand tank. What horizontal force must be applied at B for equilibrium of gate AB?

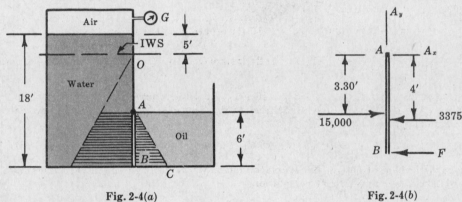

Fig. 2-4(a)   Fig. 2-4(b)

Solution:

The forces acting on the gate due to the liquids must be evaluated and located. For the right-hand side,

$$P_{\text{oil}} = wh_{cg}A = (0.750 \times 62.4)(3)(6 \times 4) = 3375 \text{ lb to the left}$$

acting $\qquad y_{cp} = \dfrac{4(6^3)/12}{3(4 \times 6)} + 3 = 4.00 \text{ ft from } A$

It should be noted that the pressure intensity acting on the right side of rectangle AB varies linearly from zero gage to a value due to 6 ft of oil ($p = wh$ is a linear equation). Loading diagram ABC indicates this fact. For a rectangular area only, the center of gravity of this loading diagram coincides with the center of pressure. The center of gravity is located $\frac{2}{3}(6) = 4$ ft from A, as above.

For the left-hand side, it is necessary to convert the negative pressure due to the air to its equivalent in feet of the liquid, water.

$$h = -\frac{p}{w} = -\frac{2.17 \times 144 \text{ lb/ft}^2}{62.4 \text{ lb/ft}^3} = -5.00 \text{ ft}$$

This negative pressure head is equivalent to having 5 ft less of water above level A. It is convenient and useful to employ an imaginary water surface (IWS) five feet below the real surface and solve the problem by direct use of basic equations. Thus

$$P_{\text{water}} = 62.4(7+3)(6 \times 4) = 15,000 \text{ lb acting to the right at the center of pressure}$$

For the submerged rectangular area, $y_{cp} = \dfrac{4(6^3)/12}{10(6 \times 4)} + 10 = 10.30 \text{ ft from } O$, or the center of pressure is $(10.30 - 7) = 3.30$ ft from A.

In Fig. 2-4(b) above, the free body diagram of gate AB shows the forces acting. The sum of the moments about A must equal zero. Taking clockwise plus,

$$+3375 \times 4 + 6F - 15,000 \times 3.30 = 0 \quad \text{and} \quad F = 6000 \text{ lb to the left}$$

7. The tank in Fig. 2-5 contains oil and water. Find the resultant force on side *ABC* which is 4 ft wide.

    Solution:

    The total force on *ABC* is equal to $(P_{AB} + P_{BC})$. Find each force, locate it and, using the principle of moments, determine the position of the total force on side *ABC*.

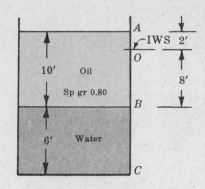

Fig. 2-5

(a) $P_{AB} = (0.800 \times 62.4)(5)(10 \times 4) = 9980$ lb acting at a point $\frac{2}{3}(10)$ ft from *A* or 6.67 ft down. The same distance can be obtained by formula, as follows:

$$y_{cp} = \frac{4(10^3)/12}{5(4 \times 10)} + 5 = 1.67 + 5 = 6.67 \text{ ft from } A$$

(b) Water is acting on area *BC* and any superimposed liquid can be converted into an equivalent depth of water. Employ an imaginary water surface (IWS) for this second calculation, locating the IWS by changing 10 ft of oil to $0.800 \times 10 = 8$ ft of water. Then

$$P_{BC} = 62.4(8+3)(6 \times 4) = 16,450 \text{ lb acting at the center of pressure}$$

$$y_{cp} = \frac{4(6^3)/12}{11(4 \times 6)} + 11 = 11.27 \text{ ft from } O \quad \text{or} \quad (2+11.27) = 13.27 \text{ ft from } A$$

The total resultant force $= 9980 + 16,450 = 26,430$ lb acting at the center of pressure for the entire area. The moment of this total force $=$ the sum of the moments of its two parts. Using *A* as a convenient axis,

$$26,430 \, Y_{cp} = 9980(6.67) + 16,450(13.27) \quad \text{and} \quad Y_{cp} = 10.76 \text{ ft from } A$$

Other methods of attack may be employed but it is believed that the method illustrated will greatly reduce mistakes in judgment and calculation.

8. In Fig. 2-6, gate *ABC* is hinged at *B* and is 4 ft long. Neglecting the weight of the gate, determine the unbalanced moment due to the water acting on the gate.

    Solution:

    $P_{AB} = 62.4(4)(9.24 \times 4) = 9220$ lb, acting $\frac{2}{3}(9.24) = 6.16$ ft from *A*.

    $P_{BC} = 62.4(8)(3 \times 4) = 6000$ lb, acting at the center of gravity of *BC* since the pressure on *BC* is uniform. Taking moments about *B*, (clockwise plus),

$$\text{Unbalanced moment} = +9220 \times 3.08 - 6000 \times 1.50$$
$$= +19,400 \text{ ft lb clockwise}$$

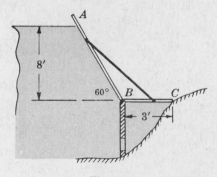

Fig. 2-6

9. Determine the resultant force due to the water acting on the vertical area shown in Fig. 2-7(a) below and locate the center of pressure in the *x* and *y* directions.

    Solution:

    Divide the area into a rectangle and a triangle. The total force acting is equal to the force P_1 acting on the rectangle plus the force P_2 acting on the triangle.

(a) $P_1 = 62.4(4)(8 \times 4) = 7980$ lb acting $\frac{2}{3}(8) = 5.33$ ft below the surface *XX*.

$$P_2 = 62.4(10)(\tfrac{1}{2} \times 6 \times 4) = 7490 \text{ lb} \quad \text{at} \quad y_{cp} = \frac{4(6^3)/36}{10(\tfrac{1}{2} \times 4 \times 6)} + 10 = 10.20 \text{ ft below } XX.$$

The resultant force $P = 7980 + 7490 = 15,470$ lb. Taking moments about axis *XX*,

$$15,470 \, Y_{cp} = 7980(5.33) + 7490(10.20) \quad \text{and} \quad Y_{cp} = 7.68 \text{ ft below surface } XX$$

(b) To locate the center of pressure in the X direction (seldom required), use the principle of moments after having located x_1 and x_2 for the rectangle and the triangle respectively. For the rectangle, the center of pressure for each horizontal strip of area dA is 2 ft from the YY-axis; therefore its center of pressure is 2 ft from that axis. For the triangle, each area dA has its center of pressure at its own center; therefore the median line contains all these centers of pressure, and the center of pressure for the entire triangle can now be calculated. Referring to Fig. 2-7(b) and using similar triangles, $x_2/2 = 3.80/6$, from which $x_2 = 1.27$ ft from YY. Taking moments,

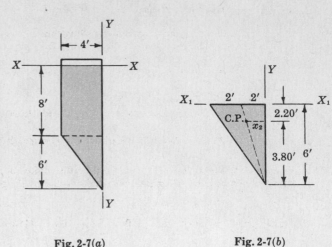

Fig. 2-7(a)   **Fig. 2-7(b)**

$$15,470\, X_{cp} = 7980(2) + 7490(1.27) \quad \text{and} \quad X_{cp} = 1.65 \text{ ft from axis } YY$$

An **alternate method** can be used to locate the center of pressure. Instead of dividing the area into two parts, calculate the center of gravity position for the *entire* area. Using the parallel-axis theorem, determine the moment of inertia and the product of inertia of the entire area about these center of gravity axes. The values of y_{cp} and x_{cp} are then calculated by formulas (2) and (4), Problems 1 and 2. Generally this alternate method has no particular advantage and may involve more arithmetic.

10. The 6 ft diameter gate AB in Fig. 2-8 swings about a horizontal pivot C located 4 in. below the center of gravity. To what depth h can the water rise without causing an unbalanced clockwise moment about the pivot C?

Solution:

If the center of pressure and axis C should coincide there would be no unbalanced moment acting on the gate. Evaluating the center of pressure distance,

$$y_{cp} = \frac{I_{cg}}{y_{cg}A} + y_{cg} = \frac{\pi d^4/64}{y_{cg}(\pi d^2/4)} + y_{cg}$$

Then $y_{cp} - y_{cg} = \dfrac{\pi 6^4/64}{(h+3)(\pi 6^2/4)} = \frac{4}{12}$ ft (given)

from which $h = 3.75$ ft above A.

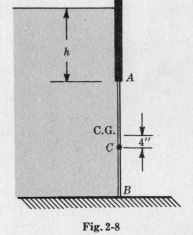

Fig. 2-8

11. Refer to Fig. 2-9 below. What is the minimum width b for the base of a dam 100 ft high if upward pressure beneath the dam is assumed to vary uniformly from full hydrostatic head at the heel to zero at the toe, and also assuming an ice thrust P_I of 12,480 lb per linear foot of dam at the top? For this study make the resultant of the reacting forces cut the base at the downstream edge of the middle third of the base (at O) and take the weight of the masonry as $2.50w$.

Solution:

Shown in the diagram are the H and V components of the reaction of the foundation, acting through O. Consider a length of one foot of dam and evaluate all the forces in terms of w and b, as follows:

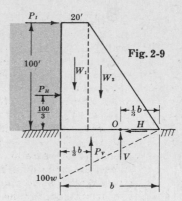

Fig. 2-9

$$P_H = w(50)(100 \times 1) = 5000w \text{ lb}$$

$$P_V = \text{area of the loading diagram}$$
$$= \tfrac{1}{2}(100w)(b \times 1) = 50wb \text{ lb}$$

$$W_1 = 2.50w(20 \times 100 \times 1) = 5000w \text{ lb}$$

$$W_2 = 2.50w[\tfrac{1}{2} \times 100(b - 20)] \times 1$$
$$= 125w(b - 20) \text{ lb} = (125wb - 2500w) \text{ lb}$$

$$P_I = 12,480 \text{ lb, as given for the ice thrust}$$

To find the value of b for equilibrium, take moments of these forces about axis O. Considering clockwise moments positive,

$$5000w\left(\frac{100}{3}\right) + 50wb\left(\frac{b}{3}\right) - 5000w\left(\frac{2}{3}b - 10\right) - (125wb - 2500w)\left[\frac{2}{3}(b - 20) - \frac{b}{3}\right] + 12,480(100) = 0$$

Simplifying and solving, $3b^2 + 100b - 24,400 = 0$ and $b = 75 \text{ ft wide.}$

12. Determine and locate the components of the force due to the water acting on the curved area AB in Fig. 2-10, per foot of its length.

Solution:

$$P_H = \text{force on vertical projection } CB = w\, h_{cg} A_{CB}$$
$$= 62.4(3)(6 \times 1) = 1123 \text{ lb} \qquad \text{acting } \tfrac{2}{3}(6) = 4 \text{ ft from } C$$

$$P_V = \text{weight of water above area } AB = 62.4(\pi 6^2/4 \times 1) = 1765 \text{ lb}$$

acting through the center of gravity of the volume of liquid. The center of gravity of a quadrant of a circle is located at a distance $4/3 \times r/\pi$ from either mutually perpendicular radius. Thus

$$x_{cp} = 4/3 \times 6/\pi = 2.55 \text{ ft to the left of line } BC$$

Note: Each force dP acts normal to the curve AB and would therefore pass through hinge C upon being extended. The total force should also pass through C. In order to confirm this statement, take moments of the components about C, as follows.

$$\Sigma M_C = -1123 \times 4 + 1765 \times 2.55 = 0 \qquad \text{(satisfied)}$$

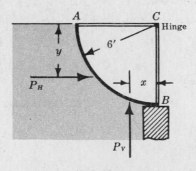

Fig. 2-10

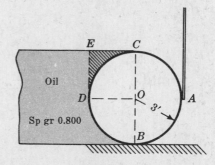

Fig. 2-11

13. The 6 ft diameter cylinder in Fig. 2-11 above weighs 5000 lb and is 5 ft long. Determine the reactions at A and B, neglecting friction.

Solution:

(a) The reaction at A is due to the horizontal component of the liquid force acting on the cylinder or

$$P_H = (0.800 \times 62.4)(3)(6 \times 5) = 4490 \text{ lb}$$

to the right. Hence the reaction at A must be 4490 lb to the left.

(b) The reaction at B is the algebraic sum of the weight of the cylinder and the net vertical component of the force due to the liquid. The curved surface CDB acted upon by the liquid consists of a concave downward part CD and a concave upward part DB. The net vertical component is the algebraic sum of the downward force and the upward force.

$$\text{Upward } P_V = \text{ weight of the liquid (real or imaginary) above curve } DB$$
$$= 0.800 \times 62.4 \times 5 \text{ (area of sector } DOB + \text{ area of square } DOCE)$$

$$\text{Downward } P_V = 0.800 \times 62.4 \times 5 \text{ (crosshatched area } DEC)$$

Noting that square $DOCE$ upward less area DEC downward equals quadrant of circle DOC, the net vertical component is

$$\text{net } P_V = 0.800 \times 62.4 \times 5 \text{ (sectors } DOB + DOC) \text{ upward}$$
$$= 0.800 \times 62.4 \times 5(\tfrac{1}{2}\pi 3^2) = 3530 \text{ lb upward}$$

Finally, $\Sigma Y = 0$, $5000 - 3530 - B = 0$ and $B = 1470$ lb upward

In this particular problem, the upward component (buoyant force) equals the weight of the displaced liquid to the left of the vertical plane COB.

14. Referring to Fig. 2-12, determine the horizontal and vertical forces due to the water acting on the 6 ft diameter cylinder per foot of its length.

Solution:

(a) Net P_H = force on CDA − force on AB.

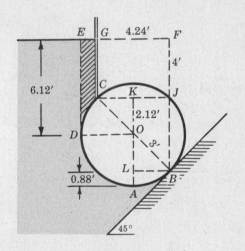

Fig. 2-12

Using the vertical projection of CDA and of AB,

$$P_H(CDA) = 62.4(4 + 2.56)(5.12 \times 1)$$
$$= 2090 \text{ lb to the right}$$

$$P_H(AB) = 62.4(4 + 4.68)(0.88 \times 1)$$
$$= 477 \text{ lb to the left}$$

Net $P_H = 2090 - 477 = 1613$ lb to the right.

(b) Net P_V = upward force on DAB
      − downward force on DC
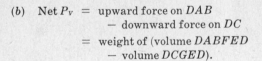
= weight of (volume $DABFED$
      − volume $DCGED$).

The crosshatched area (volume) is contained in each of the above volumes, one force being upward and the other downward. Thus they cancel and

net P_V = weight of volume $DABFGCD$

Dividing this volume into convenient geometric shapes,

$$\text{net } P_V = \text{ weight of (rectangle } GFJC + \text{ triangle } CJB + \text{ semicircle } CDAB)$$
$$= 62.4(4 \times 4.24 + \tfrac{1}{2} \times 4.24 \times 4.24 + \tfrac{1}{2}\pi 3^2)(1)$$
$$= 62.4(16.96 + 8.98 + 14.14) = 2500 \text{ lb upward}$$

Should it be desired to locate this resultant vertical component, the principle of moments would be employed. Each part of the 2500 lb resultant acts through the center of gravity of the volume it represents. By static mechanics, the centers of gravity are found and the moment equation written (see Problems 7 and 9 above).

15. In Fig. 2-13, an 8 ft diameter cylinder plugs a rectangular hole in a tank which is 3 ft long. With what force is the cylinder pressed against the bottom of the tank due to the 9 ft depth of water?

Solution:

$$\text{Net } P_V = \text{ downward force on } CDE - \text{ upward force on } CA \text{ and } BE$$
$$= 62.4 \times 3[(7 \times 8 - \tfrac{1}{2}\pi 4^2) - 2(7 \times 0.54 + \tfrac{1}{12}\pi 4^2 - \tfrac{1}{2} \times 2 \times 3.46)]$$
$$= 5760 - 1690 = 4070 \text{ lb downward}$$

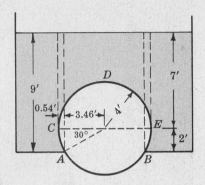

Fig. 2-13

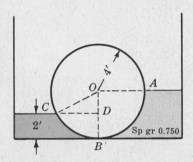

Fig. 2-14

16. In Fig. 2-14, the 8 ft diameter cylinder weighs 500 lb and rests on the bottom of a tank which is 3 ft long. Water and oil are poured into the left and right hand portions of the tank to depths of 2 and 4 ft respectively. Find the magnitudes of the horizontal and vertical components of the force which will keep the cylinder touching the tank at B.

Solution:

$$\text{Net } P_H = \text{ component on } AB \text{ to left} - \text{ component on } CB \text{ to right}$$
$$= 0.750 \times 62.4 \times 2(4 \times 3) - 62.4 \times 1(2 \times 3) = 749 \text{ lb to left}$$

$$\text{Net } P_V = \text{ component upward on } AB + \text{ component upward on } CB$$
$$= \text{ weight of quadrant of oil} + \text{ weight of (sector} - \text{triangle) of water}$$
$$= 0.750 \times 62.4 \times 3 \times \tfrac{1}{4}\pi 4^2 + 62.4 \times 3(\tfrac{1}{6}\pi 4^2 - \tfrac{1}{2} \times 2\sqrt{12}) = 2690 \text{ lb upward}$$

The components to hold the cylinder in place are 749 lb to the right and 2190 lb downward.

17. The half-conical buttress ABE shown in Fig. 2-15 is used to support a half-cylindrical tower $ABCD$. Calculate the horizontal and vertical components of the force due to water acting on the buttress ABE.

Solution:

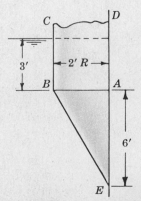

$$P_H = \text{ force on vertical projection of half-cone}$$
$$= 62.4(3 + 2)(\tfrac{1}{2} \times 6 \times 4)$$
$$= 3740 \text{ lb to the right}$$

$$P_V = \text{ weight of volume of water above curved surface (imaginary)}$$
$$= 62.4(\text{volume of half-cone} + \text{volume of half-cylinder})$$
$$= 62.4(\tfrac{1}{2} \times 6\pi 2^2/3 + \tfrac{1}{2}\pi 2^2 \times 3)$$
$$= 1965 \text{ lb upward}$$

Fig. 2-15

18. A 48 in. diameter steel pipe, $\frac{1}{4}$ in. thick, carries oil of sp gr 0.822 under a head of 400 ft of oil.   Compute (a) the stress in the steel and (b) the thickness of steel required to carry a pressure of 250 psi with an allowable stress of 18,000 psi.

Solution:

(a)
$$\sigma \text{ (stress in psi)} = \frac{p' \text{ (pressure in psi)} \times r \text{ (radius in in.)}}{t \text{ (thickness in in.)}}$$

$$= \frac{(0.822 \times 62.4 \times 400)/144 \times 24}{\frac{1}{4}} = 13,650 \text{ psi}$$

(b) $\sigma = p'r/t,$ $18,000 = 250 \times 24/t,$ $t = 0.333$ in.

19. A wooden storage vat, 20 ft in outside diameter, is filled with 24 ft of brine, sp gr 1.06.   The wood staves are bound by flat steel bands, 2 in. wide by $\frac{1}{4}$ in. thick, whose allowable stress is 16,000 psi. What is the spacing of the bands near the bottom of the vat, neglecting any initial stress?   Refer to Fig. 2-16.

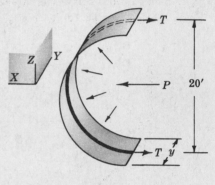

Fig. 2-16

Solution:

Force P represents the sum of all the horizontal components of small forces dP acting on length y of the vat, and forces T represent the total tension carried in a band loaded by the same length y. Since the sum of the forces in the X direction must be zero, $2T \text{ (lb)} - P \text{ (lb)} = 0$ or

$$2(\text{area steel} \times \text{stress in steel}) = p' \times Z \text{ projection of semi-cylinder}$$

Then
$$2(2 \times \tfrac{1}{4})16,000 = (1.06 \times 62.4 \times 24/144)(20 \times 12y)$$

and
$$y = 6.05 \text{ in. spacing of bands}$$

Supplementary Problems

20. For an 8 ft length of gate AB in Fig. 2-17, find the compression in strut CD due to the water pressure (B, C and D are pins).
Ans. 15,850 lb

21. A 12 ft high by 5 ft wide rectangular gate AB is vertical and is hinged at a point 6 in. below the center of gravity of the gate. The total depth of water is 20 ft. What horizontal force F must be applied at the bottom of the gate for equilibrium?
Ans. 3430 lb

22. Find the dimension z so that the total stress in rod BD in Fig. 2-18 below will be not more than 18,000 lb, using a 4 ft length perpendicular to the paper and considering BD pinned at each end.
Ans. 5.87 ft

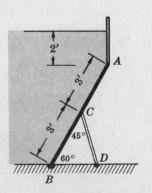

Fig. 2-17

23. Oil of specific gravity 0.800 acts on a vertical, triangular area whose apex is in the oil surface. The triangle is 9 ft high and 12 ft wide. A vertical, rectangular area 8 ft high is attached to the 12 ft base of the triangle and is acted upon by water. Find the magnitude and position of the resultant force on the entire area. *Ans.* 83,300 lb, 12.18 ft down

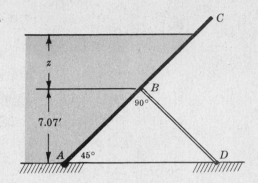

24. In Fig. 2-19 below, the gate *AB* is hinged at *B* and is 4 ft wide. What vertical force, applied at the center of gravity of the 4500 lb gate, will keep it in equilibrium? *Ans.* 12,120 lb

25. A tank is 20 ft long and of cross section shown in Fig. 2-20 below. Water is at level *AE*. Find (*a*) the total force acting on side *BC* and (*b*) the total force acting on end *ABCDE* in magnitude and position. *Ans.* 200,000 lb, 98,000 lb at 11.17 ft

Fig. 2-18

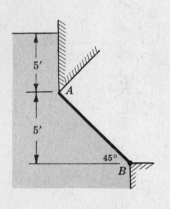

Fig. 2-19

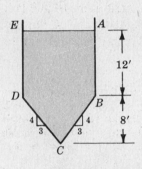

Fig. 2-20

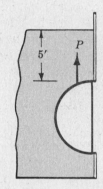

Fig. 2-21

26. In Fig. 2-21 above, the 4 ft diameter semi-cylindrical gate is 3 ft long. If the coefficient of friction between the gate and its guides is 0.100, find the force *P* required to raise the 1000 lb gate. *Ans.* 347 lb

27. A tank with vertical sides contains 3 ft of mercury and 16.5 ft of water. Find the total force on a square portion of one side, 2 ft by 2 ft in area, half of this area being below the surface of the mercury. The sides of the square are horizontal and vertical. *Ans.* 4910 lb, 16.63 ft down

28. An isosceles triangle, base 18 ft and altitude 24 ft, is immersed vertically in oil of specific gravity 0.800 with its axis of symmetry horizontal. If the head on the horizontal axis is 13 ft, determine the total force on one face of the triangle and locate the center of pressure vertically. *Ans.* 140,400 lb, 14.04 ft

29. How far below the water surface should a vertical square, 4 ft on a side with two sides horizontal, be immersed in order that the center of pressure be 3 in. below the center of gravity? What would be the total force on the square? *Ans.* 3.33 ft, 5330 lb

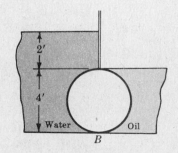

30. In Fig 2-22, the 4 ft diameter cylinder, 4 ft long, is acted upon by water on the left and oil of specific gravity 0.800 on the right. Determine (*a*) the normal force at *B* if the cylinder weighs 4000 lb and (*b*) the horizontal force due to oil and water if the oil level drops one foot. *Ans.* 1180 lb, 3100 lb to right

Fig. 2-22

31. In Fig. 2-23 below, for a length of 8 ft determine the unbalanced moment about the hinge *O* due to water at level *A*.   *Ans.* 18,000 ft lb clockwise

32. The tank, whose cross section is shown in Fig. 2-24 below, is 4 ft long and full of water under pressure. Find the components of the force required to keep the cylinder in position, neglecting the weight of the cylinder.   *Ans.* 3250 lb down, 4580 lb to left

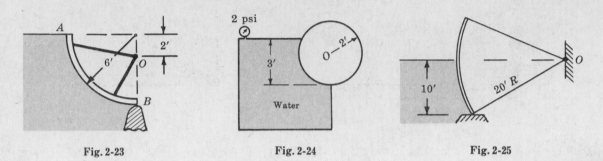

Fig. 2-23   Fig. 2-24   Fig. 2-25

33. Determine, per ft of length, the horizontal and vertical components of water pressure acting on the Tainter-type gate shown in Fig. 2-25 above.   *Ans.* 3120 and 1130 lb

34. Find the vertical force acting on the semi-cylindrical dome shown in Fig. 2-26 when gage *A* reads 8.45 psi. The dome is 6 ft long.   *Ans.* 25,400 lb

35. If the dome in Problem 34 is changed to a hemispherical dome of the same diameter, what is the vertical force acting?   *Ans.* 13,600 lb

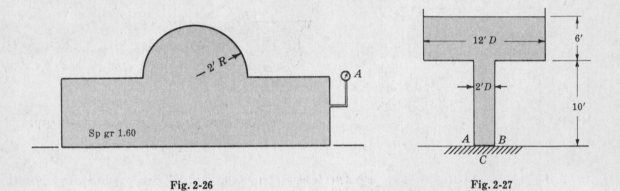

Fig. 2-26   Fig. 2-27

36. Referring to Fig. 2-27 above, determine (*a*) the force exerted by the water on the bottom plate *AB* of the 2 ft diameter riser pipe and (*b*) the total force on plane *C*.
Ans. 3140 lb,   44,400 lb

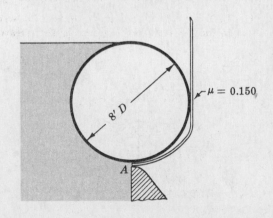

37. The cylinder shown in Fig. 2-28 is 10 ft long. Assuming a watertight condition at *A* and no rotation of the cylinder, what weight of cylinder is required for impending motion upward?   *Ans.* 12,700 lb

38. A wood stave pipe, 48 in. inside diameter, is bound by flat steel bands 4 in. wide and $\frac{3}{4}$ in. thick. For an allowable stress of 16,000 psi in the steel and an internal pressure of 160 psi, determine the spacing of the bands.   *Ans.* 12.5 in.

Fig. 2-28

39. For the parabolic sea wall shown in Fig. 2-29, what moment about A per ft of wall is created by the 10 ft depth of water ($w = 64.0$ lb/ft^3)?　　　*Ans.* 25,200 ft lb counterclockwise

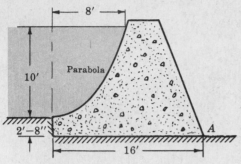

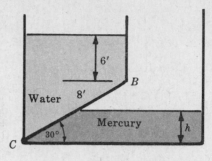

Fig. 2-29　　　　　　　　　　　　　　　　　Fig. 2-30

40. The tank shown in Fig. 2-30 is 10 ft long, and sloping bottom BC is 8' wide. What depth of mercury will cause the resultant moment about C due to the liquids to be 101,300 ft lb clockwise?　　　*Ans.* 2 ft

41. The gate shown in Fig. 2-31 is 20 ft long. What are the reactions at the hinge O due to the water? Check to see that the torque about O is zero.　　　*Ans.* 30,600 lb,　61,100 lb

42. Refer to Fig. 2-32. A flat plate hinged at C has a configuration satisfying the equation $x^2 + 1.5y = 9$. What is the force of the oil on the plate and what is the torque about hinge C due to the oil? *Ans.* 6240 lb, 16,400 ft lb

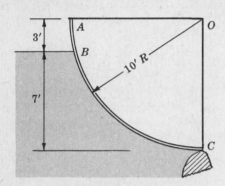

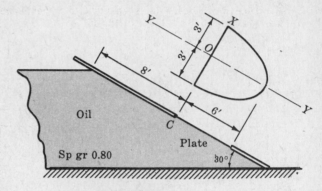

Fig. 2-31　　　　　　　　　　　　　　　　Fig. 2-32

43. In Fig. 2-33, the parabolic-shaped gate ABC is hinged at A and is acted upon by oil weighing 50 lb/ft^3. If the center of gravity of the gate is at B, what must the gate weigh per ft of length (perpendicular to the paper) in order that equilibrium exist? Vertex of parabola is at A.　　　*Ans.* 408 lb/ft

44. In Fig. 2-34, the automatic gate ABC weighs 1.50 tons/ft of length and its center of gravity is 6 ft to the right of the hinge A. Will the gate turn open due to the depth of water shown?　　　*Ans.* Yes

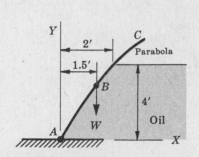

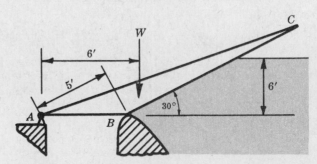

Fig. 2-33　　　　　　　　　　　　　　　　Fig. 2-34

Chapter 3

Buoyancy and Flotation

ARCHIMEDES' PRINCIPLE

Archimedes' principle has been used by man for about 2200 years. The volume of an irregular solid can be found by determining the apparent loss of weight when the body is wholly immersed in a liquid of known specific gravity. Specific gravity of liquids can be determined by means of the depth of flotation of the hydrometer. Further applications include problems of general flotation and of naval architectural design.

Any body, floating or immersed in a liquid, is acted upon by a *buoyant force* equal to the weight of the liquid displaced. The point through which this force acts is called the center of buoyancy. It is located at the center of gravity of the displaced liquid.

STABILITY of SUBMERGED and FLOATING BODIES

For stability of a submerged body the center of gravity of the body must lie directly below the center of buoyancy (gravity) of the displaced liquid. Should the two points coincide, the submerged body is in neutral equilibrium for all positions.

For stability of floating cylinders or spheres the center of gravity of the body must lie below the center of buoyancy.

Stability of other floating objects will depend upon whether a righting or overturning moment is developed when the center of gravity and center of buoyancy move out of vertical alignment due to the shifting of the position of the center of buoyancy. The center of buoyancy will shift because, if the floating object tips, the shape of the displaced liquid changes and hence its center of gravity shifts.

Solved Problems

1. A stone weighs 90 lb in air and when immersed in water it weighs 50 lb. Compute the volume of the stone and its specific gravity.

   Solution:

   All problems in engineering work can best be analyzed by the use of the free body diagram. Reference to the adjoining figure indicates the total weight of 90 lb acting downward, the tension in the cord attached to the scales of 50 lb upward, and the net buoyant force P_V acting upward. From

   $$\Sigma Y = 0$$

   we have $\qquad 90 - 50 - P_V = 0, \quad$ and $\quad P_V = 40 \text{ lb}$

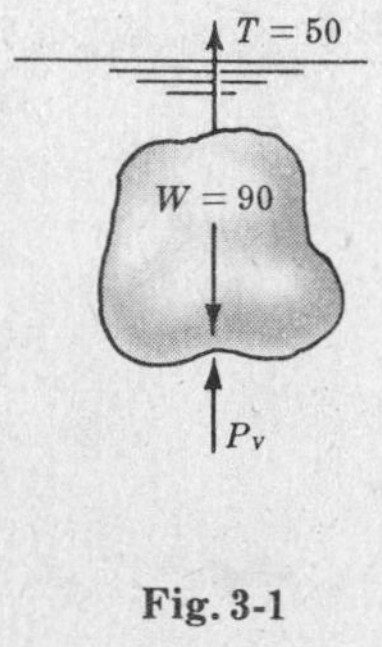

$T = 50$

$W = 90$

P_V

Fig. 3-1

Since the buoyant force = the weight of the displaced fluid,
$$40 \text{ lb} = 62.4 \text{ lb/ft}^3 \times v \qquad \text{and} \qquad v = 0.641 \text{ ft}^3$$

Specific gravity $= \dfrac{\text{weight of the stone}}{\text{weight of an equal volume of water}} = \dfrac{90 \text{ lb}}{40 \text{ lb}} = 2.25.$

2. A prismatic object 8″ thick by 8″ wide by 16″ long is weighed in water at a depth of 20 in. and found to weigh 11.0 lb. What is its weight in air and its specific gravity?

   Solution:

   Referring to the free body diagram in Fig. 3-2, $\Sigma Y = 0$; then
   $$W - \text{net } P_v - 11.0 = 0 \quad \text{or} \quad (1) \ \ W = 11.0 + \text{net } P_v$$

   Buoyant force P_v = weight of displaced liquid
   $$= 62.4(8 \times 8 \times 16)/1728 = 37.0 \text{ lb.}$$

   Therefore from (1), $W = 11 + 37 = 48 \text{ lb}$ and $\text{sp gr} = 48/37 = 1.30.$

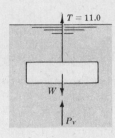

Fig. 3-2

3. A hydrometer weighs 0.00485 lb and has a stem at the upper end which is cylindrical and 0.1100 in. in diameter. How much deeper will it float in oil of sp gr 0.780 than in alcohol of sp gr 0.821?

   Solution:

   For position 1 in Fig. 3-3 in the alcohol,
   $$\text{weight of hydrometer} = \text{weight of displaced liquid}$$
   $$0.00485 = 0.821 \times 62.4 \times v_1$$
   from which $v_1 = 0.0000947 \text{ ft}^3$ (in alcohol).

   For position 2,
   $$0.00485 = 0.780 \times 62.4(v_1 + Ah)$$
   $$= 0.780 \times 62.4[0.0000947 + \tfrac{1}{4}\pi(0.1100/12)^2 h]$$
   from which $h = .0751 \text{ ft} = 0.901 \text{ in.}$

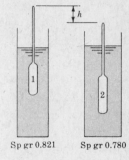

Sp gr 0.821   Sp gr 0.780

Fig. 3-3

4. A piece of wood of sp gr 0.651 is 3 in. square and 5 ft long. How many pounds of lead weighing 700 lb/ft³ must be fastened at one end of the stick so that it will float upright with 1 ft out of water?

   Solution:
   $$\text{Total weight of wood and lead} = \text{weight of displaced water}$$
   $$[0.651 \times 62.4 \times 5(3/12)^2 + 700v] = 62.4[(3/12)^2 \times 4 + v]$$
   from which $v = 0.00455 \text{ ft}^3$ and weight of lead $= 700v = 700 \times 0.00455 = 3.18 \text{ lb.}$

5. What fraction of the volume of a solid piece of metal of sp gr 7.25 floats above the surface of a container of mercury of sp gr 13.57?

   Solution:

   The free body diagram indicates that, from $\Sigma Y = 0$, $W - P_v = 0$ or
   $$\text{weight of body} = \text{buoyant force (weight of displaced mercury)}$$
   $$7.25 \times 62.4 v = 13.57 \times 62.4 v'$$
   and the ratio of the volumes is thus $v'/v = 7.25/13.57 = 0.535.$

   Hence the fraction of the volume above the mercury $= 1 - 0.535 = 0.465.$

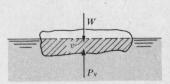

Fig. 3-4

6. A rectangular open box, 25 ft by 10 ft in plan and 12 ft deep, weighs 40 tons and is launched in fresh water. (a) How deep will it sink? (b) If the water is 12 ft deep, what weight of stone placed in the box will cause it to rest on the bottom?
    Solution:
    (a)   Weight of box = weight of displaced water
    $$40 \times 2000 = 62.4(25 \times 10 \times Y) \qquad\qquad Y = 5.12 \text{ ft submerged}$$

    (b)   Weight of box plus stone = weight of displaced water
    $$40 \times 2000 + W_s = 62.4(25 \times 10 \times 12) \qquad\qquad W_s = 107{,}200 \text{ lb stone}$$

7. A block of wood floats in water with 2 in. projecting above the water surface. When placed in glycerine of sp gr 1.35 the block projects 3 in. above the surface of that liquid. Determine the sp gr of the wood.
    Solution:

    Total weight of block is (a) $W = \text{sp gr} \times 62.4(A \times h)$, and weights of displaced water and glycerine respectively are (b) $W_w = 62.4A(h-2)/12$ and (c) $W_G = 1.35 \times 62.4A(h-3)/12$.

    Since the weight of each displaced liquid equals the total weight of the block, $(b) = (c)$ or

    $$62.4A(h-2)/12 = 1.35 \times 62.4A(h-3)/12 \qquad\qquad h = 5.86 \text{ in.}$$

    Since $(a) = (b)$, $\text{sp gr} \times 62.4A \times 5.86/12 = 62.4 \times A(5.86-2)/12 \qquad \text{sp gr} = 0.658$

8. To what depth will an 8 ft diameter log 15 ft long and of sp gr 0.425 sink in fresh water?
    Solution:
    Fig. 3-5 is drawn with center O of log above the water surface because the sp gr is less than 0.500. Had the sp gr been 0.500, then the log would be half submerged.

    Total weight of log = weight of displaced liquid

    sector − 2 triangles

    $$0.425 \times 62.4 \times \pi 4^2 \times 15 = 62.4 \times 15\left(\frac{2\theta}{360}16\pi - 2 \times \tfrac{1}{2} \times 4 \sin\theta \times 4 \cos\theta\right)$$

    Simplifying and substituting $\tfrac{1}{2}\sin 2\theta$ for $\sin\theta\cos\theta$,

    $$0.425\pi = \theta\pi/180 - \tfrac{1}{2}\sin 2\theta$$

    Solving by successive trials:

    Try $\theta = 85°$: $1.335 \overset{?}{=} 85\pi/180 - \tfrac{1}{2}(0.1737)$

    $$1.335 \neq 1.397$$

    Try $\theta = 83°$: $1.335 \overset{?}{=} 1.449 - \tfrac{1}{2}(0.242)$

    $$1.335 \neq 1.328$$

    The trial values have straddled the answer.

    Try $\theta = 83°10'$: $1.335 \overset{?}{=} 1.451 - \tfrac{1}{2}(0.236)$

    $$= 1.333 \text{ (close check)}.$$

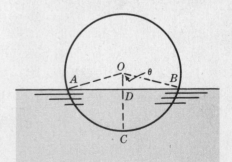

Fig. 3-5

    The depth of flotation $DC = r - OD = 4.00 - 4.00 \cos 83°10'$
    $$= 4.00(1 - 0.119) = 3.52 \text{ ft.}$$

9. (a) Neglecting the thickness of the tank walls in Fig. 3-6(a), if the tank floats in the position shown what is its weight? (b) If the tank is held so that the top is 10 ft below the surface of the water, what is the force on the inside top of the tank?

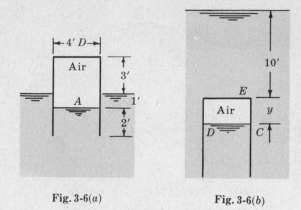

Fig. 3-6(a)   Fig. 3-6(b)

Solution:

(a) Weight of tank = weight of displaced liquid
$$= 62.4\pi 2^2(1) = 783 \text{ lb.}$$

(b) The space occupied by the air will be less at the new depth shown in Fig. 3-6(b). Assuming that the temperature of the air is constant, then for positions (a) and (b),

$$p_A v_A = p_D v_D \quad \text{(absolute pressure units must be used)}$$
$$w(34+1)(4 \times \text{area}) = w(34+10+y)(y \times \text{area})$$

which yields $y^2 + 44y - 140 = 0$ whose required positive root is $y = 3$ ft.

The pressure at $D = 13$ ft of water gage = the pressure at E. Hence the force on the inside top of the cylinder is $w h A = 62.4(13)(\pi 2^2) = 10,200$ lb.

10. A ship, with vertical sides near the water line, weighs 4000 tons and draws 22 ft in salt water $(w = 64.0 \text{ lb/ft}^3)$. Discharge of 200 tons of water ballast decreases the draft to 21 ft. What would be the draft d of the ship in fresh water?

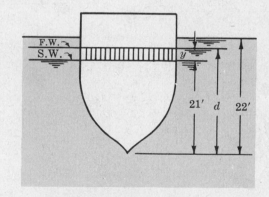

Fig. 3-7

Solution:

Because the shape of the underwater section of the ship is not known, it is best to solve the problem on the basis of volumes displaced.

A 1 ft decrease in draft was caused by a reduction in weight of 200 tons or

$$200 \times 2000 = wv = 64.0(A \times 1)$$

where v represents the volume between drafts 22 ft and 21 ft, and $(A \times 1)$ represents the water line area $\times$ 1 ft, or the same volume v. Then

$$v = A \times 1 = 200(2000)/64.0 = 6250 \text{ ft}^3/\text{ft depth}$$

Buoyant force $B = w \times$ volume of displaced liquid. Then $B/w =$ volume of displaced liquid.

From the figure, the vertically cross-hatched volume is the difference in displaced fresh water and salt water. This difference can be expressed as $\left(\dfrac{3800 \times 2000}{62.4} - \dfrac{3800 \times 2000}{64.0}\right)$ and this volume is also equal to $6250y$. Equating these values, $y = 0.486$ ft.

The draft $d = 21 + 0.49 = 21.49$ ft or 21.5 ft.

11. A barrel containing water weighs 283.5 lb. What will be the reading on the scales if a 2″ by 2″ piece of wood is held vertically in the water to a depth of 2 ft?

Solution:

For every acting force there must be an equal and opposite reacting force. The buoyant force exerted by the water upward against the bottom of the piece of wood is opposed by the $2'' \times 2''$ area of wood acting downward on the water with equal magnitude. This force will measure the increase in scale reading.

$P_V = 62.4 \times 2/12 \times 2/12 \times 2 = 3.47$ lb. The scale reading $= 283.5 + 3.5 = 287.0$ lb.

12. A block of wood 6 ft by 8 ft by 10 ft floats on oil of sp gr 0.751. A clockwise couple holds the block in the position shown in Fig. 3-8. Determine (*a*) the buoyant force acting on the block and its position, (*b*) the magnitude of the couple acting on the block, and (*c*) the location of the metacenter for the tilted position.

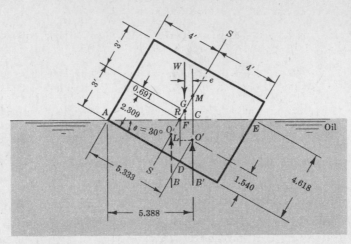

Fig. 3-8

Solution:

(*a*) Weight of block = weight of triangular prism of oil (or the buoyant force)

$$W = B' = (0.751 \times 62.4)(\tfrac{1}{2} \times 8 \times 4.618 \times 10) = 8650 \text{ lb}$$

Then $B' = 8650$ lb acting upward through the center of gravity O' of the displaced oil. The center of gravity lies 5.33 ft from A and 1.540 ft from D, as shown in the figure.

$$AC = AR + RC = AR + LO' = 5.333 \cos 30° + 1.540 \sin 30° = 5.388 \text{ ft}$$

The buoyant force of 8650 lb acts upward through the center of gravity of the displaced oil, which is 5.39 ft to the right of A.

(*b*) One method of obtaining the magnitude of the righting couple (which must equal the magnitude of the external couple for equilibrium) is to find the eccentricity e. This dimension is the distance between the two parallel, equal forces W and B' which form the righting couple.

$$e = FC = AC - AF = 5.388 - AF = 5.388 - 4.963 = 0.425 \text{ ft}$$

since $\qquad AF = AR + RF = AR + GR \sin 30° = 4.618 + 0.691(\tfrac{1}{2}) = 4.963 \text{ ft}$

The couple We or $B'e = 8650 \times 0.425 = 3670$ ft lb. Thus the moment or couple to hold the block in the position shown is 3670 ft lb clockwise.

(*c*) The point of intersection of the buoyant force and the axis of symmetry S-S is called the metacenter (point M in the figure). If the metacenter is located above the center of gravity of a floating object, the weight of the object and the buoyant force form a righting moment in tilted positions.

The metacentric distance $MG = MR - GR = \dfrac{RC}{\sin 30°} - GR = \dfrac{0.770}{\tfrac{1}{2}} - 0.691 = 0.849 \text{ ft.}$

It should be noted that the distance MG multiplied by the sine of angle θ is equal to the eccentricity e (previously calculated by another approach).

In naval architecture, an extreme angle of some 10° is taken as the limit of heel for which the metacentric distance MG can be considered constant. Formulas for locating the metacenter can be derived but such studies are beyond the scope of introductory work in fluid mechanics.

Supplementary Problems

13. An object weighs 65 lb in air and 42 lb in water. Find its volume and specific gravity.
    Ans. 0.369 ft^3, 2.83

14. An object weighs 65 lb in air and 42 lb in oil of sp gr 0.75. Find its volume and specific gravity.
    Ans. 0.491 ft^3, 2.12

15. If aluminum weighs 165 lb/ft^3, how much will a 12 in. diameter sphere weigh when immersed in water? When immersed in oil of sp gr 0.75?   *Ans.* 53.6 lb, 62.0 lb

16. A 6 in. cube of aluminum weighs 12.2 lb when immersed in water. What will be its apparent weight when immersed in a liquid of sp gr 1.25?   *Ans.* 10.25 lb

17. A stone weighs 135 lb and, when it was lowered into a square tank 2 ft on a side, the weight of the stone in the water was 72.6 lb. How much did the water rise in the tank?   *Ans.* 3 in.

18. A hollow cylinder 3 ft in diameter and 5 ft long weighs 860 lb. (a) How many pounds of lead weighing 700 lb/ft^3 must be fastened to the outside bottom to make the cylinder float vertically with 3 ft submerged? (b) How many pounds if placed inside the cylinder?   *Ans.* 510 lb, 465 lb

19. A hydrometer weighs 0.0250 lb and its stem is 0.0250 in^2 in cross sectional area. What is the difference in depth of flotation for liquids of sp gr 1.25 and 0.90?   *Ans.* 8.62 in.

20. What length of 3 in. by 12 in. timber, sp gr 0.50, will support a 100 lb boy in salt water if he stands on the timber?   *Ans.* 12.2 ft

21. An object which has a volume of 6 ft^3 requires a force of 60 lb to keep it immersed in water. If a force of 36 lb is required to keep it immersed in another liquid, what is the sp gr of that liquid?
    Ans. 0.937

22. A barge 10 ft deep has a trapezoidal cross section of 30 ft top width and 20 ft bottom width. The barge is 50 ft long and its ends are vertical. Determine (a) its weight if it draws 6 ft of water and (b) the draft if 84.5 tons of stone are placed in the barge.   *Ans.* 431,000 lb, 8 ft

23. A 4 ft diameter sphere floats half submerged in salt water ($w = 64.0$ lb/ft^3). What minimum weight of concrete ($w = 150$ lb/ft^3) used as an anchor will submerge the sphere completely?   *Ans.* 1875 lb

24. An iceberg weighing 57 lb/ft^3 floats in the ocean (64 lb/ft^3) with a volume of 21,000 ft^3 above the surface. What is the total volume of the iceberg?   *Ans.* 192,000 ft^3

25. An empty balloon and its equipment weigh 100 lb. When inflated with gas weighing 0.0345 lb/ft^3 the balloon is spherical and 20 ft in diameter. What is the maximum weight of cargo that the balloon can lift, assuming air to weigh 0.0765 lb/ft^3?   *Ans.* 76 lb

26. A cubical float, 4 ft on a side, weighs 400 lb and is anchored by means of a concrete block which weighs 1500 lb in air. Nine inches of the float are submerged when the chain connected to the concrete is taut. What rise in the water level will lift the concrete off the bottom? Concrete weighs 150 lb per cu ft.   *Ans.* 6.32 in.

27. A rectangular barge with outside dimensions of 20 ft width, 60 ft length and 10 ft height, weighs 350,000 lb. It floats in salt water ($w = 64.0$ lb/ft^3) and the center of gravity of the loaded barge is 4.50 ft from the top. (a) Locate the center of buoyancy when floating on an even keel and (b) when the barge lists at 10°, and (c) locate the metacenter for the 10° list.
    Ans. 2.28 ft from bottom on centerline, 11.28 ft to right, 4.17 ft above *CG*

28. A cube, 6″ on a side, is made of aluminum and is suspended by a string. The cube is submerged, half of it being in oil (sp gr $= 0.80$) and the other half being in water. Find the tension in the string if aluminum weighs 165 lb/ft^3.   *Ans.* 13.61 lb

29. If the cube in the preceding problem were half in air and half in oil, what would be the tension in the string?   *Ans.* 17.51 lb

Chapter 4

Translation and Rotation of Liquid Masses

INTRODUCTION

A fluid may be subjected to translation or rotation at constant accelerations without relative motion between particles. This condition is one of relative equilibrium and the fluid is free from shear. There is generally no motion between the fluid and the containing vessel. Laws of fluid statics still apply, modified to allow for the effects of acceleration.

HORIZONTAL MOTION

For horizontal motion, the surface of the liquid will become an inclined plane. The slope of the plane will be determined by

$$\tan \theta = \frac{a \text{ (linear acceleration of vessel, ft/sec}^2)}{g \text{ (gravitational acceleration, ft/sec}^2)}$$

Proof of the general equation for translation is given in Problem 4.

VERTICAL MOTION

For vertical motion, the pressure (psf) at any point in the liquid is given by

$$p = wh\left(1 \pm \frac{a}{g}\right)$$

where the positive sign is used with a constant upward acceleration and the negative sign with a constant downward acceleration.

ROTATION of FLUID MASSES — OPEN VESSELS

The form of the free surface of the liquid in a rotating vessel is that of a paraboloid of revolution. Any vertical plane through the axis of rotation which cuts the fluid will produce a parabola. The equation of the parabola is

$$y = \frac{\omega^2}{2g}x^2$$

where x and y are coordinates, in feet, of any point in the surface measured from the vertex in the axis of revolution and ω is the constant angular velocity in rad/sec. Proof of this equation is given in Problem 7.

ROTATION of FLUID MASSES — CLOSED VESSELS

The pressure in a closed vessel will be increased by rotating the vessel. (Also see

Chapter 12.)   The pressure increase between a point in the axis of rotation and a point x feet away from the axis is

$$p \text{ (psf)} = w \frac{\omega^2}{2g} x^2$$

or the increase in pressure head (ft) is

$$\frac{p}{w} = y = \frac{\omega^2}{2g} x^2$$

which equation is similar to the equation for rotating open vessels.   Since the linear velocity $V = x\omega$, the term $x^2\omega^2/2g = V^2/2g$ which we will later recognize as the velocity head, in ft units.

Solved Problems

1.   A rectangular tank 20 ft long by 6 ft deep by 7 ft wide, contains 3 ft of water.   If the linear acceleration horizontally in the direction of the tank's length is 8.05 ft/sec^2, (a) compute the total force due to the water acting on each end of the tank and (b) show that the difference between these forces equals the unbalanced force necessary to accelerate the liquid mass.   Refer to Fig. 4-1 below.

Solution:

(a) $\tan \theta = \dfrac{\text{linear acceleration}}{\text{gravitational acceleration}} = \dfrac{8.05}{32.2} = 0.250$ and $\theta = 14°2'$

From the figure, the depth d at the shallow end is $d = 3 - y = 3 - 10 \tan 14°2' = 0.500 \text{ ft}$, and the depth at the deep end is 5.50 ft.   Then

$$P_{AB} = wh_{cg}A = 62.4(5.50/2)(5.50 \times 7) = 6600 \text{ lb}$$
$$P_{CD} = wh_{cg}A = 62.4(0.500/2)(0.500 \times 7) = 54.6 \text{ lb}$$

(b)   Force needed $= \text{mass of water} \times \text{linear acceleration} = \dfrac{20 \times 7 \times 3 \times 62.4}{32.2} \times 8.05 = 6550 \text{ lb}$,

and $P_{AB} - P_{CD} = 6600 - 55 = 6545 \text{ lb}$,   which checks within slide rule accuracy.

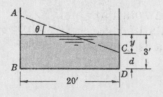

Fig. 4-1

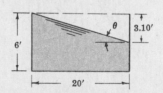

Fig. 4-2

2.   If the tank in Problem 1 is filled with water and accelerated in the direction of its length at the rate 5.00 ft/sec^2, how many gallons of water are spilled?   Refer to Fig. 4-2 above.

Solution:

Slope of surface $= \tan \theta = 5.00/32.2 = 0.155$,   and   drop in surface $= 20 \tan \theta = 3.10$ ft.

Volume spilled $= 7 \times$ triangular cross section shown in Fig. 4-2

$= 7(\frac{1}{2} \times 20.0 \times 3.10) = 213 \text{ ft}^3 = 213 \text{ ft}^3 \times 7.48 \text{ gal/ft}^3 = 1625 \text{ gal.}$

3. A tank is 5 ft square and contains 3 ft of water. How high must its sides be if no water is to be spilled when the acceleration is 12.0 ft/sec² parallel to a pair of sides?

Solution:

Slope of surface $= \tan\theta = 12.0/32.2 = 0.373$.

Rise (or fall) in surface $= 2.50 \tan\theta = 2.50 \times 0.373 = 0.932$ ft.

The tank must be at least $3 + 0.93 = 3.93$ ft deep.

4. An open vessel of water accelerates up a 30° plane at 12.0 ft/sec². What is the angle the water surface makes with the horizontal?

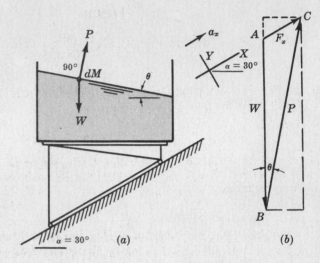

Solution:

Referring to the figure, the forces acting on each mass dM are weight W vertically downward and the force P exerted by the surrounding particles of liquid. This force P is normal to the liquid surface since no frictional component is acting. The resultant force F_x (due to W and P) for each particle of liquid must be up the plane XX at an angle of $\alpha = 30°$ with the horizontal and must cause the common acceleration a_x. Fig. (b) shows this vector relationship. The following equations may now be established:

Fig. 4-3

$$(1) \quad F_x = \frac{W}{g}a_x \quad \text{or} \quad \frac{F_x}{W} = \frac{a_x}{g}$$

$$(2) \quad F_x \sin\alpha = P\cos\theta - W$$

$$(3) \quad F_x \cos\alpha = P\sin\theta \qquad \text{from the vector diagram}$$

Multiplying (2) by $\sin\theta$ and (3) by $\cos\theta$ and solving simultaneously,

$$F_x \sin\alpha\sin\theta + W\sin\theta - F_x\cos\alpha\cos\theta = 0 \quad \text{and} \quad \frac{F_x}{W} = \frac{\sin\theta}{\cos\alpha\cos\theta - \sin\alpha\sin\theta}$$

Substituting in (1) and simplifying,

$$(4) \quad \frac{a_x}{g} = \frac{1}{\cos\alpha\cot\theta - \sin\alpha} \quad \text{from which, since } \alpha = 30°$$

$$(A) \quad \cot\theta = \tan 30° + \frac{g}{a_x\cos 30°} = 0.577 + \frac{32.2}{12 \times 0.866} = 3.68 \quad \text{and} \quad \theta = 15°12'$$

Note: For a horizontal plane, angle α becomes 0° and equation (4) above becomes $a/g = \tan\theta$, the equation given for horizontally accelerated motion. For acceleration down the plane, the sign in front of $\tan 30°$ becomes minus in equation (A).

5. A cubical tank is filled with 5 ft of oil, sp gr 0.752. Find the force acting on the side of the tank (a) when the acceleration is 16.1 ft/sec² vertically upward and (b) when the acceleration is 16.1 ft/sec² vertically downward.

Solution:

(a) Fig. 4-4 below, shows the distribution of loading on a vertical side AB. At B the intensity of pressure in psf is

$$p_B = wh\left(1 + \frac{a}{g}\right) = 0.752 \times 62.4(5)\left(1 + \frac{16.1}{32.2}\right) = 352 \text{ psf}$$

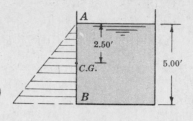

$$\text{Force } P_{AB} = \text{area of loading diagram} \times 5 \text{ ft long}$$
$$= (\tfrac{1}{2} \times 352 \times 5)(5) = 4400 \text{ lb}$$

An alternate solution is

$$P_{AB} = wh_{cg}A = p_{cg}A = \left[0.752 \times 62.4(2.5)\left(1 + \frac{16.1}{32.2}\right)\right](5 \times 5)$$
$$= 4400 \text{ lb}$$

(b) $$P_{AB} = \left[0.752 \times 62.4(2.5)\left(1 - \frac{16.1}{32.2}\right)\right](5 \times 5) = 1465 \text{ lb}$$

Fig. 4-4

6. **Determine the pressure at the bottom of the tank in Problem 5 when the acceleration is 32.2 ft/sec² vertically downward.**

   Solution:

   $$p_B = 0.752 \times 62.4(5)(1 - 32.2/32.2) = 0 \text{ psf}$$

   Hence, for a liquid mass falling freely, the pressure within the mass at any point is zero, i.e., that of the surrounding atmosphere. This conclusion is important in considering a stream of water falling through space.

7. **An open vessel partly filled with a liquid rotates about a vertical axis at constant angular velocity. Determine the equation of the free surface of the liquid after it has acquired the same angular velocity as the vessel.**

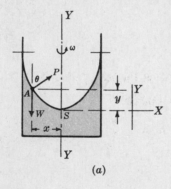

(a)   (b)

Fig. 4-5

   Solution:

   Fig. (a) represents a section through the rotating vessel, and any particle A is at a distance x from the axis of rotation. Forces acting on mass A are the weight W vertically downward and P which is normal to the surface of the liquid since no friction is acting. The acceleration of mass A is $x\omega^2$, directed toward the axis of rotation. The direction of the resultant of forces W and P must be in the direction of this acceleration, as shown in Fig. (b).

   From Newton's second law, $F_x = Ma_x$ or   (1) $P \sin\theta = \dfrac{W}{g}x\omega^2$

   From $\Sigma Y = 0$ (2) $P \cos\theta = W$

   Dividing (1) by (2),   (3) $\tan\theta = \dfrac{x\omega^2}{g}$

   Now θ is also the angle between the X-axis and a tangent drawn to the curve at A in Fig. (a). The slope of this tangent is $\tan\theta$ or dy/dx. Substituting in (3) above,

   $$\frac{dy}{dx} = \frac{x\omega^2}{g} \qquad \text{from which, by integration,} \qquad y = \frac{\omega^2}{2g}x^2 + C_1$$

   To evaluate the constant of integration, C_1: When $x = 0$, $y = 0$ and $C_1 = 0$.

8.   An open cylindrical tank, 6 ft high and 3 ft in diameter, contains 4.50 ft of water. If
the cylinder rotates about its geometric axis, (a) what constant angular velocity can
be attained without spilling any water? (b) What is the pressure at the bottom of the
tank at C and D when $\omega = 6.00$ rad/sec?

Solution:

(a)   Volume of paraboloid of revolution $= \frac{1}{2}$(volume circumscribed cylinder) $= \frac{1}{2}[\frac{1}{4}\pi 3^2(1.50 + y_1)]$

If no liquid is spilled, this volume equals the
volume above the original water level A-A, or

$$\frac{1}{2}[\frac{1}{4}\pi 3^2(1.50 + y_1)] = \frac{1}{4}\pi 3^2(1.50)$$

and $y_1 = 1.50$ ft.

To generalize, the point in the axis of rotation
drops by an amount equal to the rise of the liquid
at the walls of the vessel.

From this information, the x and y coordinates of points B are respectively 1.50 and 3.00 ft
from origin S. Then

$$y = \frac{\omega^2}{2g}x^2$$

$$3.00 = \frac{\omega^2}{2 \times 32.2}(1.50)^2$$

and $\omega = 9.26$ rad/sec.

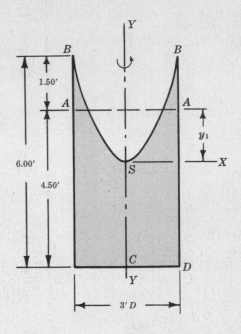

(b)   For $\omega = 6.00$ rad/sec,

$$y = \frac{\omega^2}{2g}x^2 = \frac{(6.00)^2}{2(32.2)}(1.50)^2 = 1.26 \text{ ft from } S$$

Origin S drops $\frac{1}{2}y = 0.63$ ft and S is now
$4.50 - 0.63 = 3.87$ ft from the bottom of the tank.
At the walls of the tank the depth $= 3.87 + 1.26 =$
5.13 ft (or $4.50 + 0.63 = 5.13$ ft).

At C, $p_c = wh = 62.4 \times 3.87 = 242$ psf.
At D, $p_D = wh = 62.4 \times 5.13 = 320$ psf.

Fig. 4-6

9.   Consider the tank in Problem 8 closed with the
air space subjected to a pressure of 15.5 psi.
When the angular velocity is 12.0 rad/sec, what
are the pressures in psi at points C and D in
the Fig. 4-7?

Solution:

Since there is no change in the volume of air within
the vessel,

volume above level A-A = volume of paraboloid

or   (1) $\frac{1}{4}\pi 3^2 \times 1.50 = \frac{1}{2}\pi x_2^2 y_2$

Also   (2) $y_2 = \frac{(12.0)^2}{2(32.2)}x_2^2$

Solving (1) and (2) simultaneously, $x_2^4 = 3.02$. Then $x_2 =$
1.32 ft and $y_2 = 3.89$ ft.

From the figure, S is located $6.00 - 3.89 = 2.11$ ft
above C. Then

$p_c' = 15.5 + wh/144 = 15.5 + 62.4(2.11)/144 = 16.4$ psi

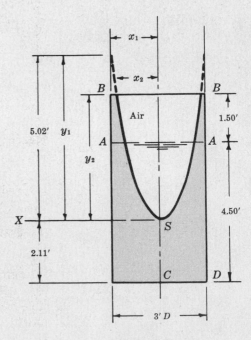

Fig. 4-7

To evaluate the pressure at D, pressure head $y_1 = \dfrac{(12.0)^2}{2 \times 32.2}(1.50)^2 = 5.02$ ft above S and

$$p_D' = 62.4(5.02 + 2.11)/144 + 15.5 = 18.6 \text{ psi}$$

10. (a) At what speed must the tank in Problem 9 be rotated in order that the center of the bottom have zero depth of water? (b) If the bottom circumferential plate is $\frac{1}{4}$ in. thick, what is the stress therein?

Solution:

(a) Origin S will now be at point C in Fig. 4-7.

Volume above liquid surface = volume of paraboloid

or

(1) $\frac{1}{4}\pi 3^2 \times 1.50 = \frac{1}{2}\pi x_2^2 (6.00)$

Also

(2) $y_2 = 6.00 = \dfrac{\omega^2}{2 \times 32.2} x_2^2$

From (1) and (2) we obtain $\omega^2 = 12(32.2)/1.125 = 343$ and $\omega = 18.6$ rad/sec.

(b) $p_D' = 15.5 + \dfrac{wh}{144}$,   where $h = y_1 = \dfrac{(18.6)^2(1.50)^2}{2 \times 32.2} = 12.1$ ft,

$= 15.5 + \dfrac{62.4 \times 12.1}{144} = 20.7$ psi.   Stress at $D = \sigma_D = \dfrac{p'r}{t} = \dfrac{20.7 \times 18}{\frac{1}{4}} = 1490$ psi.

11. A closed cylindrical tank 6 ft high and 3 ft in diameter contains 4.50 ft of water. When the angular velocity is constant at 20.0 rad/sec, how much of the bottom of the tank is uncovered?

Solution:

In order to estimate the parabolic curve to be drawn in the adjoining figure, the value of y_3 was calculated first. Now

$$y_3 = \frac{(20)^2}{2 \times 32.2}(1.50)^2 = 14.0 \text{ ft}$$

and the curved surface of the water can now be sketched, showing S below the bottom of the tank. Next

(1) $y_1 = \dfrac{(20)^2}{2 \times 32.2} x_1^2$

(2) $y_2 = 6 + y_1 = \dfrac{(20)^2}{2 \times 32.2} x_2^2$ and, since the volume of the air is constant,

(3) $\frac{1}{4}\pi 3^2 \times 1.50 =$ volume (paraboloid SAB − parab. SCD)
      $= \frac{1}{2}\pi x_2^2 y_2 - \frac{1}{2}\pi x_1^2 y_1$.

Substituting values from (1) and (2) and solving,

$x_1^2 = 0.0792$ and $x_1 = 0.282$ ft.

Hence the area uncovered $= \pi(0.282)^2 = 0.249$ ft².

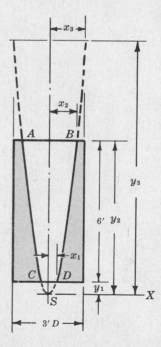

Fig. 4-8

12. A 6 ft diameter cylinder 9 ft high is completely filled with glycerin, sp gr 1.60, under a pressure of 35.2 psi at the top. The steel plates which form the cylinder are $\frac{1}{2}$ in. thick and can withstand an allowable unit stress of 12,000 psi. What maximum speed in rpm can be imposed on the cylinder?

Solution:

From the specifications for the tank and the hoop tension formula $\sigma = p'r/t$,

$$p'_A = \sigma t/r = 12{,}000(\tfrac{1}{2})/36 = 167 \text{ psi}$$

Also, $p'_A = \Sigma$ pressures (35.2 imposed + due to 9 ft glycerin + due to rotation)

or
$$167 = 35.2 + \frac{1.60 \times 62.4 \times 9}{144} + \frac{\omega^2}{2 \times 32.2} \times 3^2 \times \frac{1.60 \times 62.4}{144} \quad \text{psi}$$

Solving, $\omega = 36.0$ rad/sec or 344 rpm.

The pressure conditions are shown graphically in Fig. 4-9 below, not to scale. Line RST indicates a pressure head of 50.8 ft of glycerin above the top of the tank before rotation. The parabolic pressure curve with vertex at S is caused by the constant angular velocity of 36.0 rad/sec. Had the vessel been full but not under pressure, vertex S would coincide with the inside top of the vessel.

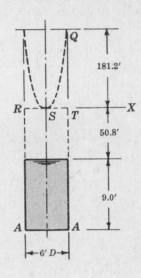

Fig. 4-9

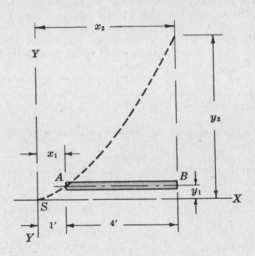

Fig. 4-10

13. A 3 in. diameter pipe, 4 ft long is just filled with oil of sp gr 0.822 and then capped. Placed in a horizontal position, it is rotated at 27.5 rad/sec about a vertical axis 12 in. from one end. What pressure in psi is developed at the far end of the pipe?

Solution:

As previously noted, the pressure throughout length AB in Fig. 4-10 above will be increased by rotation. At some speed of rotation the increased pressure would tend to compress the element of liquid and cause the pressure at A to decrease. Since liquids are practically incompressible, rotation will not lower the pressure at A nor will it increase the pressure at A. Between A and B the pressure will increase as the square of the distance from axis YY.

To evaluate the pressure at B:

$$(1)\quad y_1 = \frac{(27.5)^2}{2g} \times 1^2 = 11.8 \text{ ft} \qquad (2)\quad y_2 = \frac{(27.5)^2}{2g} \times 5^2 = 293 \text{ ft}$$

and
$$p'_B = 0.822(62.4)(293 - 11.8)/144 = 100 \text{ psi}$$

Supplementary Problems

14. A vessel partly filled with water is accelerated horizontally at a constant rate. The inclination of the water surface is 30°. What is the acceleration of the vessel? *Ans.* 18.6 ft/sec²

15. An open tank is 6 ft square, weighs 770 lb and contains 3 ft of water. It is acted upon by an unbalanced force of 2330 lb parallel to a pair of sides. What must be the height of the sides of the tank so that no water will be spilled? What is the force acting on the side where the greatest depth occurs? *Ans.* 3.93 ft, 2890 lb

16. An open tank 30 ft long by 4 ft wide by 4 ft deep is filled with 3.25 ft of oil, sp gr 0.822. It is accelerated uniformly from rest to 45.0 ft/sec. What is the shortest time in which the tank may be accelerated without spilling any oil? *Ans.* 28.0 sec

17. When an open rectangular tank, 5 ft wide, 10 ft long and 6 ft deep, containing 4 ft of water is accelerated horizontally parallel to its length at the rate 16.1 ft/sec², how much water is spilled? *Ans.* 25.0 ft³

18. At what acceleration must the tank in the preceding problem move in order that the depth at the forward edge be zero? *Ans.* 19.3 ft/sec²

19. An open tank of water accelerates down a 15° inclined plane at 16.1 ft/sec². What is the slope of the water surface? *Ans.* 29°1′

20. A vessel containing oil of sp gr 0.762 moves vertically upward with an acceleration of +8.05 ft/sec². What is the pressure at a depth of 6 ft? *Ans.* 357 psf

21. If the acceleration in Problem 20 is −8.05 ft/sec², what is the pressure at a depth of 6 ft? *Ans.* 214 psf

22. An unbalanced vertical force of 60.0 lb upward accelerates a volume of 1.55 ft³ of water. If the water is 3 ft deep in a cylindrical tank, what is the force acting on the bottom of the tank? *Ans.* 157 lb

23. An open cylindrical tank 4 ft in diameter and 6 ft deep is filled with water and rotated about its axis at 60 rpm. How much liquid is spilled and how deep is the water at the axis? *Ans.* 15.3 ft³, 3.55 ft

24. At what speed should the tank in Problem 23 be rotated in order that the center of the bottom of the tank have zero depth of water? *Ans.* 9.83 rad/sec

25. A closed vessel, 2 ft in diameter, is completely filled with water. If the vessel is rotated at 1200 rpm, what increase in pressure would occur at the top of the tank at the circumference? *Ans.* 106 psi

26. An open vessel, 18 in. in diameter and filled with water, is rotated about its vertical axis at such a velocity that the water surface 4 in. from the axis makes an angle of 40° with the horizontal. Compute the speed of rotation. *Ans.* 9.00 rad/sec

27. A U-tube with right angle bends is 12 in. wide and contains mercury which rises 9 in. in each leg when the tube is at rest. At what speed must the tube be rotated about an axis 3 in. from one leg so that there will be no mercury in that leg of the tube? *Ans.* 13.9 rad/sec

28. A 7 ft length of 2 in. diameter pipe is capped and is filled with water under 12.5 psi pressure. Placed in a horizontal position, it is rotated about a vertical axis through one end at the rate 3 rad/sec. What will be the pressure at the outer end? *Ans.* 15.5 psi

29. The 5.00 ft diameter impeller of a closed centrifugal water pump is rotated at 1500 rpm. If the casing is full of water, what pressure head is developed by rotation? *Ans.* 2390 ft

Chapter 5

Dimensional Analysis and Hydraulic Similitude

INTRODUCTION

Mathematical theory and experimental data have developed practical solutions to many hydraulic problems. Important hydraulic structures are now designed and built only after extensive model studies have been made. Application of dimensional analysis and hydraulic similitude enable the engineer to organize and simplify the experiments and to analyze the results therefrom.

DIMENSIONAL ANALYSIS

Dimensional analysis is the mathematics of dimensions of quantities and is another useful tool of modern fluid mechanics. In an equation expressing a physical relationship between quantities, absolute numerical and dimensional equality must exist. In general, all such physical relationships can be reduced to the fundamental quantities of force F, length L and time T (or mass M, length L and time T). Applications include (1) converting one system of units to another, (2) developing equations, (3) reducing the number of variables required in an experimental program and (4) establishing principles of model design.

The Buckingham Pi Theorem is outlined and illustrated in Problems 13 to 17.

HYDRAULIC MODELS

Hydraulic models, in general, may be either true models or distorted models. True models have all the significant characteristics of the prototype reproduced to scale (geometrically similar) and satisfy design restrictions (kinematic and dynamic similitude). Model-prototype comparisons have clearly shown that the correspondence of behavior is often well beyond expected limitations, as has been attested by the successful operation of many structures designed from model tests.

GEOMETRIC SIMILITUDE

Geometric similitude exists between model and prototype if the ratios of all corresponding dimensions in model and prototype are equal. Such ratios may be written

$$\frac{L_{\text{model}}}{L_{\text{prototype}}} = L_{\text{ratio}} \quad \text{or} \quad \frac{L_{\text{m}}}{L_{\text{p}}} = L_{\text{r}} \tag{1}$$

and
$$\frac{A_{\text{model}}}{A_{\text{prototype}}} = \frac{L^2_{\text{model}}}{L^2_{\text{prototype}}} = L^2_{\text{ratio}} = L^2_{\text{r}} \tag{2}$$

KINEMATIC SIMILITUDE

Kinematic similitude exists between model and prototype (1) if the paths of homologous moving particles are geometrically similar and (2) if the ratios of the velocities of homologous particles are equal. A few useful ratios follow.

Velocity:
$$\frac{V_m}{V_p} = \frac{L_m/T_m}{L_p/T_p} = \frac{L_m}{L_p} \div \frac{T_m}{T_p} = \frac{L_r}{T_r} \qquad (3)$$

Acceleration:
$$\frac{a_m}{a_p} = \frac{L_m/T_m^2}{L_p/T_p^2} = \frac{L_m}{L_p} \div \frac{T_m^2}{T_p^2} = \frac{L_r}{T_r^2} \qquad (4)$$

Discharge:
$$\frac{Q_m}{Q_p} = \frac{L_m^3/T_m}{L_p^3/T_p} = \frac{L_m^3}{L_p^3} \div \frac{T_m}{T_p} = \frac{L_r^3}{T_r} \qquad (5)$$

DYNAMIC SIMILITUDE

Dynamic similitude exists between geometrically and kinematically similar systems if the ratios of all homologous forces in model and prototype are the same.

The conditions required for complete similitude are developed from Newton's second law of motion, $\Sigma F_x = M a_x$. The forces acting may be any one, or a combination of several, of the following: viscous forces, pressure forces, gravity forces, surface tension forces and elasticity forces. The following relation between forces acting on model and prototype develops:

$$\frac{\Sigma \text{ forces (viscous} \twoheadrightarrow \text{pressure} \twoheadrightarrow \text{gravity} \twoheadrightarrow \text{surface ten.} \twoheadrightarrow \text{elasticity)}_m}{\Sigma \text{ forces (viscous} \twoheadrightarrow \text{pressure} \twoheadrightarrow \text{gravity} \twoheadrightarrow \text{surface ten.} \twoheadrightarrow \text{elasticity)}_p} = \frac{M_m a_m}{M_p a_p}$$

THE INERTIA FORCE RATIO is developed into the following form:

$$F_r = \frac{\text{force}_{\text{model}}}{\text{force}_{\text{prototype}}} = \frac{M_m a_m}{M_p a_p} = \frac{\rho_m L_m^3}{\rho_p L_p^3} \times \frac{L_r}{T_r^2} = \rho_r L_r^2 \left(\frac{L_r}{T_r}\right)^2$$

$$F_r = \rho_r L_r^2 V_r^2 = \rho_r A_r V_r^2 \qquad (6)$$

This equation expresses the general law of dynamic similarity between model and prototype and is referred to as the Newtonian equation.

INERTIA — PRESSURE FORCE RATIO (*Euler number*) gives the relationship (using $T = L/V$)

$$\frac{M a}{p A} = \frac{\rho L^3 \times L/T^2}{p L^2} = \frac{\rho L^4 (V^2/L^2)}{p L^2} = \frac{\rho L^2 V^2}{p L^2} = \frac{\rho V^2}{p} \qquad (7)$$

INERTIA — VISCOUS FORCE RATIO (*Reynolds number*) is obtained from

$$\frac{M a}{\tau A} = \frac{M a}{\mu \left(\dfrac{dV}{dy}\right) A} = \frac{\rho L^2 V^2}{\mu \left(\dfrac{V}{L}\right) L^2} = \frac{\rho V L}{\mu} \qquad (8)$$

INERTIA — GRAVITY FORCE RATIO is obtained from

$$\frac{M\,a}{M\,g} = \frac{\rho\,L^2\,V^2}{\rho\,L^3\,g} = \frac{V^2}{L\,g} \tag{9}$$

The square root of this ratio, $\dfrac{V}{\sqrt{L\,g}}$, is known as the *Froude number*.

INERTIA — ELASTICITY FORCE RATIO (*Cauchy number*) is obtained from

$$\frac{M\,a}{E\,A} = \frac{\rho\,L^2\,V^2}{E\,L^2} = \frac{\rho\,V^2}{E} \tag{10}$$

The square root of this ratio, $\dfrac{V}{\sqrt{E/\rho}}$, is known as the *Mach number*.

INERTIA — SURFACE TENSION RATIO (*Weber number*) is obtained from

$$\frac{M\,a}{\sigma\,L} = \frac{\rho\,L^2\,V^2}{\sigma\,L} = \frac{\rho\,L\,V^2}{\sigma} \tag{11}$$

In general, the engineer is concerned with the effect of the dominant force. In most fluid flow problems, gravity, viscosity and/or elasticity govern predominantly, but not necessarily simultaneously. Solutions in this book will cover cases where one predominant force influences the flow pattern, other forces causing negligible or compensating effects. If several forces jointly affect flow conditions, the problem becomes involved and is beyond the scope of this book. Problems 21 and 34 suggest the possibilities.

TIME RATIOS

The time ratios established for flow patterns governed essentially by viscosity, by gravity, by surface tension and by elasticity are respectively

$$T_r = \frac{L_r^2}{v_r} \qquad \text{(see Problem 20)} \tag{12}$$

$$T_r = \sqrt{\frac{L_r}{g_r}} \qquad \text{(see Problem 18)} \tag{13}$$

$$T_r = \sqrt{L_r^3 \times \frac{\rho_r}{\sigma_r}} \tag{14}$$

$$T_r = \frac{L_r}{\sqrt{E_r/\rho_r}} \tag{15}$$

Solved Problems

1. Express each of the following quantities (a) in terms of force F, length L and time T and (b) in terms of mass M, length L and time T.

   Solution:

Quantity	Symbol	(a) F-L-T	(b) M-L-T
(a) Area A in ft²	A	L^2	L^2
(b) Volume v in ft³	v	L^3	L^3
(c) Velocity V in ft/sec	V	$L\,T^{-1}$	$L\,T^{-1}$
(d) Acceleration a or g in ft/sec²	a, g	$L\,T^{-2}$	$L\,T^{-2}$
(e) Angular velocity ω in rad/sec	ω	T^{-1}	T^{-1}
(f) Force F in lb	F	F	$M\,L\,T^{-2}$
(g) Mass M in lb sec²/ft	M	$F\,T^2\,L^{-1}$	M
(h) Specific weight w in lb/ft³	w	$F\,L^{-3}$	$M\,L^{-2}\,T^{-2}$
(i) Density ρ in lb sec²/ft⁴	ρ	$F\,T^2\,L^{-4}$	$M\,L^{-3}$
(j) Pressure p in lb/ft²	p	$F\,L^{-2}$	$M\,L^{-1}\,T^{-2}$
(k) Absolute viscosity μ in lb sec/ft²	μ	$F\,T\,L^{-2}$	$M\,L^{-1}\,T^{-1}$
(l) Kinematic viscosity ν in ft²/sec	ν	$L^2\,T^{-1}$	$L^2\,T^{-1}$
(m) Modulus of elasticity E in lb/ft²	E	$F\,L^{-2}$	$M\,L^{-1}\,T^{-2}$
(n) Power P in ft lb/sec	P	$F\,L\,T^{-1}$	$M\,L^2\,T^{-3}$
(o) Torque T in ft lb	T	$F\,L$	$M\,L^2\,T^{-2}$
(p) Rate of flow Q in ft³/sec	Q	$L^3\,T^{-1}$	$L^3\,T^{-1}$
(q) Shearing stress τ in lb/ft²	τ	$F\,L^{-2}$	$M\,L^{-1}\,T^{-2}$
(r) Surface tension σ in lb/ft	σ	$F\,L^{-1}$	$M\,T^{-2}$
(s) Weight W in lb	W	F	$M\,L\,T^{-2}$
(t) Weight rate of flow W in lb/sec	W	$F\,T^{-1}$	$M\,L\,T^{-3}$

2. Develop an equation for the distance traveled by a freely falling body in time T, assuming the distance depends upon the weight of the body, the acceleration of gravity and the time.

   Solution:

$$\text{Distance } s = f(W, g, T)$$

   or

$$s = K\,W^a\,g^b\,T^c$$

where K is a dimensionless coefficient, generally determined experimentally.

   This equation must be dimensionally homogeneous. The exponents of each of the quantities must be the same on each side of the equation. We may write

$$F^0\,L^1\,T^0 = (F^a)\,(L^b\,T^{-2b})\,(T^c)$$

Equating exponents of F, L and T respectively, we obtain $0 = a$, $1 = b$ and $0 = -2b + c$, from which $a = 0$, $b = 1$ and $c = 2$. Substituting,

$$s = K\,W^0\,g\,T^2 \qquad \text{or} \qquad s = K\,g\,T^2$$

   Note that the exponent of the weight W is zero, signifying that the distance is independent of the weight. Factor K must be determined by physical analysis and/or experimentation.

3. The *Reynolds number* is a function of density, viscosity and velocity of a fluid, and a characteristic length.   Establish the *Reynolds number* relation by dimensional analysis.

   Solution:

   $$R_E = f(\rho, \mu, V, L)$$

   or

   $$R_E = K \rho^a \mu^b V^c L^d$$

   Then, dimensionally, $F^0 L^0 T^0 = (F^a T^{2a} L^{-4a})(F^b T^b L^{-2b})(L^c T^{-c})(L^d)$

   Equating exponents of F, L and T respectively, we obtain

   $$0 = a + b, \quad 0 = -4a - 2b + c + d, \quad 0 = 2a + b - c$$

   from which $a = -b$, $c = -b$, $d = -b$. Substituting,

   $$R_E = K \rho^{-b} \mu^b V^{-b} L^{-b} = K\left(\frac{V L \rho}{\mu}\right)^{-b}$$

   Values of K and b must be determined by physical analysis and/or experimentation. Here $K = 1$ and $b = -1$.

4. For an ideal liquid, express the flow Q through an orifice in terms of the density of the liquid, the diameter of the orifice, and the pressure difference.

   Solution:

   $$Q = f(\rho, p, d)$$

   or

   $$Q = K \rho^a p^b d^c$$

   Then, dimensionally, $F^0 L^3 T^{-1} = (F^a T^{2a} L^{-4a})(F^b L^{-2b})(L^c)$

   and $0 = a + b, \quad 3 = -4a - 2b + c, \quad -1 = 2a$

   from which $a = -\frac{1}{2}$, $b = \frac{1}{2}$, $c = 2$. Substituting,

   $$Q = K \rho^{-1/2} p^{1/2} d^2$$

   or   ideal $Q = K d^2 \sqrt{p/\rho}$

   Factor K must be obtained by physical analysis and/or experimentation.

   For an orifice in the side of a tank under head h, $p = w h$. To obtain the familiar orifice formula in Chapter 9, let $K = \sqrt{2}\,(\pi/4)$.   Then

   $$\text{ideal } Q = \sqrt{2}\,(\pi/4)\, d^2 \sqrt{wh/\rho}$$

   But $g = w/\rho$; hence   ideal $Q = \frac{1}{4}\pi\, d^2 \sqrt{2g\, h}$

5. Determine the dynamic pressure exerted by a flowing incompressible fluid on an immersed object, assuming the pressure is a function of the density and the velocity.

   Solution:

   $$p = f(\rho, V)$$

   or

   $$p = K \rho^a V^b$$

   Then, dimensionally, $F^1 L^{-2} T^0 = (F^a T^{2a} L^{-4a})(L^b T^{-b})$

   and $1 = a$, $-2 = -4a + b$, $0 = 2a - b$ from which $a = 1$, $b = 2$.   Substituting,

   $$p = K \rho V^2$$

6. Assuming the power delivered to a pump is a function of the specific weight of the fluid, the flow in cfs and the head delivered, establish an equation by dimensional analysis.
    Solution:

$$P = f(w, Q, H)$$

or

$$P = K w^a Q^b H^c$$

Then, dimensionally, $$F^1 L^1 T^{-1} = (F^a L^{-3a})(L^{3b} T^{-b})(L^c)$$

and $1 = a$, $1 = -3a + 3b + c$, $-1 = -b$ from which $a = 1$, $b = 1$, $c = 1$.   Substituting,

$$P = K w Q H$$

7. A projectile is fired at an angle θ with initial velocity V. Find the range R in the horizontal plane, assuming the range is a function of V, θ and g.
    Solution:

$$R = f(V, g, \theta) = K V^a g^b \theta^c \qquad (A)$$
Dimensionally, $$L^1 = (L^a T^{-a})(L^b T^{-2b}) \qquad (B)$$

Since θ is dimensionless, it does not appear in (B).

Solving for a and b, $a = 2$ and $b = -1$. Substituting, $R = K V^2/g$. Obviously this equation is unsatisfactory in that it is lacking in some designation of angle θ. Problem 8 will show how a solution can be attained.

8. Solve Problem 7, using a vector-directional notation.
    Solution:

In cases of two-dimensional motion, X- and Y-components may be introduced to provide a more complete analysis. Then line (A) in Problem 7 may be written

$$R_x = K V_x^a V_y^b g_y^c \theta^d \qquad (C)$$
Dimensionally, $$L_x^1 = (L_x^a T^{-a})(L_y^b T^{-b})(L_y^c T^{-2c})$$

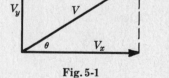

Fig. 5-1

which gives $L_x: \ 1 = a$
$T: \ 0 = -a - b - 2c$
$L_y: \ 0 = b + c$

Then $a = 1$, $b = 1$ and $c = -1$. Substituting in (C),

$$R = K \left(\frac{V_x V_y}{g} \right) \qquad (D)$$

From the vector diagram, $\cos \theta = V_x/V$, $\sin \theta = V_y/V$, and $\cos \theta \sin \theta = V_x V_y/V^2$. Substituting in (D),

$$R = K \frac{V^2 \cos \theta \sin \theta}{g} = K \frac{V^2 \sin 2\theta}{2g} \qquad (E)$$

From static mechanics, R is usually written $\dfrac{V^2 \sin 2\theta}{g}$; hence $K = 2$ in equation (E).

9. Assuming that the drag force exerted by a flowing fluid on a body is a function of the density, viscosity and velocity of the fluid, and a characteristic length of the body, develop a general equation.
    Solution:

$$F = f(\rho, \mu, L, V)$$
or

$$F = K \rho^a \mu^b L^c V^d$$

Then $$F^1 L^0 T^0 = (F^a T^{2a} L^{-4a})(F^b T^b L^{-2b})(L^c)(L^d T^{-d})$$

and $1 = a + b$, $0 = -4a - 2b + c + d$, $0 = 2a + b - d$.

It will be noted that there are more unknown exponents than equations. One method of attack is to express three of the unknowns in terms of the fourth unknown. Solving in terms of b yields

$$a = 1 - b, \quad d = 2 - b, \quad c = 2 - b$$

Substituting,
$$F = K \rho^{1-b} \mu^b L^{2-b} V^{2-b}$$

In order to express this equation in the commonly-used form, multiply by 2/2 and rearrange the terms as follows:

$$F = 2 K \rho \left(\frac{V L \rho}{\mu}\right)^{-b} L^2 \frac{V^2}{2}$$

Recognizing that $\dfrac{V L \rho}{\mu}$ is the *Reynolds number* and that L^2 represents an area, we obtain

$$F = [2K R_E^{-b}] \rho A \frac{V^2}{2} \qquad \text{or} \qquad F = C_D \rho A \frac{V^2}{2}$$

10. Develop an expression for the shear stress in a fluid flowing in a pipe assuming that the stress is a function of the diameter and the roughness of the pipe, and the density, viscosity and velocity of the fluid.

Solution:
$$\tau = f(V, d, \rho, \mu, K)$$
or
$$\tau = C V^a d^b \rho^c \mu^d K^e$$

Roughness K is usually expressed as a ratio of the size of the surface protuberances to the diameter of the pipe, ϵ/d, a dimensionless number.

Then
$$F^1 L^{-2} T^0 = (L^a T^{-a})(L^b)(F^c T^{2c} L^{-4c})(F^d T^d L^{-2d})(L^e/L^e)$$

and $1 = c + d, \ -2 = a + b - 4c - 2d + e - e, \ 0 = -a + 2c + d$. Solving in terms of d yields

$$c = 1 - d, \quad a = 2 - d, \quad b = -d$$

Substituting,
$$\tau = C V^{2-d} d^{-d} \rho^{1-d} \mu^d K^e$$

Collecting terms,
$$\tau = C \left(\frac{V d \rho}{\mu}\right)^{-d} K^e V^2 \rho$$

or
$$\tau = (C' R_E^{-d}) V^2 \rho$$

11. Develop the expression for lost head in a horizontal pipe for turbulent incompressible flow.

Solution:
For any fluid, the lost head is represented by the drop in the pressure gradient and is a measure of the resistance to flow through the pipe. The resistance is a function of the diameter of the pipe, the viscosity and density of the fluid, the length of the pipe, the velocity of the fluid, and the roughness K of the pipe. We may write

$$(p_1 - p_2) = f(d, \mu, \rho, L, V, K)$$
or
$$(p_1 - p_2) = C d^a \mu^b \rho^c L^d V^e (\epsilon/d)^f \tag{1}$$

From experiment and observation, the exponent of the length L is unity. The value of K is usually expressed as a ratio of the size of the surface protuberances ϵ to the diameter d of the pipe, a dimensionless number. We may now write

$$F^1 L^{-2} T^0 = (L^a)(F^b T^b L^{-2b})(F^c T^{2c} L^{-4c})(L^1)(L^e T^{-e})(L^f/L^f)$$

and $1 = b + c, \ -2 = a - 2b - 4c + 1 + e + f - f, \ 0 = b + 2c - e$ from which the values of a, b and c

may be determined in terms of e, or

$$c = e - 1, \quad b = 2 - e, \quad a = e - 3$$

Substituting in (1),

$$(p_1 - p_2) = C \, d^{e-3} \, \mu^{2-e} \, \rho^{e-1} \, L^1 \, V^e \, (\epsilon/d)^f$$

Dividing the left side of the equation by w and the right side by its equivalent ρg,

$$\frac{p_1 - p_2}{w} = \text{lost head} = \frac{C \, (\epsilon/d)^f \, L \, (d^{e-3} \, V^e \, \rho^{e-1} \, \mu^{2-e})}{\rho \, g}$$

which becomes (introducing 2 in numerator and in denominator)

$$\text{lost head} = 2 \, C \left(\frac{\epsilon}{d}\right)^f \frac{L}{d} \frac{V^2}{2g} \left[\frac{d^{e-2} \, V^{e-2} \, \rho^{e-2}}{\mu^{e-2}} \right]$$

$$= K' \, (R_E^{e-2}) \left(\frac{L}{d}\right) \left(\frac{V^2}{2g}\right) = f \frac{L}{d} \frac{V^2}{2g} \quad \text{(Darcy formula)}$$

12. Establish an expression for the power input to a propeller assuming that the power can be expressed in terms of the mass density of the air, the diameter, the velocity of the air stream, the rotational speed, the coefficient of viscosity and the speed of sound.

Solution:

$$\text{Power} = K \, \rho^a \, d^b \, V^c \, \omega^d \, \mu^e \, c^f$$

and, using mass, length and time as fundamental dimensions,

$$M \, L^2 \, T^{-3} = (M^a \, L^{-3a})(L^b)(L^c \, T^{-c})(T^{-d})(M^e \, L^{-e} \, T^{-e})(L^f \, T^{-f})$$

Then
$$1 = a + e$$
$$2 = -3a + b + c - e + f$$
$$-3 = -c - d - e - f$$

from which
$$a = 1 - e$$
$$b = 5 - 2e - c - f$$
$$d = 3 - c - e - f$$

Substituting,

$$\text{Power} = K \rho^{1-e} \, d^{5-2e-c-f} \, V^c \, \omega^{3-c-e-f} \, \mu^e \, c^f$$

Rearranging and collecting terms with like exponents, we obtain

$$\text{Power} = K \left[\left(\frac{\rho \, d^2 \, \omega}{\mu}\right)^{-e} \left(\frac{d \, \omega}{V}\right)^{-c} \left(\frac{d \, \omega}{c}\right)^{-f} \right] \omega^3 \, d^5 \, \rho$$

Examination of the terms in the parentheses indicates that they are all dimensionless. The first term can be written as *Reynolds number* since linear velocity $=$ radius $\times$ angular velocity. The second term is a propeller ratio and the third term, velocity to celerity, is the Mach number. Combining all these terms, the equation becomes

$$\text{Power} = C' \, \rho \, \omega^3 \, d^5$$

13. Outline the procedure to be followed when the Buckingham Pi Theorem is used.

Introduction:

Where the number of physical quantities or variables equals four or more, the Buckingham Pi Theorem provides an excellent tool by which these quantities can be organized into the smallest number of significant, dimensionless groupings, from which an equation can be evaluated. The dimensionless groupings are called Pi terms. Written in mathematical form, if there are n physical quantities q (such as velocity, density, viscosity, pressure and area) and k fundamental dimensions (such as force, length, and time or mass, length and time), then mathematically

$$f_1 (q_1, q_2, q_3, \ldots, q_n) = 0$$

This expression can be replaced by the equation

$$\phi(\pi_1, \pi_2, \pi_3, \ldots, \pi_{n-k}) = 0$$

where any one π-term depends on not more than $(k+1)$ physical quantities q and each of the π-terms are independent, dimensionless, monomial functions of the quantities q.

Procedure:

1. List the n physical quantities q entering into a particular problem, noting their dimensions and the number k of fundamental dimensions. There will be $(n-k)$ π-terms.
2. Select k of these quantities, none dimensionless and no two having the same dimensions. All fundamental dimensions must be included collectively in the quantities chosen.
3. The first π-term can be expressed as the product of the chosen quantities each to an unknown exponent, and one other quantity to a known power (usually taken as one).
4. Retain the quantities chosen in (2) as *repeating* variables and choose one of the remaining variables to establish the next π-term. Repeat this procedure for the successive π-terms.
5. For each π-term, solve for the unknown exponents by dimensional analysis.

Helpful Relationships:

a. If a quantity is dimensionless, it is a π-term without going through the above procedure.
b. If any two physical quantities have the same dimensions, their ratio will be one of the π-terms. For example, L/L is dimensionless and a π-term.
c. Any π-term may be replaced by any power of that term, including π^{-1}. For example, π_3 may be replaced by π_3^2, or π_2 by $1/\pi_2$.
d. Any π-term may be replaced by multiplying it by a numerical constant. For example, π_1 may be replaced by $3\pi_1$.
e. Any π-term may be expressed as a function of the other π-terms. For example, if there are two π-terms, $\pi_1 = \phi(\pi_2)$.

14. Solve Problem 2, using the Buckingham Pi Theorem.

Solution:

The problem can be expressed by stating that some function of distance s, weight W, gravitational acceleration g, and time T equals zero or, written mathematically,

$$f_1(s, W, g, T) = 0$$

Step 1

List the quantities and the units.

$$s = \text{length } L, \quad W = \text{force } F, \quad g = \text{acceleration } L/T^2, \quad T = \text{time } T$$

There are 4 physical quantities and 3 fundamental units, hence $(4-3)$ or one π-term.

Step 2

Choosing s, W, and T as the physical quantities provides the 3 fundamental dimensions F, L and T.

Step 3

Since physical quantities of dissimilar dimensions cannot be added or subtracted, the π-term is expressed as a product, as follows:

$$\pi_1 = (s^{x_1})(W^{y_1})(T^{z_1})(g) \tag{1}$$

Using dimensional homogenity produces

$$F^0 L^0 T^0 = (L^{x_1})(F^{y_1})(T^{z_1})(LT^{-2})$$

Equating exponents of F, L and T respectively, we obtain $0 = y_1$, $0 = x_1 + 1$, $0 = z_1 - 2$, from which $x_1 = -1$, $y_1 = 0$, $z_1 = 2$. Substituting in (1),

$$\pi_1 = s^{-1} W^0 T^2 g = \frac{W^0 T^2 g}{s}$$

Solving for s and noting that $1/\pi_1 = K$, we obtain $s = K g T^2$.

15. Solve Problem 6, using the Buckingham Pi Theorem.

Solution:

The problem can be written mathematically as

$$f(P, w, Q, H) = 0$$

The physical quantities with their dimensions in F, L, and T units are

Power $P = FLT^{-1}$ Flow $Q = L^3 T^{-1}$

Specific weight $w = FL^{-3}$ Head $H = L$

There are 4 physical quantities and 3 fundamental units, hence $(4-3)$ or one π-term.

Choosing Q, w and H as the quantities with the unknown exponents, we establish the π-term as follows:

$$\pi_1 = (Q^{x_1})(w^{y_1})(H^{z_1}) P \tag{1}$$

or

$$\pi_1 = (L^{3x_1} T^{-x_1})(F^{y_1} L^{-3y_1})(L^{z_1})(FLT^{-1})$$

Equating exponents of F, L and T respectively, we obtain $0 = y_1 + 1$, $0 = 3x_1 - 3y_1 + z_1 + 1$, $0 = -x_1 - 1$, from which $x_1 = -1$, $y_1 = -1$, $z_1 = -1$. Substituting in (1),

$$\pi_1 = Q^{-1} w^{-1} H^{-1} P = \frac{P}{wQH} \quad \text{or} \quad P = KwQH$$

16. Solve Problem 9, using the Buckingham Pi Theorem.

Solution:

The problem can be expressed as

$$\phi(F, \rho, \mu, L, V) = 0$$

The physical quantities with their dimensions in F, L, and T units are

Force $F = F$ Length $L = L$

Density $\rho = FT^2 L^{-4}$ Velocity $V = LT^{-1}$

Absolute viscosity $\mu = FTL^{-2}$

There are 5 physical quantities and 3 fundamental units, hence $(5-3)$ or two π-terms.

Choosing length L, velocity V, and density ρ as the 3 repeating variables with unknown exponents, we establish the π-terms as follows:

$$\pi_1 = (L^{a_1})(L^{b_1} T^{-b_1})(F^{c_1} T^{2c_1} L^{-4c_1})(F) \tag{1}$$

Equating exponents of F, L and T respectively, we obtain $0 = c_1 + 1$, $0 = a_1 + b_1 - 4c_1$, $0 = -b_1 + 2c_1$, from which $c_1 = -1$, $b_1 = -2$, $a_1 = -2$. Substituting in (1), $\pi_1 = F/L^2 V^2 \rho$.

To evaluate the second π-term, retain the first three physical quantities and add another quantity, in this case absolute viscosity μ. (See Problem 13, Item 4.)

$$\pi_2 = (L^{a_2})(L^{b_2} T^{-b_2})(F^{c_2} T^{2c_2} L^{-4c_2})(FTL^{-2}) \tag{2}$$

Equating exponents of F, L and T respectively, we obtain $0 = c_2 + 1$, $0 = a_2 + b_2 - 4c_2 - 2$, $0 = -b_2 + 2c_2 + 1$, from which $c_2 = -1$, $b_2 = -1$, $a_2 = -1$. Thus $\pi_2 = \mu/LV\rho$. This expression can be written $\pi_2 = LV\rho/\mu$, which we recognize as Reynolds Number.

The new relationship, written in terms of π_1 and π_2 is

$$f_1\left(\frac{F}{L^2 V^2 \rho}, \frac{LV\rho}{\mu}\right) = 0$$

or

$$\text{Force } F = (L^2 V^2 \rho) f_2\left(\frac{LV\rho}{\mu}\right)$$

which can be written

$$F = (2K R_E) \rho L^2 \frac{V^2}{2}$$

Recognizing L^2 as an area, the final equation can be stated as $F = C_D \rho A \dfrac{V^2}{2}$. (See Chapter 11.)

17. Solve Problem 11, using the Buckingham Pi Theorem.

Solution:

This problem can be written mathematically as

$$f\,(\Delta p, d, \mu, \rho, L, V, K) \;=\; 0$$

where K is the relative roughness or the ratio of the size of the surface irregularities ϵ to the diameter of the pipe d. (See Chapter 7.)

The physical quantities with their dimensions in F, L and T units are

Pressure drop $\Delta p \;=\; F\,L^{-2}$	Length $L \;=\; L$
Diameter $d \;=\; L$	Velocity $V \;=\; L\,T^{-1}$
Absolute viscosity $\mu \;=\; F\,T\,L^{-2}$	Relative roughness $K \;=\; L_1/L_2$
Density $\rho \;=\; F\,T^2\,L^{-4}$	

There are 7 physical quantities and 3 fundamental units, hence $(7-3)$ or 4 π-terms. Choosing diameter, velocity and density as the repeating variables with unknown exponents, the π-terms are

$$\pi_1 \;=\; (L^{x_1})\,(L^{y_1}\,T^{-y_1})\,(F^{z_1}\,T^{2z_1}\,L^{-4z_1})\,(F\,L^{-2})$$

$$\pi_2 \;=\; (L^{x_2})\,(L^{y_2}\,T^{-y_2})\,(F^{z_2}\,T^{2z_2}\,L^{-4z_2})\,(F\,T\,L^{-2})$$

$$\pi_3 \;=\; (L^{x_3})\,(L^{y_3}\,T^{-y_3})\,(F^{z_3}\,T^{2z_3}\,L^{-4z_3})\,(L)$$

$$\pi_4 \;=\; K \;=\; L_1/L_2$$

Evaluating the exponents, term by term, yields

π_1: $0 = z_1 + 1,\;\; 0 = x_1 + y_1 - 4z_1 - 2,\;\; 0 = -y_1 + 2z_1;$ then $x_1 = 0,\; y_1 = -2,\; z_1 = -1.$

π_2: $0 = z_2 + 1,\;\; 0 = x_2 + y_2 - 4z_2 - 2,\;\; 0 = -y_2 + 2z_2 + 1;$ then $x_2 = -1,\; y_2 = -1,\; z_2 = -1.$

π_3: $0 = z_3,\;\; 0 = x_3 + y_3 - 4z_3 + 1,\;\; 0 = -y_3 + 2z_3;$ then $x_3 = -1,\; y_3 = 0,\; z_3 = 0.$

Hence the π-terms are

$$\pi_1 \;=\; d^0\,V^{-2}\,\rho^{-1}\,\Delta p \;=\; \frac{\Delta p}{\rho\,V^2} \qquad\qquad \text{(Euler Number)}$$

$$\pi_2 \;=\; \frac{\mu}{d\,V\,\rho} \quad\text{or}\quad \frac{d\,V\,\rho}{\mu} \qquad\qquad \text{(Reynolds Number)}$$

$$\pi_3 \;=\; d^{-1}\,V^0\,\rho^0\,L \;=\; \frac{L}{d} \qquad\qquad \begin{array}{l}\text{(as might be expected;}\\ \text{see Item } \mathbf{b}, \text{ Prob. 13)}\end{array}$$

$$\pi_4 \;=\; L_1/L_2 \;=\; \frac{\epsilon}{d} \qquad\qquad \text{(see Chapter 7)}$$

The new relationship may now be written

$$f_1\Big(\frac{\Delta p}{\rho\,V^2},\, \frac{d\,V\,\rho}{\mu},\, \frac{L}{d},\, \frac{\epsilon}{d}\Big) \;=\; 0$$

Solving for Δp,

$$\Delta p \;=\; \frac{w}{g}\,V^2\,f_2\Big(R_E,\, \frac{L}{d},\, \frac{\epsilon}{d}\Big)$$

where $\rho = w/g$. Hence the pressure head drop would be

$$\frac{\Delta p}{w} \;=\; \frac{V^2}{2g}\,(2) \cdot f_2\Big(R_E,\, \frac{L}{d},\, \frac{\epsilon}{d}\Big)$$

If it were desirable to obtain the Darcy-type expression, experiment and analysis indicate that the pressure drop is a function of L/d to the first power; hence

$$\frac{\Delta p}{w} = \frac{V^2}{2g} \cdot \frac{L}{d} \cdot 2 \cdot f_3\left(R_E, \frac{\epsilon}{d}\right)$$

which can be expressed as

$$\frac{\Delta p}{w} = (\text{factor } f)\left(\frac{L}{d}\right)\left(\frac{V^2}{2g}\right)$$

Note 1

If the flow were compressible, another physical quantity, bulk modulus E, would be included and the fifth π-term would yield the dimensionless ratio $\frac{E}{\rho V^2}$. This is usually rewritten in the form $\frac{V}{\sqrt{E/\rho}}$, which is the Mach Number.

Note 2

If gravity should enter the general flow problem, the gravitational force would be another physical quantity and a sixth π-term would yield the dimensionless ratio $\frac{V^2}{gL}$. This term is recognized as the Froude Number.

Note 3

If the surface tension σ should enter into the general flow problem, another physical quantity is added which yields a seventh π-term. This π-term would take the form of $\frac{V^2 L \rho}{\sigma}$, which is the Weber Number.

18. For model and prototype, show that, when gravity and inertia are the only influences, the ratio of flows Q is equal to the ratio of the length dimension to the five-halves power.

Solution:

$$\frac{Q_m}{Q_p} = \frac{L_m^3/T_m}{L_p^3/T_p} = \frac{L_r^3}{T_r}$$

The time ratio must be established for the conditions influencing the flow. Expressions can be written for the gravitation and inertia forces, as follows.

Gravity:
$$\frac{F_m}{F_p} = \frac{W_m}{W_p} = \frac{w_m}{w_p} \times \frac{L_m^3}{L_p^3} = w_r L_r^3$$

Inertia:
$$\frac{F_m}{F_p} = \frac{M_m a_m}{M_p a_p} = \frac{\rho_m}{\rho_p} \times \frac{L_m^3}{L_p^3} \times \frac{L_r}{T_r^2} = \rho_r L_r^3 \times \frac{L_r}{T_r^2}$$

Equating the force ratios,

$$w_r L_r^3 = \rho_r L_r^3 \times \frac{L_r}{T_r^2}$$

which, when solved for the time ratio, yields

$$T_r^2 = L_r \times \frac{\rho_r}{w_r} = \frac{L_r}{g_r} \tag{1}$$

Recognizing that the value of g_r is unity, substitution in the flow ratio expression gives

$$Q_r = \frac{Q_m}{Q_p} = \frac{L_r^3}{L_r^{1/2}} = L_r^{5/2} \tag{2}$$

19. For the conditions laid down in the preceding problem, establish (a) the velocity ratio and (b) the pressure ratio and force ratio.

Solution:

(a) Dividing both sides of equation (1) of the preceding problem by L_r^2 gives

$$\frac{T_r^2}{L_r^2} = \frac{L_r}{L_r^2 g_r} \qquad \text{or, since} \quad V = \frac{L}{T}, \qquad V_r^2 = L_r g_r$$

But the value of g_r may be considered unity. This means that, for model and prototype, $V_r^2 = L_r$, which may be called the Froude model law for velocity ratios.

(b)   The force ratio for pressure forces $= \dfrac{p_m L_m^2}{p_p L_p^2} = p_r L_r^2.$

The force ratio for inertia forces $= \dfrac{\rho_r L_r^4}{T_r^2} = w_r L_r^3.$

Equating these, we obtain $\qquad\qquad p_r L_r^2 = w_r L_r^3$

$$p_r = w_r L_r \tag{1}$$

For model studies with a free surface the Froude numbers of model and prototype are the same. The Euler numbers of model and prototype are also the same.

Using $V_r^2 = L_r$ we may write equation (1) as

$$p_r = w_r V_r^2$$

and, since force $F = pA$, $\qquad\qquad F_r = p_r L_r^2 = w_r L_r^3 \tag{2}$

20. Develop the Reynolds model law for time and velocity ratios for incompressible liquids.
Solution:
For flow patterns subjected to inertia and viscous forces only (other effects negligible), these forces for model and prototype must be evaluated.

For inertia: $\qquad \dfrac{F_m}{F_p} = \rho_r L_r^3 \times \dfrac{L_r}{T_r^2}$ (from preceding problem)

For viscosity: $\qquad \dfrac{F_m}{F_p} = \dfrac{\tau_m A_m}{\tau_p A_p} = \dfrac{\mu_m (dV/dy)_m A_m}{\mu_p (dV/dy)_p A_p} = \dfrac{\mu_m (L_m/T_m \times 1/L_m) L_m^2}{\mu_p (L_p/T_p \times 1/L_p) L_p^2}$

$$= \frac{\mu_m L_m^2/T_m}{\mu_p L_p^2/T_p} = \frac{\mu_r L_r^2}{T_r}$$

Equating the two force ratios, we obtain $\rho_r \dfrac{L_r^4}{T_r^2} = \dfrac{\mu_r L_r^2}{T_r}$ from which $T_r = \dfrac{\rho_r L_r^2}{\mu_r}.$

Since $\nu = \dfrac{\mu}{\rho}$, we may write $\qquad\qquad T_r = \dfrac{L_r^2}{\nu_r} \tag{1}$

The velocity ratio $\qquad\qquad V_r = \dfrac{L_r}{T_r} = \dfrac{L_r}{L_r^2} \nu_r = \dfrac{\nu_r}{L_r} \tag{2}$

Writing these ratio values in terms of model and prototype, we obtain from (2)

$$\frac{V_m}{V_p} = \frac{\nu_m}{\nu_p} \times \frac{L_p}{L_m}$$

Collecting the terms for model and prototype yields $V_m L_m/\nu_m = V_p L_p/\nu_p$ which the reader will recognize as: Reynolds number for model = Reynolds number for prototype.

21. Oil of kinematic viscosity 50×10^{-5} ft²/sec is to be used in a prototype in which both viscous and gravity forces dominate. A model scale of 1:5 is also desired. What viscosity of model liquid is necessary to make both the Froude number and the Reynolds number the same in model and prototype?

Solution:

Using the scale ratios for velocity for the Froude and Reynolds laws (see Problems 19 and 20), we equate

$$(L_r\, g_r)^{1/2} \;=\; \nu_r/L_r$$

Since $g_r = 1$, $L_r^{3/2} = \nu_r$ and $\nu_r = (1/5)^{3/2} = .0894$.

This means that $\dfrac{\nu_m}{\nu_p} \;=\; .0894 \;=\; \dfrac{\nu_m}{50 \times 10^{-5}}$ and thus $\nu_m = 4.47 \times 10^{-5}$ ft²/sec.

Using the time, acceleration and discharge scale ratios will produce the same results. For example, the time ratios equated (Problems 18 and 20) yield

$$\frac{L_r^{1/2}}{g_r^{1/2}} \;=\; \frac{\rho_r\, L_r^2}{\mu_r} \quad \text{or, since } g_r = 1, \quad \frac{\mu_r}{\rho_r} \;=\; \nu_r \;=\; L_r^{3/2}, \text{ as before.}$$

22. Water at 60°F flows at 12.0 ft/sec in a 6″ pipe.   At what velocity must medium fuel oil at 90°F flow in a 3″ pipe for the two flows to be dynamically similar?

Solution:

Since the flow pattern in pipes is subject to viscous and inertia forces only, Reynolds number is the criterion for similarity. Other properties of the fluid flowing, such as elasticity and surface tension, as well as the gravity forces, do not affect the flow picture. Thus, for dynamic similarity,

$$\text{Reynolds number for the water} \;=\; \text{Reynolds number for the oil}$$

$$\frac{V\,d}{\nu} \;=\; \frac{V'\,d'}{\nu'}$$

Obtaining values of kinematic viscosity from Table 2 in the Appendix and substituting,

$$\frac{12.0 \times 6/12}{1.217 \times 10^{-5}} \;=\; \frac{V' \times 3/12}{3.19 \times 10^{-5}}$$

and $V' = 63.0$ ft/sec for the oil.

23. Air at 68°F is to flow through a 24″ pipe at an average velocity of 6.00 ft/sec.   For dynamic similarity, what size pipe carrying water at 60°F at 3.65 ft/sec should be used?

Solution:

Equate the two Reynolds numbers: $\dfrac{6.00 \times 2.0}{16.0 \times 10^{-5}} \;=\; \dfrac{3.65 \times d}{1.217 \times 10^{-5}},$ $d = 0.250$ ft $= 3.0$ in.

24. A 1:15 model of a submarine is to be tested in a towing tank containing salt water. If the submarine moves at 12.0 mph, at what velocity should the model be towed for dynamic similarity?

Solution:

Equate the Reynolds numbers for prototype and model: $\dfrac{12.0 \times L}{\nu} \;=\; \dfrac{V \times L/15}{\nu},$ $V = 180$ mph.

25. A 1:80 model of an airplane is tested in 68°F air which has a velocity of 150 ft/sec.
   (a) At what speed should the model be towed when fully submerged in 80°F water?
   (b) What prototype drag in air would a model resistance of 1.25 lb in water represent?

Solution:

(a) Equating the Reynolds numbers, $\dfrac{150 \times L}{16.0 \times 10^{-5}} = \dfrac{V \times L}{0.930 \times 10^{-5}}$ or $V = 8.71$ ft/sec in water.

(b) Since p varies with ρV^2, equating the Euler numbers will produce

$$\frac{\rho_m V_m^2}{p_m} = \frac{\rho_p V_p^2}{p_p} \quad \text{or} \quad \frac{p_m}{p_p} = \frac{\rho_m V_m^2}{\rho_p V_p^2}$$

But forces acting are (pressure × area), or $p L^2$; hence

$$\frac{F_m}{F_p} = \frac{p_m L_m^2}{p_p L_p^2} = \frac{\rho_m V_m^2 L_m^2}{\rho_p V_p^2 L_p^2}$$

or $\qquad F_r = \rho_r V_r^2 L_r^2 \qquad$ (equation (6), Page 51).

To obtain the velocity of the prototype in air, equate the Reynolds numbers. We obtain

$$\frac{V_m L_m}{\nu_{air}} = \frac{V_p L_p}{\nu_{air}} \quad \text{or} \quad \frac{150 \times L_p/80}{\nu_{air}} = \frac{V_p L_p}{\nu_{air}} \quad \text{and} \quad V_p = 1.875 \text{ ft/sec}$$

Then $\qquad \dfrac{1.25}{F_p} = \left(\dfrac{1.94}{.00233}\right)\left(\dfrac{8.71}{1.875}\right)^2\left(\dfrac{1}{80}\right)^2 \quad$ and $\quad F_p = 0.444$ lb

26. A model of a torpedo is tested in a towing tank at a velocity of 80.0 ft/sec. The
   prototype is expected to attain a velocity of 20.0 ft/sec in 60°F water. (a) What model
   scale has been used? (b) What would be the model speed if tested in a wind tunnel
   under a pressure of 20 atmospheres and at constant temperature 80°F?

Solution:

(a) Equating the Reynolds numbers for prototype and model, $\dfrac{20.0 \times L}{\nu} = \dfrac{80.0 \times L/x}{\nu}$ or $x = 4$.
   The model scale is 1:4.

(b) For the air, from Table 1B the absolute viscosity is 3.85×10^{-7} lb sec/ft² and the density

$$\rho = \frac{w}{g} = \frac{p}{g\,R\,T} = \frac{20 \times 14.7 \times 144}{32.2(53.3)(460 + 80)} = .0456 \text{ slug/ft}^3 \text{ (or } \rho = 20 \text{ times value in Table 1B for}$$

80°F $= 20 \times .00228 = .0456$). Then

$$\frac{20.0 \times L}{1.217 \times 10^{-5}} = \frac{V \times L/4}{3.85 \times 10^{-7}/.0456} \quad \text{and} \quad V = 55.6 \text{ ft/sec}$$

27. A centrifugal pump pumps medium lubricating oil at 60°F while rotating at 1200 rpm.
   A model pump, using 68°F air, is to be tested. If the diameter of the model is 3 times
   the diameter of the prototype, at what speed should the model run?

Solution:

   Using the peripheral speeds (which equal radius times angular velocity in radians/sec) as the
velocities in Reynolds number, we obtain

$$\frac{(d/2)\,\omega_p\,(d)}{188 \times 10^{-5}} = \frac{(3d/2)\,\omega_m\,(3d)}{16.0 \times 10^{-5}}$$

Hence $\omega_p = 106\omega_m$ and model speed $= 1200/106 = 11.3$ rpm.

28. An airplane wing of 3 ft chord is to move at 90 mph in air.   A model of 3 in. chord is to be tested in a wind tunnel with air velocity at 108 mph.   For air temperature of 68°F in each case, what should be the pressure in the wind tunnel?

Solution:

Equate the Reynolds numbers, model and prototype, using velocities in identical units.

$$\frac{V_m L_m}{\nu_m} = \frac{V_p L_p}{\nu_p}, \qquad \frac{108 \times 3/12}{\nu_{tunnel}} = \frac{90 \times 3}{16.0 \times 10^{-5}}, \qquad \nu_{tunnel} = 1.6 \times 10^{-5} \text{ ft}^2/\text{sec}$$

The pressure which produces this kinematic viscosity of 68°F air can be found by remembering that the absolute viscosity is not affected by pressure changes.   The kinematic viscosity equals absolute viscosity divided by density.   But density increases with pressure (temperature constant); then

$$\nu = \frac{\mu}{\rho} \quad \text{and} \quad \frac{\nu_m}{\nu_p} = \frac{16.0 \times 10^{-5}}{1.6 \times 10^{-5}} = 10.0$$

Thus the density of the air in the tunnel must be ten times standard (68°F) air and the resulting pressure in the tunnel must be ten atmospheres.

29. A ship whose hull length is 460 ft is to travel at 25.0 ft/sec.   (a) Compute the Froude number N_F.   (b) For dynamic similarity, at what velocity should a 1:30 model be towed through water?

Solution:

(a)
$$N_F = \frac{V}{\sqrt{g L}} = \frac{25.0}{\sqrt{32.2 \times 460}} = 0.206$$

(b)   When two flow patterns with geometrically similar boundaries are influenced by inertia and gravity forces, the Froude number is the significant ratio in model studies.   Then

$$\text{Froude number of prototype} \quad = \quad \text{Froude number of model}$$

or
$$\frac{V}{\sqrt{g L}} = \frac{V'}{\sqrt{g' L'}}$$

Since $g = g'$ in practically all cases, we may write

$$\frac{V}{\sqrt{L}} = \frac{V'}{\sqrt{L'}}, \qquad \frac{25.0}{\sqrt{460}} = \frac{V'}{\sqrt{460/30}}, \qquad V' = 4.56 \text{ ft/sec for the model}$$

30. A spillway model is to be built to a scale of 1:25 across a flume which is 2 ft wide.   The prototype is 37.5 ft high and the maximum head expected is 5.0 ft.   (a) What height of model and what head on the model should be used?   (b) If the flow over the model at 0.20 ft head is 0.70 cfs, what flow per ft of prototype may be expected?   (c) If the model shows a measured hydraulic jump of 1.0 in., how high is the jump in the prototype?   (d) If the energy dissipated in the model at the hydraulic jump is 0.15 horsepower, what would be the energy dissipation in the prototype?

Solution:

(a)   Since $\dfrac{\text{lengths in model}}{\text{lengths in prototype}} = \dfrac{1}{25}$,   height of model $= \dfrac{1}{25} \times 37.5 = 1.5$ ft   and

$$\text{head on model} = \frac{1}{25} \times 5.0 = 0.20 \text{ ft.}$$

(b)   From Problem 18, $Q_r = L_r^{5/2}$, since gravity forces predominate; then

$$Q_p = \frac{Q_m}{L_r^{5/2}} = 0.70(25 \times 25 \times 5) = 2188 \text{ cfs}$$

This quantity may be expected over 2×25 or 50 ft of length of prototype. Thus the flow per ft of prototype $= 2188/50 = 43.8$ cfs.

(c) $$\frac{h_m}{h_p} = L_r \quad \text{or} \quad h_p = \frac{h_m}{L_r} = \frac{1.0}{1/25} = 25 \text{ in. (height of jump)}$$

(d) Power ratio $P_r = (\text{ft lb/sec})_r = \frac{F_r L_r}{T_r} = \frac{w_r L_r^3 L_r}{\sqrt{L_r/g_r}}$. But $g_r = 1$ and $w_r = 1$. Then

$$\frac{P_m}{P_p} = L_r^{7/2} = \left(\frac{1}{25}\right)^{7/2} \quad \text{and} \quad P_p = P_m(25)^{7/2} = 0.15(25)^{7/2} = 11,700 \text{ hp}$$

31. A model of a reservoir is drained in 4 minutes by opening the sluice gate. The model scale is 1:225. How long should it take to empty the prototype?

Solution:

Since gravity is the dominant force, the Q ratio, from Problem 18, is equal to $L_r^{5/2}$.

Also, $Q_r = \dfrac{Q_m}{Q_p} = \dfrac{L_m^3}{L_p^3} \div \dfrac{T_m}{T_p}$. Then $L_r^{5/2} = L_r^3 \times \dfrac{T_p}{T_m}$ and $T_p = T_m/L_r^{1/2} = 4(225)^{1/2} = 60$ minutes.

32. A rectangular pier in a river is 4 ft wide by 12 ft long and the average depth of water is 9.0 ft. A model is built to a scale of 1:16. The velocity of flow of 2.50 ft/sec is maintained in the model and the force acting on the model is 0.90 lb. (a) What are the values of velocity in and force on the prototype? (b) If a standing wave in the model is 0.16 ft high, what height of wave should be expected at the nose of the pier? (c) What is the coefficient of drag resistance?

Solution:

(a) Since gravity forces predominate, from Problem 19, we obtain

$$\frac{V_m}{V_p} = \sqrt{L_r} \quad \text{and} \quad V_p = \frac{2.50}{(1/16)^{1/2}} = 10.0 \text{ ft/sec}$$

Also, $$\frac{F_m}{F_p} = w_r L_r^3 \quad \text{and} \quad F_p = \frac{0.90}{1.0(1/16)^3} = 3690 \text{ lb}$$

(b) Since $\dfrac{V_m}{V_p} = \dfrac{\sqrt{L_m}}{\sqrt{L_p}}$, $\sqrt{h_p} = \sqrt{0.16} \times \dfrac{10.0}{2.5}$ and $h_p = 2.56$ ft wave height.

(c) Drag force $= C_D \rho A \dfrac{V^2}{2}$, $0.90 = C_D (1.94)\left(\dfrac{4}{16} \times \dfrac{9}{16}\right)\dfrac{(2.50)^2}{2}$ and $C_D = 1.06$.

Had the prototype values been used for this calculation, we would have the following:

$$3690 = C_D (1.94)(4 \times 9)\frac{(10.0)^2}{2} \quad \text{and} \quad C_D = 1.06, \text{ as expected}$$

33. The measured resistance in fresh water of an 8 ft ship model moving at 6.50 ft/sec was 9.60 lb. (a) What would be the velocity of the 128 ft prototype? (b) What force would be required to drive the prototype at this speed in salt water?

Solution:

(a) Since gravity forces predominate, we obtain

$$\frac{V_m}{V_p} = \sqrt{L_r} = \sqrt{8/128} \quad \text{and} \quad V_p = \frac{6.50}{(1/16)^{1/2}} = 26.0 \text{ ft/sec}$$

(b) $$\frac{F_m}{F_p} = w_r L_r^3 \quad \text{and} \quad F_p = \frac{9.60}{(62.4/64.0)(1/16)^3} = 40{,}300 \text{ lb}$$

This latter value may be obtained by using the drag formula: Drag force $= C_f \rho \dfrac{A}{2} V^2$.

For model, $9.60 = C_f \dfrac{62.4}{2g} \dfrac{A}{(16)^2} (6.50)^2 \quad \text{and} \quad \dfrac{C_f A}{2g} = \dfrac{9.60(16)^2}{62.4(6.50)^2}$ (1)

For prototype, $\text{force} = C_f \dfrac{64.0}{2g} A (26.0)^2 \quad \text{and} \quad \dfrac{C_f A}{2g} = \dfrac{\text{force}}{64.0(26.0)^2}$ (2)

Equating (1) and (2) since the value of C_f is the same for model and prototype, we obtain

$$\frac{9.6(16)^2}{62.4(6.50)^2} = \frac{\text{force}}{64.0(26.0)^2} \quad \text{from which} \quad \text{force} = 40{,}300 \text{ lb, as before}$$

34. (a) Evaluate the model scale when both viscous and gravity forces are necessary to secure similitude. (b) What should be the model scale if oil of viscosity 100×10^{-5} ft²/sec is used in the model tests and if the prototype liquid has a viscosity of 800×10^{-5} ft²/sec? (c) What would be the velocity and flow ratios for these liquids for the 1:4 model-prototype scale?

Solution:

(a) For this situation, both the Reynolds number and the Froude number must be satisfied simultaneously. We shall equate the velocity ratios for each model law. Using information from Problems 19 and 20,

$$\text{Reynolds number } V_r = \text{Froude number } V_r$$

$$(\nu/L)_r = \sqrt{L_r g_r}$$

Since $g_r = 1$, we obtain $L_r = \nu_r^{2/3}$.

(b) Using the above length ratio, $L_r = \left(\dfrac{100 \times 10^{-5}}{800 \times 10^{-5}}\right)^{2/3} = \dfrac{1}{4}$. Model scale is 1:4.

(c) Using the Froude model laws (see Problems 18 and 19),

$$V_r = \sqrt{L_r g_r} = \sqrt{L_r} = \sqrt{\tfrac{1}{4}} = \tfrac{1}{2} \quad \text{and} \quad Q_r = L_r^{5/2} = (\tfrac{1}{4})^{5/2} = \tfrac{1}{32}$$

Or, using the Reynolds model laws (see Problem 20),

$$V_r = \frac{\nu_r}{L_r} = \frac{100/800}{1/4} = \frac{1}{2} \quad \text{and} \quad Q_r = A_r V_r = L_r^2 \times \frac{\nu_r}{L_r} = L_r \nu_r = \frac{1}{4}\left(\frac{100}{800}\right) = \frac{1}{32}$$

Supplementary Problems

35. Check the expression $\tau = \mu \, (dV/dy)$ dimensionally.

36. Show that the kinetic energy of a body equals $K M V^2$, using methods of dimensional analysis.

37. Using methods of dimensional analysis, prove that centrifugal force equals $K M V^2/r$.

38. A body falls freely for distance s, from rest. Develop an equation for velocity.   *Ans.* $V = K \sqrt{s g}$

39. A body falls freely from rest for time T. Develop an equation for velocity.   *Ans.* $V = K g T$

40. Develop an expression for the frequency of a simple pendulum, assuming it is a function of the length and mass of the pendulum, and the gravitational acceleration.   *Ans.* Frequency $= K \sqrt{g/L}$

41. Assuming that flow Q over a rectangular weir varies directly with length L and is a function of head H and the acceleration of gravity g, establish a weir formula.   *Ans.* $Q = K L H^{3/2} g^{1/2}$

42. Develop the relationship for the distance s traveled by a freely falling body, assuming the distance depends upon the initial velocity V, the time T and the acceleration of gravity g.
    Ans. $s = K V T (g T/V)^b$

43. Establish the expression for the Froude number if it is a function of velocity V, gravitational acceleration g and length L.   *Ans.* $N_F = K (V^2/L g)^{-c}$

44. Establish the expression for the Weber number if it is a function of velocity V, density ρ, length L and surface tension σ.   *Ans.* $N_W = K (\rho L V^2/\sigma)^{-d}$

45. Establish a dimensionless number if it is a function of gravitational acceleration g, surface tension σ, absolute viscosity μ and density ρ.   *Ans.* Number $= K (\sigma^3 \rho/g \mu^4)^d$

46. Assuming that the drag force of a ship is a function of the absolute viscosity μ and density ρ of the fluid, velocity V, gravitational acceleration g and the size (length factor L) of the ship, establish a drag force formula.   *Ans.* Force $= K (R_E^{-a} N_F^{-d} \rho V^2 L^2)$

47. Solve Problem 9 including the compressibility effect by adding the variable *celerity* c, the speed of sound.   *Ans.* Force $= K' R_E^{-b} N_M^{-e} \rho A V^2/2$

48. Show that, for geometrically similar orifices, the velocity ratio is essentially the square root of the head ratio.

49. Show that the time and velocity ratios when surface tension is the dominant force are

$$T_r = \sqrt{L_r^3 \times \frac{\rho_r}{\sigma_r}} \quad \text{and} \quad V_r = \sqrt{\frac{\sigma_r}{L_r \rho_r}} \quad \text{respectively}$$

50. Show that the time and velocity ratios when elasticity is the dominant force are

$$T_r = \frac{L_r}{\sqrt{E_r/\rho_r}} \quad \text{and} \quad V_r = \sqrt{\frac{E_r}{\rho_r}} \quad \text{respectively}$$

51. The model of a spillway is built to a scale of 1:36. If the model velocity and discharge are 1.25 ft/sec and 2.50 cfs respectively, what are the corresponding values for the prototype?
    Ans. 7.50 ft/sec, 19,440 cfs

52. At what velocity should tests be run in a wind tunnel on a model of an airplane wing of 6 in. chord in order that the Reynolds number shall be the same as that of the prototype of 3 ft chord moving at 90 mph? Air is under atmospheric pressure in the wind tunnel.   *Ans.* 540 mph

53. Oil ($\nu = 6.09 \times 10^{-5}$ ft²/sec) is to flow at 12.0 ft/sec in a 6 in. pipe. At what velocity should 60°F water flow in a 12 in. pipe in order that the Reynolds numbers shall be the same?   *Ans.* 1.20 ft/sec

54. Gasoline at $60°F$ flows in a 4 in. pipe at 10.0 ft/sec. What size pipe should be used to carry $60°F$ water at 5.00 ft/sec in order that the Reynolds numbers shall be the same?   *Ans.* 13.3 in.

55. Water at $60°F$ flows at 12.0 ft/sec in a 6 in. pipe. For dynamic similarity, (*a*) what should be the velocity of medium fuel oil at $90°F$ in a 12 in. pipe? (*b*) What size pipe should be used if the oil has a velocity of 63.0 ft/sec?   *Ans.* 15.75 ft/sec, $d = 3.0$ in.

56. A model is tested in standard ($68°F$) air at a speed of 90.0 ft/sec. At what speed should it be tested when fully submerged in water at $60°F$ in a towing tank to obtain similar dynamic conditions? *Ans.* 6.84 ft/sec

57. A surface vessel 512 ft long is to move at the rate of 22.4 ft/sec. At what velocity should a geometrically similar model 8 ft long be tested?   *Ans.* 2.80 ft/sec

58. What force would be exerted against a sea wall if a 1:36 model which was 3 ft long experienced a wave force of 27.0 lb?   *Ans.* 11,700 lb/ft

59. A submerged object is anchored in fresh water ($60°F$) which flows at the rate of 8.0 ft/sec. The resistance of a 1:5 model in a wind tunnel at standard conditions is 4.5 lb. What force acts on the prototype under dynamically similar conditions?   *Ans.* 21.7 lb

60. For flow with viscous and pressure forces dominant, evaluate an expression for velocity ratio and lost head ratio for model and prototype.   *Ans.* $V_r = p_r L_r / \mu_r$ and $L H_r = V_r \mu_r / w_r L_r$

61. Develop the relation for friction factor f if this coefficient depends upon pipe diameter d, average velocity V, fluid density ρ, fluid viscosity μ and absolute pipe roughness ϵ. Use the Buckingham π-theorem.   *Ans.* $f = \phi (R_E, \epsilon/d)$

Chapter 6

Fundamentals of Fluid Flow

INTRODUCTION

Chapters 1 through 4 have considered fluids at rest in which fluid weight was the only property of significance. This chapter will outline additional concepts required for the study of fluids in motion. Fluid flow is complex and not always subject to exact mathematical analysis. Unlike solids, the elements of a flowing fluid may move at different velocities and may be subject to different accelerations. Three significant concepts in fluid flow are:

(a) the principle of conservation of mass, from which the equation of continuity is developed,

(b) the principle of kinetic energy, from which certain flow equations are derived, and

(c) the principle of momentum, from which equations evaluating dynamic forces exerted by flowing fluids may be established (see Chapters 11 and 12).

FLUID FLOW

Fluid flow may be steady or unsteady; uniform or non-uniform; laminar or turbulent (Chapter 7); one-dimensional, two-dimensional or three-dimensional; and rotational or irrotational.

True one-dimensional flow of an incompressible fluid occurs when the direction and magnitude of the velocity at all points are identical. However, one-dimensional flow analysis is acceptable when the single dimension is taken along the central streamline of the flow, and when velocities and accelerations normal to the streamline are negligible. In such cases, average values of velocity, pressure, and elevation are considered to represent the flow as a whole, and minor variations can be neglected. For example, flow in curved pipelines is analyzed by means of one-dimensional flow principles despite the fact that the structure has three dimensions and that the velocity varies across any cross section normal to the flow.

Two-dimensional flow occurs when the fluid particles move in planes or parallel planes and the streamline patterns are identical in each plane.

For an ideal fluid in which no shear stresses occur and hence no torques exist, rotational motion of fluid particles about their own mass centers cannot exist. Such ideal flow, which can be represented by a flow net, is called irrotational flow.

In Chapter 4, the liquid in the rotating tanks illustrates rotational flow where the velocity of each particle varies directly as the distance from the center of rotation.

STEADY FLOW

Steady flow occurs if, at any point, the velocity of successive fluid particles is the same at successive periods of time. Thus, the velocity is constant with respect to time or $\partial V/\partial t = 0$, but it may vary at different points or with respect to distance. This statement infers that other fluid variables will not vary with time, or $\partial p/\partial t = 0$, $\partial \rho/\partial t = 0$, $\partial Q/\partial t = 0$, etc. Most practical engineering flow problems involve steady flow conditions. For example, pipelines carrying liquids under constant head conditions or orifices flowing under constant heads illustrate steady flow. These flows may be uniform or non-uniform.

The complexities of unsteady flow are beyond the scope of a book on introductory fluid mechanics. Flow is unsteady when conditions at any point in a fluid change with time, or $\partial V/\partial t \neq 0$. Problem 7 will develop a general equation for unsteady flow, and Chapter 9 will present a few simple problems in which head and flow vary with time.

UNIFORM FLOW

Uniform flow occurs when the magnitude and direction of the velocity do not change from point to point in the fluid, or $\partial V/\partial s = 0$. This statement implies that other fluid variables do not change with distance, or $\partial y/\partial s = 0$, $\partial \rho/\partial s = 0$, $\partial p/\partial s = 0$, etc. Flow of liquids under pressure through long pipelines of constant diameter is uniform flow whether the flow is steady or unsteady.

Non-uniform flow occurs when velocity, depth, pressure, etc., change from point to point in the fluid flow, or $\partial V/\partial s \neq 0$, etc. (See Chapter 10.)

STREAMLINES

Streamlines are imaginary curves drawn through a fluid to indicate the direction of motion in various sections of the flow of the fluid system. A tangent at any point on the curve represents the instantaneous direction of the velocity of the fluid particles at that point. The average direction of velocity may likewise be represented by tangents to streamlines. Since the velocity vector has a zero component normal to the streamline, it should be apparent that there can be no flow across a streamline at any point.

STREAMTUBES

A streamtube represents elementary portions of a flowing fluid bounded by a group of streamlines which confine the flow. If the cross sectional area of the streamtube is sufficiently small, the velocity of the midpoint of any cross section may be taken as the average velocity for the section as a whole. The streamtube will be used to derive the equation of continuity for steady one-dimensional, incompressible flow (Problem 1).

EQUATION of CONTINUITY

The equation of continuity results from the principle of conservation of mass. For steady flow, the mass of fluid passing all sections in a stream of fluid per unit of time is the same. This may be evaluated as

$$\rho_1 A_1 V_1 = \rho_2 A_2 V_2 = \text{constant} \tag{1}$$

or

$$w_1 A_1 V_1 = w_2 A_2 V_2 \qquad \text{(in lb/sec)} \tag{2}$$

For *incompressible* fluids and where $w_1 = w_2$ for all practical purposes, the equation becomes

$$Q = A_1 V_1 = A_2 V_2 = \text{constant} \qquad \text{(in ft}^3\text{/sec)} \tag{3}$$

where A_1 and V_1 are respectively the cross sectional area in ft^2 and the average velocity of the stream in ft/sec at Section 1, with similar terms for Section 2. (See Problem 1.) Units of flow commonly used are cubic feet per second (cfs), although gallons per minute (gpm) and million gallons per day (mgd) are used in water supply work.

The equation of continuity for steady, two-dimensional, incompressible flow is

$$A_{n_1} V_1 = A_{n_2} V_2 = A_{n_3} V_3 = \text{constant} \tag{4}$$

where the A_n terms represent the areas normal to the respective velocity vectors (see Problems 10 and 11).

The equation of continuity for three-dimensional flow is derived in Problem 7, for steady and unsteady flow. The general equation is reduced to steady flow conditions for two- and one-dimensional flows also.

FLOW NETS

Flow nets are drawn to indicate flow patterns in cases of two-dimensional flow, or even three-dimensional flow. The flow net consists of (*a*) a system of streamlines so spaced that the rate of flow q is the same between each successive pair of lines, and (*b*) another system of lines normal to the streamlines and so spaced that the distance between the normal lines equals the distance between adjacent streamlines. An infinite number of streamlines are required to describe completely the flow under given boundary conditions. However, it is usual practice to use a small number of such streamlines, as long as acceptable accuracy is obtained.

While the technique of drawing flow nets is beyond the scope of introductory fluid mechanics, the significance of flow nets is important (see Problems 13 and 14). After a flow net for a particular boundary configuration has been obtained, it may be used for all other irrotational flows as long as the boundaries are geometrically similar.

ENERGY EQUATION

The energy equation results from application of the principle of conservation of energy to fluid flow. The energy possessed by a flowing fluid consists of internal energy and energies due to pressure, velocity and position. In the direction of flow, the energy principle is summarized by a general equation as follows:

$$\begin{matrix} \text{Energy at} \\ \text{Section 1} \end{matrix} + \begin{matrix} \text{Energy} \\ \text{Added} \end{matrix} - \begin{matrix} \text{Energy} \\ \text{Lost} \end{matrix} - \begin{matrix} \text{Energy} \\ \text{Extracted} \end{matrix} = \begin{matrix} \text{Energy at} \\ \text{Section 2} \end{matrix}$$

This equation, for steady flow of incompressible fluids in which the change in internal energy is negligible, simplifies to

$$\left(\frac{p_1}{w} + \frac{V_1^2}{2g} + z_1\right) + H_A - H_L - H_E = \left(\frac{p_2}{w} + \frac{V_2^2}{2g} + z_2\right) \tag{5}$$

This equation is known as the *Bernoulli theorem*. Proof of equation (5) and its modifications for compressible fluids will be found in Problem 20.

The units used are ft lb/lb of fluid or feet of the fluid. Practically all problems dealing with flow of liquids utilize this equation as the basis of solution. Flow of gases, in many instances, involves principles of thermodynamics and heat transfer which are beyond the scope of this book.

VELOCITY HEAD

The velocity head represents the kinetic energy per unit weight that exists at a particular point. If the velocity at a cross section were uniform, then the velocity head calculated with this uniform or average velocity would be the true kinetic energy per unit weight of fluid. But, in general, the velocity distribution is not uniform. The true kinetic energy is found by integrating the differential kinetic energies from streamline to streamline (see Problem 16). The kinetic-energy correction factor α to be applied to the $V_{av}^2/2g$ term is given by the expression

$$\alpha = \frac{1}{A} \int_A \left(\frac{v}{V}\right)^3 dA \tag{6}$$

where 　V = average velocity in the cross section
　　　　v = velocity at any point in the cross section
　　　　A = area of the cross section.

Studies indicate that $\alpha = 1.0$ for uniform distribution of velocity, $\alpha = 1.02$ to 1.15 for turbulent flows, and $\alpha = 2.00$ for laminar flow. In most fluid mechanics computations, α is taken as 1.0, without serious error being introduced into the result, since the velocity head is generally a small percentage of the total head (energy).

APPLICATION of the BERNOULLI THEOREM

Application of the Bernoulli Theorem should be rational and systematic. Suggested procedure is as follows.

(1) Draw a sketch of the system, choosing and labeling all cross sections of the stream under consideration.

(2) Apply the Bernoulli equation in the direction of flow. Select a datum plane for each equation written. The low point is logical in that minus signs are avoided and mistakes reduced in number.

(3) Evaluate the energy upstream at Section 1. The energy is in ft lb/lb units which reduce to feet of fluid units. For liquids the pressure head may be expressed in gage or absolute units, but the same basis must be used for the pressure head at Section 2. Gage units are simpler for liquids and will be used throughout this book. Absolute pressure head units must be used where specific weight w is not constant. As in the equation of continuity, V_1 is taken as the average velocity at the section, without loss of acceptable accuracy.

(4) Add, in feet of the fluid, any energy contributed by mechanical devices, such as pumps.

(5) Subtract, in feet of the fluid, any energy lost during flow.

(6) Subtract, in feet of the fluid, any energy extracted by mechanical devices, such as turbines.

(7) Equate this summation of energy to the sum of the pressure head, velocity head and elevation head at Section 2.

(8) If the two velocity heads are unknown, relate them to each other by means of the equation of continuity.

ENERGY LINE

The energy line is a graphical representation of the energy at each section. With respect to a chosen datum, the total energy (as a linear value in feet of fluid) can be plotted at each representative section, and the line so obtained is a valuable tool in many flow problems. The energy line will slope (drop) in the direction of flow except where energy is added by mechanical devices.

HYDRAULIC GRADE LINE

The hydraulic grade line (gradient) lies below the energy line by an amount equal to the velocity head at the section. The two lines are parallel for all sections of equal cross sectional area. The ordinate between the center of the stream and the hydraulic grade line is the pressure head at the section.

POWER

Power is calculated by multiplying the number of lb of fluid flowing per sec (wQ) by the energy H in ft lb/lb. There results the equation

$$\text{Power } P = wQH = \text{lb/ft}^3 \times \text{ft}^3\text{/sec} \times \text{ft lb/lb} = \text{ft lb/sec}$$
$$\text{Horsepower} = wQH/550.$$

Solved Problems

1. Develop the equation of continuity for steady flow of (a) a compressible fluid and (b) an incompressible fluid.

   Solution:

   (a) Consider flow through the streamtube in which Sections 1 and 2 are normal to the streamlines composing the tube. For values of mass density ρ_1 and normal velocity V_1, the mass per unit of time passing section 1 is $\rho_1 V_1 dA_1$, since $V_1 dA_1$ is the volume per unit of time. Similarly, the mass passing section 2 is $\rho_2 V_2 dA_2$. Since for steady flow the mass cannot change with respect to time, and since no flow can pass through the boundaries of a streamtube, the mass flowing through the streamtube is constant. Therefore,

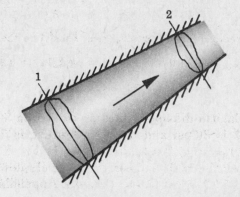

Fig. 6-1

$$\rho_1 V_1 dA_1 = \rho_2 V_2 dA_2 \tag{A}$$

The mass densities ρ_1 and ρ_2 are constant over any cross section dA, and velocities V_1 and V_2 represent the velocities of the streamtube at Sections 1 and 2 respectively. Then

$$\rho_1 V_1 \int_{A_1} dA_1 = \rho_2 V_2 \int_{A_2} dA_2$$

Integrating, $\rho_1 V_1 A_1 = \rho_2 V_2 A_2$ or $w_1 V_1 A_1 = w_2 V_2 A_2$ (B)

(b) For incompressible fluids (and for some cases of compressible flow) the mass density is constant, or $\rho_1 = \rho_2$. Therefore,

$$Q = A_1 V_1 = A_2 V_2 = \text{constant} \qquad \text{(in ft}^3\text{/sec)} \qquad (C)$$

Thus the discharge is constant along a collection of streamtubes. In many cases of fluid flow, average velocity at a cross section can be utilized in the equations of continuity (B) and (C).

2. When 500 gallons per minute flow through a 12″ pipe which later reduces to a 6″ pipe, calculate the average velocities in the two pipes.

Solution:

$$Q \text{ in cfs} = \frac{500}{60} \text{ gal/sec} \times \frac{1}{7.48} \text{ ft}^3\text{/gal} = 1.113 \text{ cfs}$$

$$V_{12} = \frac{Q \text{ in cfs}}{A \text{ in ft}^2} = \frac{1.113}{\frac{1}{4}\pi(1)^2} = 1.42 \text{ ft/sec} \quad \text{and} \quad V_6 = \frac{1.113}{\frac{1}{4}\pi(\frac{1}{2})^2} = 5.68 \text{ ft/sec}$$

3. If the velocity in a 12″ pipe is 1.65 ft/sec, what is the velocity in a 3″ diameter jet issuing from a nozzle attached to the pipe?

Solution:

$Q = A_{12} V_{12} = A_3 V_3$, or, since areas vary as the diameters squared, $(12)^2 V_{12} = (3)^2 V_3$.

Then $V_3 = (12/3)^2 V_{12} = 16 \times 1.65 = 26.4$ ft/sec.

4. Air flows in a 6″ pipe at a pressure of 30.0 psi gage and a temperature of 100°F. If barometric pressure is 14.7 psi and velocity 10.5 ft/sec, how many pounds of air per second are flowing?

Solution:

The gas laws require absolute units for temperature and pressure (psf). Thus

$$w_{\text{air}} = \frac{p}{RT} = \frac{(30.0 + 14.7) \times 144}{53.3(100 + 460)} = 0.215 \text{ lb/ft}^3$$

where $R = 53.3$ for air, from Table 1 in the Appendix.

$$W \text{ in lb/sec} = wQ = wA_6 V_6 = 0.215 \text{ lb/ft}^3 \times \tfrac{1}{4}\pi(\tfrac{1}{2})^2 \text{ ft}^2 \times 10.5 \text{ ft/sec} = 0.443 \text{ lb/sec}$$

5. Carbon dioxide passes point A in a 3″ pipe at a velocity of 15 ft/sec. The pressure at A is 30 psi and the temperature is 70°F. At a point B downstream the pressure is 20 psi and the temperature 90°F. For a barometric reading of 14.7 psi, calculate the velocity at B and compare the cfs flows at A and B. The value of R for carbon dioxide is 34.9, from Table 1 in the Appendix.

Solution:

$$w_A = \frac{p_A}{RT} = \frac{44.7 \times 144}{34.9 \times 530} = 0.348 \text{ lb/ft}^3, \qquad w_B = \frac{34.7 \times 144}{34.9 \times 550} = 0.260 \text{ lb/ft}^3$$

(a) W in lb/sec $= w_A A_A V_A = w_B A_B V_B$. But since $A_A = A_B$, then

$$w_A V_A = w_B V_B = 0.348 \times 15.0 = 0.260 V_B \quad \text{and} \quad V_B = 20.1 \text{ ft/sec}$$

(b) The number of pounds per second flowing is constant but the flows in cfs will differ because the specific weight is not constant.

$$Q_A = A_A V_A = \tfrac{1}{4}\pi(\tfrac{1}{4})^2 \times 15.0 = 0.736 \text{ cfs}, \quad Q_B = A_B V_B = \tfrac{1}{4}\pi(\tfrac{1}{4})^2 \times 20.1 = 0.986 \text{ cfs}$$

6. What minimum diameter of pipe is necessary to carry 0.500 lb/sec of air with a maximum velocity of 18.5 ft/sec? The air is at 80°F and under an absolute pressure of 34.0 psi.

Solution:

$$w_{air} = \frac{p}{RT} = \frac{34.0 \times 144}{53.3(80 + 460)} = 0.170 \text{ lb/ft}^3$$

$$W = 0.500 \text{ lb/sec} = wQ \quad \text{or} \quad Q = \frac{W}{w} = \frac{0.500 \text{ lb/sec}}{0.170 \text{ lb/ft}^3} = 2.94 \text{ ft}^3/\text{sec (cfs)}.$$

$$\text{Minimum area } A \text{ needed} = \frac{\text{flow } Q \text{ in cfs}}{\text{average velocity } V} = \frac{2.94}{18.5} = 0.159 \text{ ft}^2.$$

Hence, minimum diameter = 0.450 ft (or 5.40 in.).

7. Develop the general equation of continuity for three-dimensional flow of a compressible fluid (*a*) for unsteady flow, and (*b*) for steady flow.

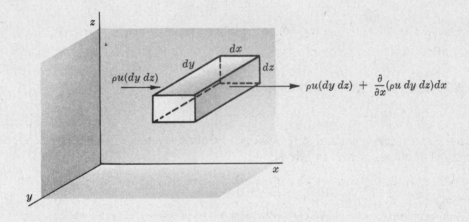

Fig. 6-2

Solution:

(*a*) Let the components of velocity in the *x*, *y* and *z* directions be *u*, *v* and *w* respectively. Consider the flow through the parallelepiped whose dimensions are *dx*, *dy*, and *dz*. The mass of fluid flowing into any face of this volume in unit time is the density of the fluid times the cross sectional area of the face times the velocity normal to the face, or in the *x* direction, $\rho u(dy\,dz)$. In the *x* direction the approximate flows are (see Fig. 6-2).

$$\textit{Inflow} \quad \rho u(dy\,dz) \quad \text{and} \quad \textit{Outflow} \quad \rho u(dy\,dz) + \frac{\partial}{\partial x}(\rho u\,dy\,dz)dx,$$

or the approximate net inflow is $-\dfrac{\partial}{\partial x}(\rho u\,dy\,dz)dx$ or $-\dfrac{\partial}{\partial x}(\rho u\,dx\,dy\,dz)$.

If we write similar expressions for the net inflow terms in the *y* and *z* directions, and sum these net inflow values, we obtain

$$-\left[\frac{\partial}{\partial x}\rho u + \frac{\partial}{\partial y}\rho v + \frac{\partial}{\partial z}\rho w\right] dx\,dy\,dz$$

These quantities become more accurate as *dx*, *dy* and *dz* approach zero.

The positive *rate of change* of mass within the parallelepiped is

$$\frac{\partial}{\partial t}(\rho\,dx\,dy\,dz) \quad \text{or} \quad \frac{\partial\rho}{\partial t}(dx\,dy\,dz)$$

where $\partial\rho/\partial t$ is the rate of change of density within the volume with respect to time. Since the net inflow is identical with the rate of change of mass, we obtain

$$-\left[\frac{\partial}{\partial x}\rho u + \frac{\partial}{\partial y}\rho v + \frac{\partial}{\partial z}\rho w\right]dx\,dy\,dz \;=\; \frac{\partial \rho}{\partial t}(dx\,dy\,dz)$$

Thus the continuity equation for three-dimensional, unsteady flow of a compressible fluid becomes

$$-\left[\frac{\partial}{\partial x}\rho u + \frac{\partial}{\partial y}\rho v + \frac{\partial}{\partial z}\rho w\right] \;=\; \frac{\partial \rho}{\partial t} \tag{A}$$

(b) For steady flow, the fluid properties do not change with respect to time, or $\partial \rho/\partial t = 0$. The continuity equation for steady, compressible flow is

$$\left[\frac{\partial}{\partial x}\rho u + \frac{\partial}{\partial y}\rho v + \frac{\partial}{\partial z}\rho w\right] \;=\; 0 \tag{B}$$

Furthermore, for steady, incompressible flow ($\rho = $ constant) the three-dimensional equation becomes

$$\frac{\partial u}{\partial x} + \frac{\partial v}{\partial y} + \frac{\partial w}{\partial z} \;=\; 0 \tag{C}$$

Should $\partial w/\partial z = 0$, the steady flow is two-dimensional and

$$\frac{\partial u}{\partial x} + \frac{\partial v}{\partial y} \;=\; 0 \tag{D}$$

Should both $\partial w/\partial z$ and $\partial v/\partial y = 0$, the steady flow is one-dimensional and

$$\frac{\partial u}{\partial x} \;=\; 0 \tag{E}$$

This expression represents uniform flow.

8. Is the continuity equation for steady, incompressible flow satisfied if the following velocity components are involved:

$$u = 2x^2 - xy + z^2, \quad v = x^2 - 4xy + y^2, \quad w = -2xy - yz + y^2$$

Solution:

Differentiating each component with respect to the appropriate dimension,

$$\partial u/\partial x = 4x - y, \quad \partial v/\partial y = -4x + 2y, \quad \partial w/\partial z = -y$$

Substituting in equation (C) above, $(4x - y) + (-4x + 2y) + (-y) = 0$. Satisfied.

9. The velocity components for steady, incompressible flow are $u = (2x - 3y)t$, $v = (x - 2y)t$ and $w = 0$. Is the equation of continuity satisfied?

Solution:

Differentiating each component with respect to the appropriate dimension,

$$\partial u/\partial x = 2t, \quad \partial v/\partial y = -2t, \quad \partial w/\partial z = 0$$

Substituting in equation (C) of Problem 7 gives 0. Satisfied.

10. For steady, incompressible flow, are the following values of u and v possible?

$$(a)\ u = 4xy + y^2,\ v = 6xy + 3x \qquad (b)\ u = 2x^2 + y^2,\ v = -4xy$$

Solution:

For the two-dimensional flows indicated, equation (D) of Problem 7 must be satisfied.

(a) $\partial u/\partial x = 4y$, $\partial v/\partial y = 6x$, $4y + 6x \neq 0$ (b) $\partial u/\partial x = 4x$, $\partial v/\partial y = -4x$, $4x - 4x = 0$

　　　Flow is not possible.　　　　　　　　　　　　Flow is possible.

11. A fluid flows between two converging plates
which are 18″ wide, and the velocity varies
according to the expression

$$\frac{v}{v_{max}} = 2\,\frac{n}{n_0}\left(1 - \frac{n}{n_0}\right)$$

For values of $n_0 = 2''$ and $v_{max} = 1.0$ ft/sec,
determine (a) the total flow in cfs, (b) the
mean velocity for the section, and (c) the
mean velocity for the section where $n = 0.75$ in.

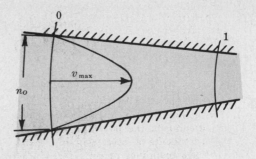

Solution:

Fig. 6-3

(a) The flow/unit of width perpendicular to the
paper is

$$q = \int_0^{n_0} v\,dn = \frac{2v_{max}}{n_0}\int_0^{n_0}(n - n^2/n_0)\,dn = \frac{1}{3}v_{max}\,n_0 = \frac{1}{18} \text{ cfs/ft width}$$

and the total flow $Q = \frac{1}{18}(18/12) = \frac{1}{12}$ cfs.

(b) The mean velocity $V_0 = q/n_0 = \frac{1}{3}$ ft/sec, where $n_0 = \frac{1}{6}$ ft. Or, $V_0 = Q/A = \frac{1}{3}$ ft/sec.

(c) Using equation (4), $V_0 A_{n_0} = V_1 A_{n_1}$, $\frac{1}{3}(2/12)(18/12) = V_1(0.75/12)(18/12)$, and $V_1 = \frac{8}{9}$ ft/sec.

12. If magnitudes and directions of velocities are measured in a vertical plane YY at distances Δy apart, show that the flow q per unit of width can be expressed as $\Sigma v_x \Delta y$.

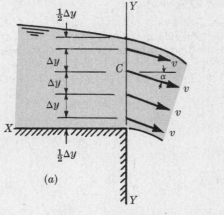

Fig. 6-4

Solution:

Flow per unit width $= q = \Sigma\,\Delta q$, where each Δq can be expressed as $v(\Delta A_n)$.
From Fig. 6-4(b), $A'B' = \Delta A_n = \Delta y\cos\alpha$. Then $q = \Sigma v(\Delta y\cos\alpha) = \Sigma v_x\Delta y$ per unit width.

13. (a) Outline the procedure for drawing the flow net for two-dimensional steady flow of
an ideal fluid between the boundaries indicated in Fig. 6-5 below.

(b) If the uniform velocities at Section 2 equal 30.0 ft/sec and the values of Δn_2 are
each 0.10 ft, determine flow q and the uniform velocities at Section 1, where the
values of Δn_1 are 0.30 ft.

Solution:

(a) The procedure of preparing the flow net for this case can be applied to more complex cases. For an *ideal* fluid we proceed as follows:

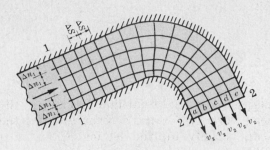

Fig. 6-5

1. Divide the flow at a cross section between parallel boundaries into a number of strips of equal width Δn (using a unit thickness perpendicular to the paper). Each strip represents a streamtube bounded by streamlines or by a streamline and a boundary. The flow between the boundaries is equally divided among the tubes and $\Delta q \cong v(\Delta n) \cong$ constant, where Δn is measured normal to the local velocity. Since $\Delta q \cong v_1 \Delta n_1 \cong v_2 \Delta n_2$, then $v_1/v_2 \cong \Delta n_2/\Delta n_1 \cong \Delta S_2/\Delta S_1$. The smaller the values of Δn and ΔS, the closer the approximation is to an exact relationship. A sufficient number of streamlines should be chosen so that the accuracy will be acceptable, without unnecessary refinement and detail.

2. In order to estimate the *direction* of the streamlines, normal or equipotential lines are drawn. These lines are spaced such that $\Delta S = \Delta n$. The equipotential lines must be perpendicular to the streamlines at each intersection, and to the boundaries since boundaries are streamlines. Thus the diagram will resemble a group of (approximate) squares throughout the entire flow net.

3. At or near changes in boundary shape it is impossible to draw good squares. Changes in initial sketching will be necessary, and a useful check is to draw diagonal lines through all the "squares". These diagonal lines, drawn in both directions, should also form approximate squares.

4. The boundaries themselves usually represent actual streamlines. If this is not so, the flow net will not represent the true flow pattern. For example, where the flow "separates" from the boundary, the boundary itself in this region cannot be used as a streamline. In general where divergent flows occur, separation zones may develop.

Mathematical solutions for irrotational flows are based upon the definition of the *streamfunction*, which definition includes the continuity principle and the properties of a streamline. Flow rate ψ of a streamline is a constant (since no flow can cross the streamline), and if ψ can be established as a function of x and y, the streamline may be plotted. Similarly, the equipotential lines can be defined by $\phi(x, y) =$ a constant. From these expressions we may deduce that

$$u = \partial\psi/\partial y \quad \text{and} \quad v = -\partial\psi/\partial x \qquad \text{for the streamlines}$$

and

$$u = -\partial\phi/\partial x \quad \text{and} \quad v = -\partial\phi/\partial y \qquad \text{for the equipotential lines}$$

These equations must satisfy the Laplace Equation, i.e.,

$$\frac{\partial^2 \psi}{\partial x^2} + \frac{\partial^2 \psi}{\partial y^2} = 0 \qquad \text{or} \qquad \frac{\partial^2 \phi}{\partial x^2} + \frac{\partial^2 \phi}{\partial y^2} = 0$$

and the continuity equation

$$\frac{\partial u}{\partial x} + \frac{\partial v}{\partial y} = 0$$

In general, equipotential functions are evaluated and plotted. Then the orthogonal streamlines are sketched to indicate the flow.

These exact solutions are presented in texts on Advanced Fluid Mechanics, Hydrodynamics, and Complex Variables.

(b) Flow/unit width $= q = \Sigma \Delta q = q_a + q_b + q_c + q_d + q_e = 5(v_2)(A_{n_2})$.

For 1 unit width, $A_{n_2} = 1(\Delta n_2)$ and $q = 5(30.0)(1 \times 0.10) = 15.0$ cfs/unit width.

Then for $\Delta n_1 = 0.30$ ft, $5v_1(0.30 \times 1) = 15.0$ or $v_1 = 10.0$ ft/sec.

v_1 may also be found from: $v_1/v_2 \cong \Delta n_2/\Delta n_1$, $v_1/30.0 \cong 0.10/0.30$, $v_1 = 10.0$ ft/sec.

14. Sketch the streamlines and equipotential lines for the boundary conditions shown in Fig. 6-6. (The area left unmarked is reserved for the reader to use.)

Solution:

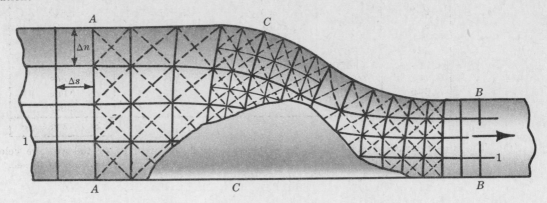

Fig. 6-6

1. Where the flow occurs between parallel boundaries, divide each width into 4 equal parts or stream-tubes (at AA and BB). Try to picture the path of a particle along one of these streamlines, sketching line 1-1, for example (see preceding problem). Proceed in the same way with the other two streamlines.

2. The equipotential lines must be perpendicular to the streamlines at all points and to the boundaries also. They should be so located as to form approximate squares. Starting at the center section, sketch these orthogonal lines in each direction. Frequent use of the eraser will be necessary before a satisfactory flow net is obtained.

3. Draw diagonals (dotted) to check on the reasonableness of the flow net. These diagonals should produce squares.

4. In the above sketch, area C was divided into 8 streamtubes. It will be seen that the smaller quadrangles are more nearly squares than the larger quadrangles. The more numerous the stream-tubes, the more satisfactory will be the "squares" obtained in the flow net.

15. Fig. 6-7 depicts a streamline for two-dimensional flow and its associated equipotential lines 1 to 10, each normal to the streamline. The spacing between the equipotential lines is given in the second column of the table below. If the average velocity between 1 and 2 is 0.500 ft/sec, calculate (a) the average velocities of the streamline between each equipotential line and (b) the time it will take for a particle to move from 1 to 10 along the streamline.

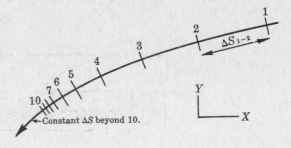

Fig. 6-7

Solution:

(a) Referring to the relation between velocity and Δn in Problem 13,

$$V_{1-2}\,\Delta n_{1-2} \;=\; V_{2-3}\,\Delta n_{2-3} \;=\; V_{3-4}\,\Delta n_{3-4} \;=\; \ldots$$

Also $\Delta S_{1-2} \cong \Delta n_{1-2}, \qquad \Delta S_{2-3} \cong \Delta n_{2-3}, \qquad \ldots$

Thus $V_{2-3} \cong V_{1-2}(\Delta S_{1-2}/\Delta S_{2-3}) = 0.500(0.500/0.400) = 0.625$ ft/sec. Similarly, $V_{3-4} = 0.500(0.500/0.300) = 0.833$ ft/sec, etc. The values of the average velocities are tabulated below.

Position	ΔS (ft)	$\Delta S_{1-2}/\Delta S$	$V = 0.500(0.500/\Delta S)$ ft/sec	$t = (\Delta S)/V$ sec
1—2	0.500	1.000	0.500	1.000
2—3	0.400	1.250	0.625	0.640
3—4	0.300	1.667	0.833	0.360
4—5	0.200	2.500	1.250	0.160
5—6	0.100	5.000	2.500	0.040
6—7	0.0700	7.143	3.571	0.020
7—8	0.0450	11.11	5.56	0.008
8—9	0.0300	16.67	8.33	0.004
9—10	0.0208	24.00	12.00	0.002
				$\Sigma = 2.234$ sec

(b) The time to travel from 1 to 2 equals the distance 1 to 2 divided by the average velocity from 1 to 2, or $t_{1-2} = (0.500/0.500) = 1.000$ sec. Similarly, $t_{2-3} = 0.400/0.625 = 0.640$ sec. The total time to travel from 1 to 10 is the sum of the last column, 2.234 sec.

16. Derive the expression for the kinetic-energy correction factor α for steady, incompressible flow.

Solution:

The true kinetic energy of a particle is $\frac{1}{2} dM\, v^2$, and so the total kinetic energy of a flowing fluid is

$$\frac{1}{2} \int_A (dM)\, v^2 \;=\; \frac{1}{2} \int_A \frac{w}{g}(dQ)v^2 \;=\; \frac{w}{2g} \int_A (v\, dA)v^2$$

To evaluate this expression it must be integrated over the area A.

The kinetic energy calculated by means of the average velocity at a cross section is $\frac{1}{2}(wQ/g)V_{av}^2 = \frac{1}{2}(wA/g)V_{av}^3$. Applying a correction factor α to this expression and equating it to the true kinetic energy, we obtain

$$\alpha\left(\frac{wA}{2g}\right)(V_{av}^3) \;=\; \frac{w}{2g} \int_A (v\, dA)v^2 \qquad \text{or} \qquad \alpha \;=\; \frac{1}{A} \int_A \left(\frac{v}{V_{av}}\right)^3 dA$$

17. A liquid flows through a circular pipe. For a velocity profile satisfying the equation $v = v_{max}(r_o^2 - r^2)/r_o^2$, evaluate the kinetic-energy correction factor α.

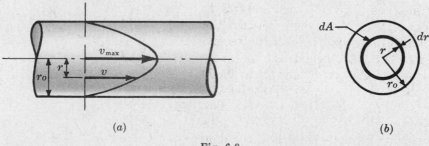

(a)   (b)

Fig. 6-8

Solution:

Evaluate the average velocity so that the equation in Problem 16 may be utilized. From the equation of continuity,

$$V_{av} \;=\; \frac{Q}{A} \;=\; \frac{\int v\, dA}{\pi r_o^2} \;=\; \frac{\int (v_{max}/r_o^2)(r_o^2 - r^2)(2\pi r\, dr)}{\pi r_o^2} \;=\; \frac{2v_{max}}{r_o^4} \int_0^{r_o} (r_o^2 r - r^3)\, dr \;=\; \frac{v_{max}}{2}$$

This value may also be obtained by considering that the given equation represents a parabola and that the volume under the generated paraboloid is half the volume of the circumscribed cylinder. Thus

$$V_{av} \;=\; \frac{\text{volume/sec}}{\text{area of base}} \;=\; \frac{\frac{1}{2}(\pi r_o^2)v_{max}}{\pi r_o^2} \;=\; \frac{v_{max}}{2}$$

Using the value of the average velocity in the equation for α,

$$\alpha \;=\; \frac{1}{A}\int_A\Big(\frac{v}{V_{av}}\Big)^3 dA \;=\; \frac{1}{\pi r_0^2}\int_0^{r_0}\Big(\frac{v_{max}(r_0^2-r^2)/r_0^2}{\tfrac{1}{2}v_{max}}\Big)^3 2\pi r\,dr \;=\; 2.00$$

(See Laminar Flow in Chapter 7.)

18. Oil of sp gr 0.750 is flowing through a 6″ pipe under a pressure of 15.0 psi. If the total energy relative to a datum plane 8.00 ft below the center of the pipe is 58.6 ft lb/lb, determine the flow of oil in cfs.

Solution:

$$\text{Energy per lb of oil} \;=\; \underset{\text{energy}}{\text{pressure}} \;+\; \underset{\text{energy}}{\text{kinetic (velocity head)}} \;+\; \underset{\text{energy}}{\text{potential}}$$

$$58.6 \;=\; \frac{15.0\times144}{0.750\times62.4} + \frac{V_6^2}{2g} + 8.00$$

from which $V_6 = 16.65$ ft/sec. Thus $Q = A_6 V_6 = \tfrac{1}{4}\pi(\tfrac{1}{2})^2 \times 16.65 = 3.26$ cfs.

19. A turbine is rated at 600 horsepower when the flow of water through it is 21.5 cfs. Assuming an efficiency of 87%, what head is acting on the turbine?

Solution:

$$\text{Rated horsepower} \;=\; \text{extracted horsepower} \times \text{efficiency} \;=\; (wQH_T/550) \times \text{efficiency}$$
$$600 \;=\; (62.4\times21.5\times H_T/550)(0.87) \quad\text{and}\quad H_T = 283 \text{ ft.}$$

20. Derive the equations of motion for steady flow of any fluid.

Solution:

Consider as a free body the elementary mass of fluid dM shown in Fig. 6-9(a) and (b). Motion is in the plane of the paper and the x-axis is chosen in the direction of motion. The forces normal to the direction of motion have not been shown acting on the free body dM. Forces acting in the x-direction are due to (1) the pressures acting on the end areas, (2) the component of the weight and (3) the shearing forces (dF_s in pounds) exerted by the adjacent fluid particles.

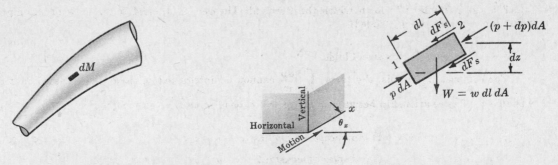

Fig. 6-9(a)   Fig. 6-9(b)

From the equation of motion $\Sigma F_x = Ma_x$, we obtain

$$[+\,p\,dA - (p+dp)dA - w\,dA\,dl\sin\theta_x - dF_s] \;=\; \frac{w\,dA\,dl}{g}\Big(\frac{dV}{dt}\Big) \tag{1}$$

Dividing (1) by $w\,dA$ and replacing dl/dt by velocity V,

$$\left[\frac{p}{w} - \frac{p}{w} - \frac{dp}{w} - dl\sin\theta_x - \frac{dF_s}{w\,dA}\right] \;=\; \frac{V\,dV}{g} \tag{2}$$

The term $\dfrac{dF_s}{w\,dA}$ represents the resistance to flow in length dl. The shearing forces dF_s can be replaced by the intensity of the shear τ times the area over which it acts (perimeter $\times$ length), or $dF_s = \tau\,dP\,dl$.

Then $\dfrac{dF_s}{w\,dA} = \dfrac{\tau\,dP\,dl}{w\,dA} = \dfrac{\tau\,dl}{wR}$, where R is called the hydraulic radius which is defined as the cross sectional area divided by the wetted perimeter or, in this case, dA/dP. The sum of all the shearing forces is the measure of energy lost due to the flow, and, in ft lb/lb, is

$$\text{lost head } dh_L = \frac{\tau\,dl}{wR} = \frac{\text{lb/ft}^2 \times \text{ft}}{\text{lb/ft}^3 \times \text{ft}^2/\text{ft}} = \text{ft}$$

For future reference, $$\tau = wR\left(\frac{dh_L}{dl}\right) \tag{3}$$

Returning to expression (2) above, since $dl \sin \theta_x = dz$, it is written in final form as

$$\frac{dp}{w} + \frac{V\,dV}{g} + dz + dh_L = 0 \tag{4}$$

This expression is known as Euler's Equation when applied to an ideal fluid (lost head $= 0$). When integrated for fluids of constant density, it is known as the Bernoulli Equation. This differential equation (4) for steady flow is a fundamental fluid flow equation.

CASE 1. Flow of Incompressible Fluids.

For *incompressible* fluids, the integration is simple, as follows:

$$\int_{p_1}^{p_2} \frac{dp}{w} + \int_{V_1}^{V_2} \frac{V\,dV}{g} + \int_{z_1}^{z_2} dz + \int_1^2 dh_L = 0 \tag{A}$$

Methods of evaluating the last term will be discussed in a subsequent chapter. The total lost head term will be called H_L. Integrating and substituting limits,

$$\left(\frac{p_2}{w} - \frac{p_1}{w}\right) + \left(\frac{V_2^2}{2g} - \frac{V_1^2}{2g}\right) + (z_2 - z_1) + H_L = 0$$

$$\left(\frac{p_1}{w} + \frac{V_1^2}{2g} + z_1\right) - H_L = \left(\frac{p_2}{w} + \frac{V_2^2}{2g} + z_2\right)$$

which is the customary form in which the Bernoulli Theorem is applied to the flow of incompressible fluids (no external energy added).

CASE 2. Flow of Compressible Fluids.

For compressible fluids, the term $\displaystyle\int_{p_1}^{p_2} \frac{dp}{w}$ cannot be integrated until w is expressed in terms of variable p. The relationship between w and p will depend upon the thermodynamic conditions involved.

(a) For *isothermal* (constant temperature) conditions, the general gas law may be expressed as

$$p_1/w_1 = p/w = \text{constant} \qquad \text{or} \qquad w = (w_1/p_1)p$$

where w_1/p_1 is a constant and p must be in lb/ft^2 *absolute*. Substituting in (A) above,

$$\int_{p_1}^{p_2} \frac{dp}{(w_1/p_1)p} + \int_{V_1}^{V_2} \frac{V\,dV}{g} + \int_{z_1}^{z_2} dz + \int_1^2 dh_L = 0$$

Integrating and substituting limits, $\dfrac{p_1}{w_1}\log_e \dfrac{p_2}{p_1} + \left(\dfrac{V_2^2}{2g} - \dfrac{V_1^2}{2g}\right) + (z_2 - z_1) + H_L = 0$ or, converting to the customary form,

$$\frac{p_1}{w_1}\log_e p_1 + \frac{V_1^2}{2g} + z_1 - H_L = \frac{p_1}{w_1}\log_e p_2 + \frac{V_2^2}{2g} + z_2. \tag{B}$$

Combining this equation with the equation of continuity and the gas law for isothermal conditions yields an expression with only one unknown velocity. Thus, for steady flow,

$$w_1 A_1 V_1 = w_2 A_2 V_2 \quad \text{and} \quad \frac{p_1}{w_1} = \frac{p_2}{w_2} = RT \quad \text{from which} \quad V_1 = \frac{w_2 A_2 V_2}{(w_2/p_2) p_1 A_1} = \frac{A_2}{A_1}\Big(\frac{p_2}{p_1}\Big) V_2$$

Substituting in the Bernoulli form (B) above,

$$\left[\frac{p_1}{w_1} \log_e p_1 + \Big(\frac{A_2}{A_1}\Big)^2 \Big(\frac{p_2}{p_1}\Big)^2 \frac{V_2^2}{2g} + z_1 \right] - H_L = \left[\frac{p_1}{w_1} \log_e p_2 + \frac{V_2^2}{2g} + z_2 \right] \tag{C}$$

(b) For *adiabatic* (no heat gained or lost) conditions, the general gas law reduces to

$$\Big(\frac{w}{w_1}\Big)^k = \frac{p}{p_1} \quad \text{or} \quad \frac{p_1^{1/k}}{w_1} = \frac{p^{1/k}}{w} = \text{constant}, \quad \text{and thus} \quad w = w_1 \Big(\frac{p}{p_1}\Big)^{1/k}$$

where k is the adiabatic exponent.

Establishing and integrating the term dp/w separately, we obtain

$$\int_{p_1}^{p_2} \frac{dp}{w_1 (p/p_1)^{1/k}} = \frac{p_1^{1/k}}{w_1} \int_{p_1}^{p_2} \frac{dp}{p^{1/k}} = \Big(\frac{k}{k-1}\Big) \times \frac{p_1}{w_1}\left[\Big(\frac{p_2}{p_1}\Big)^{(k-1)/k} - 1 \right]$$

and the Bernoulli Equation in customary form becomes

$$\left[\Big(\frac{k}{k-1}\Big) \frac{p_1}{w_1} + \frac{V_1^2}{2g} + z_1 \right] - H_L = \left[\Big(\frac{k}{k-1}\Big)\Big(\frac{p_1}{w_1}\Big)\Big(\frac{p_2}{p_1}\Big)^{(k-1)/k} + \frac{V_2^2}{2g} + z_2 \right] \tag{D}$$

Combining this equation with the equation of continuity and the gas law for adiabatic conditions yields an expression with only one unknown velocity.

Using $w_1 A_1 V_1 = w_2 A_2 V_2$ and $\dfrac{p_1^{1/k}}{w_1} = \dfrac{p_2^{1/k}}{w_2} = \text{constant}$, $V_1 = \dfrac{w_2 A_2 V_2}{w_1 A_1} = \Big(\dfrac{p_2}{p_1}\Big)^{1/k}\Big(\dfrac{A_2}{A_1}\Big) V_2$

and the Bernoulli Equation becomes

$$\left[\Big(\frac{k}{k-1}\Big) \frac{p_1}{w_1} + \Big(\frac{p_2}{p_1}\Big)^{2/k}\Big(\frac{A_2}{A_1}\Big)^2 \frac{V_2^2}{2g} + z_1 \right] - H_L = \left[\Big(\frac{k}{k-1}\Big)\Big(\frac{p_1}{w_1}\Big)\Big(\frac{p_2}{p_1}\Big)^{(k-1)/k} + \frac{V_2^2}{2g} + z_2 \right] \tag{E}$$

21. In Fig. 6-10, water flows from A to B at the rate of 13.2 cfs and the pressure head at A is 22.1 ft. Considering no loss of energy from A to B, find the pressure head at B. Draw the energy line.

Solution:

Apply the Bernoulli equation — A to B, datum A.

Energy at A + energy added − energy lost = energy at B

$$\Big(\frac{p_A}{w} + \frac{V_{12}^2}{2g} + z_A\Big) + 0 - 0 = \Big(\frac{p_B}{w} + \frac{V_{24}^2}{2g} + z_B\Big)$$

where $V_{12} = Q/A_{12} = 13.2/(\frac{1}{4}\pi 1^2) = 16.8$ and
$V_{24} = (\frac{1}{2})^2 (16.8) = 4.20$ ft/sec. Substituting,

$$\Big(22.1 + \frac{(16.8)^2}{2g} + 0\Big) - 0 = \Big(\frac{p_B}{w} + \frac{(4.20)^2}{2g} + 15.0\Big) \quad \text{and} \quad \frac{p_B}{w} = 11.2 \text{ ft water}$$

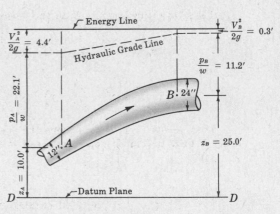

Fig. 6-10

Total energy at any section can be plotted above a chosen datum plane. Using D-D in this case,

Energy at A = $p_A/w + V_{12}^2/2g + z_A$ = $22.1 + 4.4 + 10.0 = 36.5$ ft
Energy at B = $p_B/w + V_{24}^2/2g + z_B$ = $11.2 + 0.3 + 25.0 = 36.5$ ft

Note that transformation from one form of energy to another occurs during flow. In this case a portion of both the pressure energy and the kinetic energy at A is transformed to potential energy at B.

22. For the Venturi meter shown in Fig. 6-11, the deflection of the mercury in the differential gage is 14.3 inches. Determine the flow of water through the meter if no energy is lost between A and B.

Solution:

Apply the Bernoulli equation $-A$ to B, datum A.

$$\left(\frac{p_A}{w} + \frac{V_{12}^2}{2g} + 0\right) - 0 = \left(\frac{p_B}{w} + \frac{V_6^2}{2g} + 2.50\right)$$

and $\left(\dfrac{p_A}{w} - \dfrac{p_B}{w}\right) = \left(\dfrac{V_6^2}{2g} - \dfrac{V_{12}^2}{2g} + 2.50\right)$ *(1)*

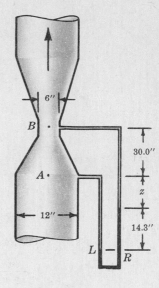

The equation of continuity yields $A_{12}V_{12} = A_6 V_6$ or $V_{12} = (\frac{6}{12})^2 V_6 = \frac{1}{4}V_6$, and $V_{12}^2 = \frac{1}{16}V_6^2$. For the gage,

pressure head at L = pressure head at R (ft of water)

$p_A/w + z + 14.3/12 = p_B/w + 2.50 + z + (14.3/12)(13.6)$

from which $(p_A/w - p_B/w) = 17.5$ ft water. Substituting in *(1)*, we obtain $V_6 = 32.1$ ft/sec and $Q = \frac{1}{4}\pi(\frac{1}{2})^2 \times 32.1 = 6.30$ cfs.

Fig. 6-11

23. A pipe carrying oil of sp gr 0.877 changes in size from 6″ at Section E to 18″ at Section R. Section E is 12 ft lower than R and the pressures are 13.2 psi and 8.75 psi respectively. If the discharge is 5.17 cfs, determine the lost head and the direction of flow.

Solution:

The average velocity at each cross section $= Q/A$. Then

$$V_6 = \frac{5.17}{\frac{1}{4}\pi(1/2)^2} = 26.3 \text{ ft/sec} \quad \text{and} \quad V_{18} = \frac{5.17}{\frac{1}{4}\pi(3/2)^2} = 2.92 \text{ ft/sec}$$

Using the lower section E as the datum plane, the energy at each section is:

$$\text{at } E, \quad \left(\frac{p}{w} + \frac{V_6^2}{2g} + z\right) = \frac{13.2 \times 144}{0.877 \times 62.4} + \frac{(26.3)^2}{2g} + 0 = 45.5 \text{ ft lb/lb}$$

$$\text{at } R, \quad \left(\frac{p}{w} + \frac{V_{18}^2}{2g} + z\right) = \frac{8.75 \times 144}{0.877 \times 62.4} + \frac{(2.92)^2}{2g} + 12.0 = 35.1 \text{ ft lb/lb}$$

Flow occurs from E to R since the energy at E exceeds that at R. The lost head can be found, using E to R, datum E: $45.5 -$ lost head $= 35.1$ or lost head $= 10.4$ ft, E to R.

24. For the meter in Problem 22 consider air at $80°F$ flowing with the pressure at A equal to 37.5 psi gage. Consider a deflection of the gage of 14.3 inches of water. Assuming that the specific weight of air does not change between A and B and that the energy loss is negligible, determine the amount of air flowing in lb/sec.

Solution:

Using A to B, datum A, as in Problem 22 we obtain $\left(\dfrac{p_A}{w} - \dfrac{p_B}{w}\right) = \dfrac{15}{16}\dfrac{V_6^2}{2g} + 2.50.$ *(1)*

To obtain the pressure head of fluid flowing, the specific weight of the air must be calculated.

$$w = \frac{p}{RT} = \frac{(37.5 + 14.7)144}{53.3(80 + 460)} = 0.261 \text{ lb/ft}^3$$

For the differential gage, $p_L = p_R$ (in psf gage)

or $p_A + 0.261(z + 14.3/12) = p_B + 0.261(2.50 + z) + 62.4(14.3/12)$

and $(p_A - p_B) = 74.7$ psf.   Substituting in (1), we obtain $V_6 = 140.0$ ft/sec and

$$W = wQ = 0.261[\tfrac{1}{4}\pi(\tfrac{1}{2})^2 \times 140.0] = 7.17 \text{ lb/sec of air}$$

25. A horizontal air duct reduces in cross sectional area from 0.75 ft² to 0.20 ft². Assuming no losses, what pressure change will occur when 1·50 lb/sec of air flows? Use $w = 0.200$ lb/ft³ for pressure and temperature conditions involved.)

Solution:

$$Q = \frac{1.50 \text{ lb/sec}}{0.200 \text{ lb/ft}^3} = 7.50 \text{ cfs}, \quad V_1 = \frac{Q}{A_1} = \frac{7.50}{0.75} = 10.0 \text{ ft/sec}, \quad V_2 = \frac{Q}{A_2} = \frac{7.50}{0.20} = 37.5 \text{ ft/sec}.$$

Applying the Bernoulli theorem, Section 1 to Section 2, gives

$$\left(\frac{p_1}{w} + \frac{(10.0)^2}{2g} + 0\right) - 0 = \left(\frac{p_2}{w} + \frac{(37.5)^2}{2g} + 0\right) \quad \text{or} \quad \left(\frac{p_1}{w} - \frac{p_2}{w}\right) = 20.3 \text{ ft air}$$

and $p_1' - p_2' = (20.3 \times 0.200)/144 = 0.0282$ psi change. This small pressure change justifies the assumption of constant density of the fluid.

26. A 6″ pipe 600 ft long carries water from A at elevation 80.0 to B at elevation 120.0. The frictional stress between the liquid and the pipe walls is 0.62 lb/ft². Determine the pressure change in the pipe and the lost head.

Solution:

(a) The forces acting on the mass of water are the same as those shown in Fig. (b) of Prob. 20.

Using $P_1 = p_1 A_6$, $P_2 = p_2 A_6$ we obtain, from $\Sigma F_x = 0$,

$$p_1 A_6 - p_2 A_6 - W \sin \theta_x - \tau(\pi d)L = 0$$

Now $W = w(\text{volume}) = 62.4[\tfrac{1}{4}\pi(1/2)^2 \times 600]$ and $\sin \theta_x = (120 - 80)/600$. Then

$$p_1[\tfrac{1}{4}\pi(\tfrac{1}{2})^2] - p_2[\tfrac{1}{4}\pi(\tfrac{1}{2})^2] - 62.4[\tfrac{1}{4}\pi(\tfrac{1}{2})^2 \times 600] \times 40/600 - 0.62(\pi \times \tfrac{1}{2} \times 600) = 0$$

from which $p_1 - p_2 = 5480$ psf $= 38.0$ psi.

(b) Using the energy equation, datum at A,

$$\text{energy at } A - \text{lost head} = \text{energy at } B$$

$$\left(\frac{p_A}{w} + \frac{V_A^2}{2g} + 0\right) - \text{lost head} = \left(\frac{p_B}{w} + \frac{V_B^2}{2g} + 40\right)$$

or lost head $= (p_A/w - p_B/w) - 40 = 5480/62.4 - 40 = 47.8$ ft.

Another method:

Using (3) of Problem 20, lost head $= \dfrac{\tau L}{wR} = \dfrac{0.62(600)}{62.4(0.50/4)} = 47.8$ ft.

27. Water at 90°F is to be lifted from a sump at a velocity of 6.50 ft/sec through the suction pipe of a pump. Calculate the theoretical maximum height of the pump setting under the following conditions: atmospheric pressure = 14.25 psia, vapor pressure = 0.70 psia (see Table 1C), and lost head in the suction pipe = 3 velocity heads.

Solution:

The specific weight of water at 90°F from Table 1C is 62.1 lb/ft³. The minimum pressure at entrance to the pump cannot exceed the vapor pressure of the liquid. The energy equation will be applied from the surface of the water outside the suction pipe to the entrance to the pump, using absolute pressure heads.

$$\text{Energy at water surface} - \text{lost head} = \text{energy at entrance to pump}$$

$$\left(\frac{14.25 \times 144}{62.1} + 0 + 0\right) - \frac{3(6.50)^2}{2g} = \left(\frac{0.70 \times 144}{62.1} + \frac{(6.50)^2}{2g} + z\right)$$

from which $z = 28.8$ ft above water surface.

Under these conditions serious damage due to cavitation will probably occur. See Chapter 12.

28. For the system shown in Fig. 6-12, pump *BC* must deliver 5.62 cfs of oil, sp gr = 0.762 to reservoir *D*. Assuming that the energy lost from *A* to *B* is 8.25 ft lb/lb and from *C* to *D* is 21.75 ft lb/lb, (*a*) how many horsepower units must the pump supply to the system? (*b*) Plot the energy line.

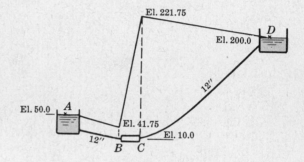

Fig. 6-12

Solution:

(*a*) The velocity of particles at *A* and *D* will be very small and hence the velocity head terms will be negligible.

    A to *D*, datum *BC* (*A* would be satisfactory also),

$$\left(\frac{p_A}{w} + \frac{V_A^2}{2g} + z_A\right) + H_{pump} - H_{lost} = \left(\frac{p_D}{w} + \frac{V_D^2}{2g} + z_D\right)$$

$$(0 + negl + 40.0) + H_{pump} - (8.25 + 21.75) = (0 + negl + 190.0)$$

and $H_{pump} = 180.0$ ft (or ft lb/lb).

    Horsepower = $wQH_{pump}/550 = (0.762 \times 62.4)(5.62)(180.0)/550 = 87.4$ delivered to system.

Note that the pump has supplied a head sufficient to raise the oil 150.0 ft and also has overcome 30.0 ft of losses in the piping. Thus 180 ft is delivered to the system.

(*b*) The energy line at *A* is at Elevation 50.0 above datum zero. From *A* to *B* the energy loss is 8.25 ft and the energy line drops by this amount, the elevation at *B* being 41.75. The pump adds 180.0 ft of energy and the elevation at *C* is thus 221.75. Finally, the loss of energy between *C* and *D* being 21.75 ft, the elevation at *D* = 221.75 − 21.75 = 200.0. These data have been shown graphically on the above figure.

29. Water flows through the turbine in Fig. 6-13 at the rate of 7.55 cfs and the pressures at *A* and *B* respectively are 21.4 psi and −5.00 psi. Determine the horsepower delivered to the turbine by the water.

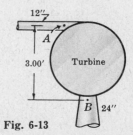

Solution:

Using *A* to *B* (datum *B*), with

$$V_{12} = 7.55/A_{12} = 9.60 \quad \text{and} \quad V_{24} = 9.60/4 = 2.40 \text{ ft/sec,}$$

Fig. 6-13

$$\left(\frac{p_A}{w} + \frac{V_{12}^2}{2g} + z_A\right) + 0 - H_{\text{Turbine}} = \left(\frac{p_B}{w} + \frac{V_{24}^2}{2g} + z_B\right)$$

$$\left(\frac{21.4(144)}{62.4} + \frac{9.60^2}{2g} + 3.00\right) - H_T = \left(\frac{-5.00(144)}{62.4} + \frac{2.40^2}{2g} + 0\right) \quad \text{and} \quad H_T = 65.4 \text{ ft.}$$

Horsepower $= wQH_T/550 = 62.4(7.55)(65.4)/550 = 56.0$ to turbine.

30. For the turbine of Problem 29, if 65.5 horsepower is extracted while the pressure gages at A and B are reading 20.5 psi and -4.8 psi respectively, how much water is flowing?

Solution:

Using A to B (datum B), $\quad \left(\frac{20.5(144)}{62.4} + \frac{V_{12}^2}{2g} + 3.00\right) - H_T = \left(\frac{-4.8(144)}{62.4} + \frac{V_{24}^2}{2g} + 0\right) \quad$ and

(a) $\qquad H_T = \left(\frac{25.3(144)}{62.4} + 3.00 + \frac{V_{12}^2}{2g} - \frac{V_{24}^2}{2g}\right)$

(b) $\qquad A_{12}V_{12} = A_{24}V_{24} \quad \text{or} \quad \frac{V_{24}^2}{2g} = \left(\frac{1}{2}\right)^4 \frac{V_{12}^2}{2g} = \frac{1}{16}\frac{V_{12}^2}{2g}$

(c) $\qquad 65.5 \text{ HP} = \frac{wQH_T}{550} = \frac{62.4 \times \frac{1}{4}\pi(1)^2 V_{12} \times H_T}{550} \quad \text{or} \quad H_T = \frac{735}{V_{12}}$

Equating (a) and (c) (with velocity head substitution), $\quad 735/V_{12} = 61.5 + \frac{15}{16}(V_{12}^2/2g) \quad$ or

$$61.5 V_{12} + 0.0146 V_{12}^3 = 735$$

Solving this equation by successive trials:

Try $V_{12} = 10$ ft/sec, $\qquad 615 + 15 \neq 735 \quad$ (must increase V)
Try $V_{12} = 12$ ft/sec, $\qquad 738 + 25 \neq 735 \quad$ (answer straddled)
Try $V_{12} = 11.6$ ft/sec, $\qquad 712 + 23 = 735 \quad$ (good result)

The flow $Q = A_{12}V_{12} = \frac{1}{4}\pi(1)^2 \times 11.6 = 9.10$ cfs.

31. Oil of specific gravity 0.761 flows from tank A to tank E as shown in Fig. 6-14. Lost head items may be assumed to be as follows:

A to $B = 0.60 \dfrac{V_{12}^2}{2g} \qquad C$ to $D = 0.40 \dfrac{V_6^2}{2g}$

B to $C = 9.0 \dfrac{V_{12}^2}{2g} \qquad D$ to $E = 9.0 \dfrac{V_6^2}{2g}$

Find (a) the flow Q in cfs,
     (b) the pressure at C in psi and
     (c) the horsepower at C, datum E.

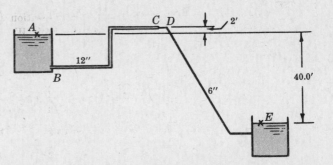

Fig. 6-14

Solution:

(a) Using A to E, datum E,

at A		A to B	B to C	C to D	D to E		at E

$$(0 + \text{negl} + 40.0) - \left[\left(0.60\frac{V_{12}^2}{2g} + 9.0\frac{V_{12}^2}{2g}\right) + \left(0.40\frac{V_6^2}{2g} + 9.0\frac{V_6^2}{2g}\right)\right] = (0 + \text{negl} + 0)$$

or $40.0 = 9.6(V_{12}^2/2g) + 9.4(V_6^2/2g)$. Also, $V_{12}^2 = (\frac{1}{2})^4 V_6^2 = \frac{1}{16}V_6^2$. Substituting and solving,

$$V_6^2/2g = 4 \text{ ft}, \quad V_6 = 16.1 \text{ ft/sec} \quad \text{and} \quad Q = \frac{1}{4}\pi(\frac{1}{2})^2 \times 16.1 = 3.16 \text{ cfs}$$

(b) Using A to C, datum A,

$$(0 + \text{negl} + 0) - (0.60 + 9.0)\frac{V_{12}^2}{2g} = \left(\frac{p_c}{w} + \frac{V_{12}^2}{2g} + 2\right) \quad \text{and} \quad \frac{V_{12}^2}{2g} = \frac{1}{16}\frac{V_6^2}{2g} = \frac{1}{16}(4) = \frac{1}{4} \text{ ft}$$

Then $p_c/w = -4.65$ ft of oil gage and $p_c' = (0.761 \times 62.4)(-4.65)/144 = -1.54$ psi gage.

The Bernoulli Equation could have been applied from C to E with equally satisfactory results. The two equations obtained by the two choices would *not* be independent, simultaneous equations.

(c) Horsepower at $C = \dfrac{wQH_c}{550} = \dfrac{(0.761 \times 62.4)(3.16)(-4.65 + 0.25 + 42.0)}{550} = 10.25$, datum E.

32. The head extracted by turbine CR in the figure is 200 ft and the pressure at T is 72.7 psi. For losses of $2.0(V_{24}^2/2g)$ between W and R and $3.0(V_{12}^2/2g)$ between C and T, determine (a) how much water is flowing and (b) the pressure head at R. Draw the energy line.

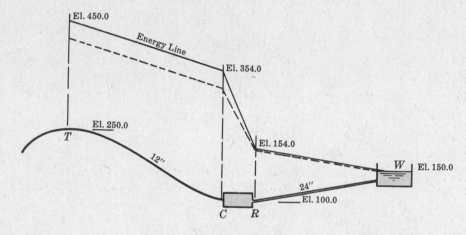

Fig. 6-15

Solution:

Because the energy line at T is at Elevation $\left(250.0 + \dfrac{72.7 \times 144}{62.4} + \dfrac{V_{12}^2}{2g}\right)$ and well above the elevation at W, the water flows into the reservoir W.

(a) Using T to W, datum zero,

$$\overset{\text{at } T}{\left(\frac{72.7 \times 144}{62.4} + \frac{V_{12}^2}{2g} + 250\right)} - \overset{T \text{ to } C \quad R \text{ to } W}{\left[3.0\frac{V_{12}^2}{2g} + 2.0\frac{V_{24}^2}{2g}\right]} \overset{H_T}{- 200} = \overset{\text{at } W}{(0 + \text{negl} + 150)}$$

Substituting $V_{24}^2 = \frac{1}{16}V_{12}^2$ and solving, $V_{12}^2/2g = 32.0$ ft or $V_{12} = 45.4$ ft/sec. Then

$$Q = \tfrac{1}{4}\pi(1)^2 \times 45.4 = 35.6 \text{ cfs}$$

(b) Using R to W, datum R, $(p_R/w + \frac{1}{16} \times 32.0 + 0) - 2(\frac{1}{16} \times 32.0) = (0 + \text{negl} + 50)$ and $p_R/w = 52.0$ ft. The reader may check this pressure head by applying the Bernoulli Equation between T and R.

To plot the energy line in the figure, evaluate the energy at the 4 sections indicated.

$$\begin{aligned}
\text{Elevation of energy line at } T &= 168.0 + 32.0 + 250.0 &&= 450.0 \\
\text{at } C &= 450.0 - 3 \times 32.0 &&= 354.0 \\
\text{at } R &= 354.0 - 200.0 &&= 154.0 \\
\text{at } W &= 154.0 - 2 \times \tfrac{1}{16} \times 32 &&= 150.0
\end{aligned}$$

It will be shown in the following chapter that the energy line is a straight line for steady flow in a pipe of constant diameter. The hydraulic grade line will be parallel to the energy line and $V^2/2g$ below it (shown dotted).

33. (a) What is the pressure on the nose of a torpedo moving in salt water at 100 ft/sec at a depth of 30.0 ft?   (b) If the pressure at a point C on the side of the torpedo at the same elevation as the nose is 10.0 psi gage, what is the relative velocity at that point?

Solution:

(a) In this case, greater clarity in the application of the Bernoulli equation may be attained by considering the relative motion of a stream of water past the stationary torpedo. The velocity at the nose of the torpedo will then be zero. Assuming no lost head in the streamtube from a point A in the undisturbed water just ahead of the torpedo to a point B on the nose of the torpedo, the Bernoulli equation becomes

$$\left(\frac{p_A}{w} + \frac{V_A^2}{2g} + z_A\right) - 0 = \left(\frac{p_B}{w} + \frac{V_B^2}{2g} + z_B\right) \quad \text{or} \quad 30.0 + \frac{(100)^2}{2g} + 0 = \frac{p_B}{w} + 0 + 0$$

Then $p_B/w = 185$ ft of salt water, and $p'_B = wh/144 = 64.0(185)/144 = 82.2$ psi gage.

This pressure is called the stagnation pressure and may be expressed as $p_s = p_0 + \frac{1}{2}\rho V_0^2$ in psf units. For further discussion, see Chapters 9 and 11.

(b) The Bernoulli equation may be applied either from point A to point C or from point B to point C. Considering A and C,

$$\left(\frac{p_A}{w} + \frac{V_A^2}{2g} + z_A\right) - 0 = \left(\frac{p_C}{w} + \frac{V_C^2}{2g} + z_C\right) \quad \text{or} \quad \left(30.0 + \frac{(100)^2}{2g} + 0\right) = \left(\frac{10.0 \times 144}{64.0} + \frac{V_C^2}{2g} + 0\right)$$

from which $V_C = 102.3$ ft/sec.

34. A sphere is placed in an air stream which is at atmospheric pressure and is moving at 100.0 ft/sec. Using the density of air constant at 0.00238 slug/ft³, (a) calculate the stagnation pressure and (b) calculate the pressure on the surface of the sphere at a point B, 75° from the stagnation point, if the velocity there is 220.0 ft/sec.

Solution:

(a) Applying the expression given in the preceding problem, we obtain

$$p_s = p_0 + \frac{1}{2}\rho V_0^2 = 14.7(144) + \frac{1}{2}(0.00238)(100.0)^2 = 2117 + 11.9 = 2129 \text{ psf}$$

(b) The specific weight of air $= \rho g = 0.00238(32.2) = 0.0765$ lb/ft³.

Applying the Bernoulli equation, stagnation point to point B, produces

$$\left(\frac{p_s}{w} + \frac{V_s^2}{2g} + 0\right) - 0 = \left(\frac{p_B}{w} + \frac{V_B^2}{2g} + 0\right) \quad \text{or} \quad \left(\frac{2129}{0.0765} + 0 + 0\right) = \left(\frac{p_B}{w} + \frac{(220)^2}{2g} + 0\right)$$

from which $p_B/w = 27,050$ ft of air, and $p'_B = wh/144 = 0.0765(27050)/144 = 14.4$ psi.

35. A large closed tank is filled with ammonia under a pressure of 5.30 psi gage and at 65°F. The ammonia discharges into the atmosphere through a small opening in the side of the tank. Neglecting friction losses, calculate the velocity of the ammonia leaving the tank (a) assuming constant density and (b) assuming adiabatic flow conditions.

Solution:

(a)   Apply the Bernoulli equation – tank to atmosphere.

$$\left(\frac{5.30 \times 144}{w_1} + 0 + 0\right) \; = \; \left(0 + \frac{V^2}{2g} + 0\right) \quad \text{where} \quad w_1 \; = \; \frac{p_1}{RT} \; = \; \frac{(5.30 + 14.7)144}{89.5(460 + 65)} \; = \; 0.0613 \text{ lb/ft}^3$$

Substituting and solving, $V = 895$ ft/sec.

For a constant specific weight w, either gage or absolute pressure heads may be used. Absolute pressure head *must* be used for cases where w is not constant.

(b)   For $V_1 = 0$ and $z_1 = z_2$, the adiabatic expression (D) in Problem 20 may be written

$$\left(\frac{k}{k-1}\right)\frac{p_1}{w_1}\left[1 - \left(\frac{p_2}{p_1}\right)^{(k-1)/k}\right] \; = \; \frac{V_2^2}{2g}$$

For ammonia, $k = 1.32$ from Table 1 in the Appendix, and

$$\frac{1.32}{0.32} \times \frac{20.0 \times 144}{0.0613}\left[1 - \left(\frac{14.7 \times 144}{20.0 \times 144}\right)^{0.242}\right] \; = \; \frac{V_2^2}{2g} \; = \; 13,900 \quad \text{or} \quad V_2 \; = \; 945 \text{ ft/sec}$$

The error in using the velocity based upon the constant density assumption is about 5.3%. The specific weight of the ammonia in the jet is calculated by using the expression

$$\frac{p_1}{p_2} \; = \; \left(\frac{w_1}{w_2}\right)^k \quad \text{or} \quad \frac{20.0}{14.7} \; = \; \left(\frac{0.0613}{w_2}\right)^{1.32} \quad \text{and} \quad w_2 \; = \; 0.0486 \text{ lb/ft}^3$$

Despite this 20.7% change in density, the error in the velocity was but 5.3%.

36. Compare the velocities in (a) and (b) of Problem 35 for a pressure of 15.3 psi gage in the tank.

Solution:

(a) $w_1 = \dfrac{p_1}{RT} = \dfrac{30.0 \times 144}{89.5 \times 525} = 0.0922$ lb/ft³ and, from the preceding problem,

$$\frac{15.3 \times 144}{0.0922} \; = \; \frac{V^2}{2g} \quad \text{and} \quad V \; = \; 1240 \text{ ft/sec}$$

(b)   Using the adiabatic expression given in the preceding problem,

$$\frac{V^2}{2g} \; = \; \frac{1.32}{0.32} \times \frac{30.0 \times 144}{0.0922}\left[1 - \left(\frac{14.7 \times 144}{30.0 \times 144}\right)^{0.242}\right] \; = \; 30,700 \quad \text{and} \quad V \; = \; 1410 \text{ ft/sec}$$

The error in using the velocity based upon the constant density assumption is about 12%. The change in density in this case is about 42%.

Limitations on the magnitude of velocity will be discussed in Chapter 11. It should be pointed out that the limiting velocity for the temperature involved is 1412 ft/sec.

37. Nitrogen flows from a 2″ pipe in which the temperature and pressure are 40°F and 40 psi gage respectively into a 1″ pipe in which the pressure is 21.3 psi gage. Calculate the velocity in each pipe, assuming isothermal conditions apply and no losses.

Solution:

Referring to Problem 20, the equation (C) for isothermal conditions may be solved for V_2, noting that $z_1 = z_2$,

$$\frac{V_2^2}{2g}\left[1 - \left(\frac{A_2 p_2}{A_1 p_1}\right)^2\right] = \frac{p_1}{w_1}\log_e\left(\frac{p_1}{p_2}\right) = RT\log_e\left(\frac{p_1}{p_2}\right) \quad \text{or} \quad V_2 = \sqrt{2g \times \frac{RT\log_e(p_1/p_2)}{1 - (A_2 p_2/A_1 p_1)^2}}$$

Substituting herein, using $R = 55.1$ for nitrogen from Table 1 in the Appendix,

$$V_2 = \sqrt{2g \times \frac{55.1 \times 500 \log_e(54.7 \times 144)/(36.0 \times 144)}{1 - (\frac{1}{2})^4[(36.0 \times 144)/(54.7 \times 144)]^2}} = 875 \text{ ft/sec}$$

Also $V_1 = (A_2/A_1)(p_2/p_1)V_2 = (\frac{1}{2})^2(36.0/54.7)(875) = 144$ ft/sec.

38. In Problem 37, had the pressure, velocity and temperature in the 2″ pipe been 38.0 psi gage, 143 ft/sec and 32°F respectively, calculate the velocity and pressure in the 1″ pipe. Assume no lost head and isothermal conditions.

Solution:

From Problem 20 for isothermal conditions using equation (C) in terms of V_1 instead of V_2,

$$(a) \quad \frac{(143)^2}{2g}\left[1 - \left(\frac{4}{1}\right)^2\left(\frac{52.7 \times 144}{p_2' \times 144}\right)^2\right] = 55.1 \times 492 \log_e\frac{p_2' \times 144}{52.7 \times 144}$$

Only one unknown appears, yet a direct solution is difficult. The method of successive trials seems indicated, assuming a value for p_2' in the denominator of the bracket.

(1) Assume $p_2' = 52.7$ psi absolute, and solve for p_2' in the right-hand side of the equation.

$$318[1 - 16(1)^2] = 27,100 \log_e(p_2'/52.7)$$

from which $p_2' = 44.4$ psi absolute.

(2) The effect of using $p_2' = 44.4$ psi in (a) would result in an inequality. Anticipating the result, assume the new value of p_2' at 35.0 psi and solve.

$$318[1 - 16(52.7/35.0)^2] = 27,100 \log_e p_2'/52.7$$

from which $p_2' = 35.2$ psi absolute (slide rule, close enough). For velocity,

$$V_2 = \frac{w_1 A_1}{w_2 A_2}V_1 \quad \text{or} \quad V_2 = \frac{p_1}{p_2}\left(\frac{A_1}{A_2}\right)V_1 = \frac{52.7 \times 144}{35.2 \times 144}\left(\frac{2}{1}\right)^2 \times 143 = 855 \text{ ft/sec}$$

Supplementary Problems

39. What average velocity in a 6″ pipe will produce a flow of 1.0 mgd of water? *Ans.* 7.87 ft/sec

40. What size pipe can carry 70.7 cfs at an average velocity of 10.0 ft/sec? *Ans.* 3 ft

41. A 12″ pipe carrying 3.93 cfs connects to a 6″ pipe. Find the velocity head in the 6″ pipe. *Ans.* 6.21 ft

42. A 6″ pipe carries 2.87 cfs of water. The pipe branches into two pipes, one 2″ in diameter and the other 4″ in diameter. If the velocity in the 2″ pipe is 40 ft/sec, what is the velocity in the 4″ pipe? *Ans.* 22.9 ft/sec

43. Determine if the following expressions for velocity components satisfy the conditions for steady, incompressible flow.

(a) $u = 3xy^2 + 2x + y^2$
    $v = x^2 - 2y - y^3$

(b) $u = 2x^2 + 3y^2$
    $v = -3xy$

Ans. (a) Yes   (b) No

44. A 12″ diameter pipe carries oil with a velocity distribution of $v = 9(r_o^2 - r^2)$. Determine the average velocity and the value of the kinetic-energy correction factor.   *Ans.* $\alpha = 2.00$, $V_{av} = 1.13$ ft/sec

45. Show that the equation of continuity may be written in the form $1 = \dfrac{1}{A} \displaystyle\int_A \left(\dfrac{v}{V_{av}}\right) dA$.

46. A 12″ pipe carries oil of sp gr 0.812 at a rate of 3.93 cfs and the pressure at a point A is 2.67 psi gage. If point A is 6.20 ft above the datum plane, calculate the energy at A in ft lb/lb.
    Ans. 14.2 ft lb/lb

47. How many lb/sec of carbon dioxide are flowing through a 6″ pipe when the pressure is 25.0 psi gage, the temperature 80°F and the average velocity 8.00 ft/sec?   *Ans.* 0.476 lb/sec

48. An 8″ pipe carries air at 80.0 ft/sec, 21.5 psi absolute and 80°F. How many pounds of air are flowing? The 8″ pipe reduces to a 4″ pipe and the pressure and temperature in the 4″ pipe are 19.0 psi absolute and 52°F respectively. Find the velocity in the 4″ pipe and compare the flows in cfs in the two pipes.
    Ans. 3.00 lb/sec, 343 ft/sec, 27.9 cfs, 29.9 cfs

49. Air flows with a velocity of 16.0 ft/sec in a 4″ pipe. A pressure gage measures 30 psi and the temperature is 60°F. At another point downstream a gage measures 20 psi and the temperature is 80°F. For a standard barometric reading, calculate the velocity downstream and compare the cfs rate of flow at each section.   *Ans.* 21.4 ft/sec, 1.40 cfs, 1.87 cfs

50. Sulfur dioxide flows through a 12″ discharge duct which reduces to 6″ in diameter at discharge into a stack. The pressures in the duct and discharge stream are respectively 20.0 psi absolute and atmospheric (14.7 psi). The velocity in the duct is 50.0 ft/sec and the temperature is 80°F. Calculate the velocity in the discharge stream if the temperature of the gas is 23°F.   *Ans.* 244 ft/sec

51. Water flows through a horizontal 6″ pipe under a pressure of 60 psi. Assuming no losses, what is the flow in cfs if the pressure at a 3″ diameter reduction is 20 psi?   *Ans.* $Q = 3.91$ cfs

52. If oil of specific gravity 0.752 flows in Problem 51, find flow Q.   *Ans.* 4.51 cfs

53. If carbon tetrachloride (sp gr 1.594) flows in Problem 51, find flow Q.   *Ans.* 3.09 cfs

54. Water flows upward in a vertical 12″ pipe at the rate of 7.85 cfs. At point A in the pipe the pressure is 30.5 psi. At B, 15 ft above A, the diameter is 24″ and the lost head A to B equals 6 ft. Determine the pressure at B in psi.   *Ans.* 22.1 psi

55. A 12″ pipe contains a short section in which the diameter is gradually reduced to 6″ and then enlarged again to 12″. The 6″ section is 2 ft below Section A in the 12″ where the pressure is 75 psi. If a differential gage containing mercury is attached to the 12″ and 6″ sections, what is the deflection of the gage when the flow of water is 4.25 cfs downward? Assume no lost head.   *Ans.* 6.46 in.

56. A 12″ pipe line carries oil of specific gravity 0.811 at a veloctiy of 80.0 ft/sec. At points A and B measurements of pressure and elevation were respectively 52.6 psi and 42.0 psi and 100.0 ft and 110.0 ft. For steady flow, find the lost head between A and B.   *Ans.* 20.2 ft

57. A stream of water, 3″ in diameter, discharges into the atmosphere at a velocity 80.0 ft/sec. Find the horsepower in the jet using the datum plane through the center of the jet.   *Ans.* 44.3 HP

58. A reservoir supplies water to a horizontal 6″ pipe 800 ft long. The pipe flows full and discharges into the atmosphere at the rate of 2.23 cfs. What is the pressure in psi midway in the pipe, assuming the only lost head is 6.20 ft in each 100 ft of length?   *Ans.* 10.7 psi

59. Oil of specific gravity 0.750 is pumped from a tank over a hill through a 24″ pipe with a pressure at the top of the hill maintained at 25.5 psi. The summit is 250 ft above the surface of the oil in the tank and oil is pumped at the rate of 22.0 cfs. If the lost head from tank to summit is 15.7 ft, what horsepower must the pump supply to the liquid?   *Ans.* 645 HP

60. A pump draws water from a sump through a vertical 6″ pipe. The pump has a horizontal discharge pipe 4″ in diameter which is 10.6 ft above the water level in the sump. While pumping 1.25 cfs, gages near the pump at entrance and discharge read −4.6 psi and +25.6 psi respectively. The discharge gage is 3.0 ft above the suction gage. Compute the horsepower output of the pump and the head lost in the 6″ suction pipe.   *Ans.* 10.7 HP, 2.4 ft

61. Compute the lost head in a 6″ pipe if it is necessary to maintain a pressure of 33.5 psi at a point upstream and 6 ft below where the pipe discharges water into the atmosphere at the rate of 1.963 cfs. *Ans.* 71.3 ft

62. A large tank is partly filled with water, the air space above being under pressure. A 2″ hose connected to the tank discharges on the roof of a building 50 ft above the level in the tank. The friction loss is 18 ft. What air pressure must be maintained in the tank to deliver 0.436 cfs on the roof? *Ans.* 32.1 psi

63. Water is pumped from reservoir A at elevation 750.0 to reservoir E at elevation 800.0 through a 12″ pipe line. The pressure in the 12″ pipe at point D, at elevation 650.0, is 80.0 psi. The lost heads are: A to pump suction $B = 2.0$ ft, pump discharge C to $D = 38V^2/2g$, and D to $E = 40V^2/2g$. Find discharge Q and horsepower supplied by pump BC.   *Ans.* 5.95 cfs, 82 HP

64. A horizontal Venturi meter has diameters at inlet and throat of 24″ and 18″ respectively. A differential gage connected to inlet and throat contains water which is deflected 4″ when air flows through the meter. Considering the specific weight of air to be constant at 0.0800 lb/ft³ and neglecting friction, determine the flow in cfs.   *Ans.* 276 cfs

65. Water is to be siphoned from a tank at the rate of 3.15 cfs. The flowing end of the siphon pipe must be 14 ft below the water surface. The lost head terms are $1.50V^2/2g$ from tank to summit of siphon and $1.00V^2/2g$ from summit to end of siphon. The summit is 5 ft above the water surface. Find the size pipe needed and the pressure at the summit.   *Ans.* 6″, −6.50 psi

66. A horizontal 24″ pipe line carries oil of sp gr 0.825 flowing at a rate of 15.7 cfs. Each of 4 pumps required along the line is the same, i.e., the pressure on the suction side and on the discharge side will be −8.0 psi and +350 psi respectively. If the lost head at the discharge stated is 6.0 ft for each 1000 ft of pipe, how far apart may the pumps be placed?   *Ans.* 167,000 ft

67. A large closed tank is filled with air under a pressure of 5.3 psi gage and a temperature of 65°F. The air discharges into the atmosphere (14.7 psi) through a small opening in the side of the tank. Neglecting friction losses, compute the velocity of the air leaving the tank assuming (a) constant density of the air and (b) adiabatic flow conditions.   *Ans.* 692 ft/sec, 728 ft/sec

68. For Problem 67, had the pressure been 10.0 psi what would be the velocity for (a) and for (b)? *Ans.* 855 ft/sec, 934 ft/sec

69. Carbon dioxide flows from a 1″ pipe where the pressure is 60 psi gage and the temperature is 40°F into a ½″ pipe at the rate of 0.0600 lb/sec. Neglecting friction and assuming isothermal conditions, determine the pressure in the ½″ pipe.   *Ans.* 2.79 psi absolute

70. An air blower is to deliver 40,000 ft³/min. Two U-tube gages measure suction and discharge pressures. The suction gage reads a negative 2″ of water. The discharge gage, located 3 ft above the point to which the suction gage is attached, reads +3″ of water. The discharge and suction ducts are of equal diameter. What size motor should be used to drive the blower, if the overall efficiency is 68% ($w = 0.0750$ lb/ft³ for air).   *Ans.* 46.7 HP

71. A 12″ pipe is being tested to evaluate the lost head. When the flow of water is 6.31 cfs, the pressure at point A in the pipe is 40 psi. A differential gage is attached to the pipe at A and at a point B downstream which is 10 ft higher than A. The deflection of mercury in the gage is 3.33 ft, indicating a greater pressure at A. What is the lost head A to B?   *Ans.* 42.0 ft

72. Prandtl has suggested that the velocity distribution for turbulent flow in conduits may be approximated by using $v = v_{max}(y/r_o)^{1/7}$, where r_o is the pipe radius and y is the distance from the pipe wall. Determine the expression for average velocity in the pipe in terms of the center velocity v_{max}. *Ans.* $V = 0.817v_{max}$

73. What is the kinetic-energy correction factor for the velocity distribution of Prob. 72?   *Ans.* $\alpha = 1.06$

74. Two large plates are spaced 1″ apart. Show that $\alpha = 1.54$ if the velocity profile is represented by $v = v_{max}(1 - 576r^2)$ where r is measured from the centerline between the plates.

75. Air flows isentropically through a duct whose area varies. For steady flow, show that the velocity V_2 at any section downstream from section 1 may be written

$$V_2 = V_1(p_1/p_2)^{1/k}(A_1/A_2) \text{ for any shaped duct, and } V_2 = V_1(p_1/p_2)^{1/k}(D_1/D_2)^2 \text{ for a circular duct}$$

76. The pressure inside the pipe at S must not fall below 3.46 psia. Neglecting losses, how high above water level A may point S be located? Refer to Fig. 6-16 below.   *Ans.* 22 ft

77. Pump B delivers a head of 140.6 feet to the water flowing to E as shown in Fig. 6-17 below. If the pressure at C is -2.0 psi and the lost head from D to E is $8.0(V^2/2g)$, what flow occurs?
Ans. 8.91 cfs

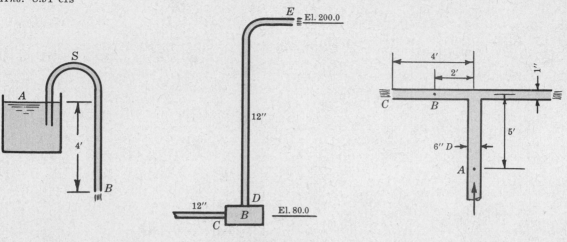

Fig. 6-16   Fig. 6-17   Fig. 6-18

78. Water flows radially between the two flanges at the end of a 6″ diameter pipe as shown in Fig. 6-18 above. Neglecting losses, if the pressure head at A is -1.0 ft, find the pressure head at B and the flow in cfs.   *Ans.* -0.15 ft, 3.88 cfs

79. Show that the average velocity V in a circular pipe of radius r_o equals $2v_{max}\left[\dfrac{1}{(K+1)(K+2)}\right]$ for a velocity distribution which can be expressed as $v = v_{max}(1 - r/r_o)^K$.

80. Find the kinetic-energy correction factor α for Problem 79.   *Ans.* $\alpha = \dfrac{(K+1)^3(K+2)^3}{4(3K+1)(3K+2)}$

Chapter 7

Fluid Flow in Pipes

INTRODUCTION

The energy principle is applied to the solution of practical pipe flow problems in different branches of engineering practice. The flow of a real fluid is more complex than the flow of an ideal fluid. Shear forces between fluid particles and the boundary walls and between the fluid particles themselves result from the viscosity of the real fluid. The partial differential equations which might evaluate the flow (Euler equations) have no general solution. The results of experimentation and semi-empirical methods must be utilized in order to solve flow problems.

Two types of steady flow of real fluids exist, and must be understood and considered. They are called laminar flow and turbulent flow. Different laws govern the two types of flow.

LAMINAR FLOW

In laminar flow the fluid particles move along straight, parallel paths in layers or laminae. The magnitudes of the velocities of adjacent laminae are not the same. Laminar flow is governed by the law relating shear stress to rate of angular deformation, i.e., the product of the viscosity of the fluid and velocity gradient or $\tau = \mu \, dv/dy$ (see Chapter 1). The viscosity of the fluid is dominant and thus suppresses any tendency to turbulent conditions.

CRITICAL VELOCITY

The critical velocity of practical interest to the engineer is the velocity below which all turbulence is damped out by the viscosity of the fluid. It is found that the upper limit of laminar flow of practical interest is represented by a Reynolds Number of about 2000.

REYNOLDS NUMBER

The Reynolds number, which is dimensionless, represents the ratio of the inertia forces to the viscous forces (see Chapter 5 on Dynamic Similitude).

For circular pipes flowing full,

$$\text{Reynolds Number } R_E \;=\; \frac{Vd\rho}{\mu} \quad \text{or} \quad \frac{Vd}{v} \;=\; \frac{V(2r_o)}{v} \tag{1a}$$

where
V = mean velocity in ft/sec
d = diameter of pipe in ft, r_o = radius of the pipe in ft
v = kinematic viscosity of the fluid in ft²/sec
ρ = mass density of fluid in slugs/ft³ or lb sec²/ft⁴
μ = absolute viscosity in lb sec/ft²

For non-circular cross sections, the ratio of the cross sectional area to the wetted perimeter, called the hydraulic radius R (in feet), is used in the Reynolds number. The expression becomes

$$R_E \;=\; \frac{V(4R)}{v} \tag{1b}$$

TURBULENT FLOW

In turbulent flow the particles of the fluid move in a haphazard fashion in all directions. It is impossible to trace the motion of an individual particle.

The shear stress for turbulent flow can be expressed as

$$\tau = \left(\mu + \eta\right)\frac{dv}{dy} \tag{2a}$$

where η (eta) = a factor depending upon the density of the fluid and the fluid motion. The first factor (μ) represents the effects of viscous action and the second factor (η) represents the effects of turbulent action.

The results of experimentation provide means by which the solution for shear stress in turbulent flow may be accomplished. Prandtl suggested that

$$\tau = \rho l^2 \left(\frac{dv}{dy}\right)^2 \tag{2b}$$

was a valid equation for shear stress in turbulent flow. This expression has the disadvantage that the mixing length l is a function of y. The greater the distance y from the pipe wall the greater the value of l. Later, von Karman suggested that

$$\tau = \tau_o\left(1 - \frac{y}{r_o}\right) = \rho k^2 \frac{(dv/dy)^4}{(d^2v/dy^2)^2} \tag{2c}$$

While k is not precisely constant, this dimensionless number is approximately 0.40. Integration of this expression leads to formulas of the type shown as (7b) below.

SHEARING STRESS at a PIPE WALL

Shearing stress at a pipe wall, as developed in Problem 5, is

$$\tau_o = f\rho V^2/8 \quad \text{in psf units} \tag{3}$$

where f is a dimensionless frictional factor (described in a subsequent paragraph).

It will be shown in Problem 4 that the shear variation at a cross section is linear and that

$$\tau = \frac{(p_1 - p_2)}{2L}r \quad \text{or} \quad \tau = \left(\frac{wh_L}{2L}\right)r \tag{4}$$

The term $\sqrt{\tau_o/\rho}$ is called shear velocity or friction velocity, and is designated by the symbol v_*. From expression (3) we obtain

$$v_* = \sqrt{\tau_o/\rho} = V\sqrt{f/8} \tag{5}$$

VELOCITY DISTRIBUTION

Velocity distribution at a cross section will follow a parabolic law of variation for *laminar* flow. The maximum velocity is at the center of the pipe and is twice the average velocity. The equation of the velocity profile for laminar flow (see Problem 6) may be expressed as

$$v = v_c - \left(\frac{wh_L}{4\mu L}\right)r^2 \tag{6}$$

For *turbulent* flows, more uniform velocity distribution results. From experiments of Nikuradse and others, equations of velocity profiles in terms of center velocity v_c or shear velocity v_* follow.

(a) An empirical formula is

$$v = v_c(y/r_o)^n \tag{7a}$$

where $n = \frac{1}{7}$, for smooth tubes, up to $R_E = 100{,}000$
        $n = \frac{1}{8}$, for smooth tubes for R_E from 100,000 to 400,000

(b) For *smooth* pipes,

$$v = v_*(5.5 + 5.75 \log y v_*/\nu) \tag{7b}$$

For the $y v_*/\nu$ term, see Problem 8, part (e)

(c) For *smooth* pipes (for $5000 < R_E < 3,000,000$) and for pipes in the wholly rough zone,

$$(v_c - v) = -2.5\sqrt{v_o/\rho} \ln y/r_o = -2.5 v_* \ln y/r_o \tag{7c}$$

In terms of the average velocity V, Vennard suggests that V/v_c may be written

$$\frac{V}{v_c} = \frac{1}{1 + 4.07\sqrt{f/8}} \tag{8}$$

(d) For *rough* pipes,

$$v = v_*(8.5 + 5.75 \log y/\epsilon) \tag{9a}$$

where ϵ is the absolute roughness of the boundary.

(e) For *rough or smooth* boundaries,

$$\frac{v - V}{V\sqrt{f}} = 2 \log y/r_o + 1.32 \tag{9b}$$

Also
$$v_c/V = 1.43\sqrt{f} + 1 \tag{9c}$$

LOSS of HEAD for LAMINAR FLOW

Loss of head for laminar flow is expressed by the Hagen-Poiseuille equation, developed in Problem 6. The expression is

$$\text{Lost head (ft)} = \frac{32 \text{ (viscosity } \mu)(\text{length } L \text{ ft})(\text{average velocity } V)}{(\text{specific weight } w)(\text{diameter } d \text{ ft})^2}$$

$$= \frac{32 \mu LV}{w d^2} \tag{10a}$$

In terms of kinematic viscosity, we obtain, since $\mu/w = \nu/g$,

$$\text{Lost head} = \frac{32\nu LV}{gd^2} \tag{10b}$$

DARCY-WEISBACH FORMULA

The Darcy-Weisbach formula, as developed in Chapter 5, Problem 11, is the basis of evaluating the lost head for fluid flow in pipes and conduits. The equation is

$$\text{Lost head (ft)} = \text{friction factor } f \times \frac{\text{length } L \text{ (ft)}}{\text{diameter } d \text{ (ft)}} \times \text{velocity head } \frac{V^2}{2g} \text{ (ft)}$$

$$= f \frac{L}{d} \frac{V^2}{2g} \tag{11}$$

As noted in Chapter 6, the exact velocity head at a cross section is obtained by multiplying the average velocity squared $(Q/A)^2$ by a coefficient α and dividing by $2g$. For turbulent flow in pipes and conduits, α may be considered as unity without causing appreciable error in the results.

FRICTION FACTOR

Friction factor f can be derived mathematically for laminar flow, but no simple mathematical relation for the variation of f with Reynolds Number is available for turbulent flow. Furthermore, Nikuradse and others found that the relative roughness of the pipe (the ratio of the size of the surface imperfections ϵ to the inside diameter of the pipe) affects the value of f also.

(a) For *laminar* flow, equation (*10b*) above may be rearranged as follows:

$$\text{Lost head} = 64 \frac{v}{Vd} \frac{L}{d} \frac{V^2}{2g} = \frac{64}{R_E} \frac{L}{d} \frac{V^2}{2g} \tag{12a}$$

Thus, *for laminar flow in all pipes for all fluids*, the value of f is

$$f = 64/R_E \tag{12b}$$

R_E has a practical maximum value of 2000 for laminar flow.

(b) For *turbulent flow*, many hydraulicans have endeavored to evaluate f from the results of their own experiments and from those of others.

(1) For turbulent flow in *smooth and rough pipes,* universal resistance laws can be derived from

$$f = 8\tau_o/\rho V^2 = 8V_*^2/V^2 \tag{13}$$

(2) For *smooth* pipes Blasius suggests, for Reynolds Numbers between 3000 and 100,000,

$$f = 0.316/R_E^{0.25} \tag{14}$$

For values of R_E up to about 3,000,000, von Karman's equation modified by Prandtl is

$$1/\sqrt{f} = 2 \log{(R_E\sqrt{f})} - 0.8 \tag{15}$$

(3) For *rough* pipes,

$$1/\sqrt{f} = 2 \log{r_o/\epsilon} + 1.74 \tag{16}$$

(4) For *all* pipes, the Hydraulic Institute and many engineers consider the Colebrook equation reliable when evaluating f. The equation is

$$\frac{1}{\sqrt{f}} = -2 \log{\left[\frac{\epsilon}{3.7d} + \frac{2.51}{R_E\sqrt{f}}\right]} \tag{17}$$

Inasmuch as this equation (*17*) is awkward to solve, diagrams are available to give the relation between friction factor f, Reynolds Number R_E, and the relative roughness ϵ/d. Two such diagrams are included in the Appendix. Diagram A-1 (the Moody Diagram, published through the courtesy of the American Society of Mechanical Engineers) is usually used when flow Q is known, and Diagram A-2 is used when the flow is to be evaluated. The latter form was first suggested by S. P. Johnson and by Hunter Rouse.

It should be observed that for smooth pipes where the value of ϵ/d is very small, the first term in the bracket of (*17*) can be neglected; then (*17*) and (*15*) are similar. Likewise, should Reynolds Number R_E be very large, the second term in the bracket of (*17*) can be neglected; in such cases the effect of viscosity is negligible, and f depends upon the relative roughness of the pipe. This statement is indicated graphically in Diagram A-1 by the fact that the curves become horizontal at high Reynolds Numbers.

Before formulas or diagrams can be utilized, the engineer must estimate the relative roughness of the pipe from his own experience and/or that of others. Suggested values of the size of surface imperfections ϵ for new surfaces are included on Diagrams A-1 and A-2.

OTHER LOSSES of HEAD

Other losses of head, as in pipe fittings, are generally expressed as

$$\text{Lost head (ft)} = K(V^2/2g) \tag{18}$$

A tabulation of loss of head items commonly required is given in Tables 4 and 5 of the Appendix.

Solved Problems

1. Determine the critical velocity for (a) medium fuel oil at 60°F flowing through a 6″ pipe and (b) water at 60°F flowing in the 6″ pipe.

   Solution:

   (a) For laminar flow, the maximum value of Reynolds Number is 2000. From Table 2 in the Appendix, the kinematic viscosity at 60°F is 4.75×10^{-5} ft²/sec.

   $$2000 = R_E = V_c d/\nu = V_c(\tfrac{1}{2})/(4.75 \times 10^{-5}) \qquad V_c = 0.190 \text{ ft/sec}$$

   (b) From Table 2, $\nu = 1.217 \times 10^{-5}$ ft²/sec for 60°F water.
   $$2000 = V_c(\tfrac{1}{2})/(1.217 \times 10^{-5}), \qquad V_c = 0.049 \text{ ft/sec}$$

2. Determine the type of flow occurring in a 12″ pipe when (a) water at 60°F flows at a velocity of 3.50 ft/sec and (b) heavy fuel oil at 60°F flows at the same velocity.

   Solution:

   (a) $R_E = Vd/\nu = 3.50(1)/(1.217 \times 10^{-5}) = 288,000 > 2000$.   The flow is turbulent.

   (b) From Table 2 in the Appendix, $\nu = 221 \times 10^{-5}$ ft²/sec.
   $R_E = Vd/\nu = 3.50(1)/(221 \times 10^{-5}) = 1580 < 2000$.   The flow is laminar.

3. For laminar flow conditions, what size pipe will deliver 90 gpm of medium fuel oil at 40°F? ($\nu = 6.55 \times 10^{-5}$ ft²/sec)

   Solution:
   $$Q = 90/(7.48 \times 60) = 0.201 \text{ cfs.}\quad V = Q/A = Q/\tfrac{1}{4}\pi d^2 = 4Q/\pi d^2 = 0.804/\pi d^2 \text{ cfs.}$$

   $$R_E = \frac{Vd}{\nu}, \quad 2000 = \frac{0.804}{\pi d^2}\left(\frac{d}{6.55 \times 10^{-5}}\right), \quad d = 1.95 \text{ ft.}\quad \text{Use a standard 2 ft diameter pipe.}$$

4. Determine the nature of the distribution of shear stress at a cross section in a horizontal, circular pipe under steady flow conditions.

   Solution:

   (a) For the free body in Fig. 7-1(a) below, since the flow is steady, each particle moves to the right without acceleration. Hence the summation of the forces in the x-direction must equal zero.

   $$p_1(\pi r^2) - p_2(\pi r^2) - \tau(2\pi rL) = 0 \qquad \text{or} \qquad \tau = \frac{(p_1 - p_2)r}{2L} \tag{A}$$

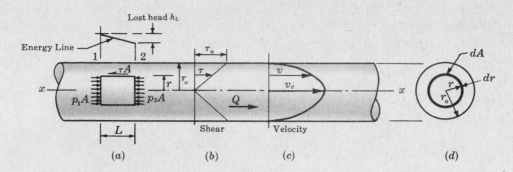

Fig. 7-1

When $r = 0$, the shear stress τ is zero; and when $r = r_o$, the stress τ_o at the wall is a maximum. The variation is linear and is so indicated in Fig. 7-1(b) above. This equation (A) holds for laminar and turbulent flows as no limitations concerning flow were imposed in the derivation.

Since $(p_1 - p_2)/w$ represents the drop in the energy line, or the lost head h_L, multiplying equation (A) by w/w yields

$$\tau = \frac{wr}{2L}\left(\frac{p_1 - p_2}{w}\right) \quad \text{or} \quad \tau = \frac{wh_L}{2L}r \qquad (B)$$

5. Develop the expression for shear stress at a pipe wall.

Solution:

From Problem 4, $h_L = \dfrac{2\tau_o L}{w r_o} = \dfrac{4\tau_o L}{wd}$. The Darcy-Weisbach formula is $h_L = f\dfrac{L}{d}\dfrac{V^2}{2g}$.

Equating these expressions, $\dfrac{4\tau_o L}{wd} = f\dfrac{L}{d}\dfrac{V^2}{2g}$ and $\tau_o = f\dfrac{w}{g}\dfrac{V^2}{8} = f\rho V^2/8$ in psf.

6. For steady, laminar flow (a) what is the relation between the velocity at a point in the cross section and the velocity at the center of the pipe, and (b) what is the equation for velocity distribution?

Solution:

(a) For laminar flow, the shear stress (see Chap. 1) is $\tau = -\mu(dv/dr)$. Equating this with the value of τ in equation (A) of Problem 4, we obtain

$$-\mu\frac{dv}{dr} = \frac{(p_1 - p_2)r}{2L}$$

Since $(p_1 - p_2)/L$ is not a function of r,

$$-\int_{v_c}^{v} dv = \frac{p_1 - p_2}{2\mu L}\int_{0}^{r} r\,dr \quad \text{and} \quad -(v - v_c) = \frac{(p_1 - p_2)r^2}{4\mu L}$$

or

$$v = v_c - \frac{(p_1 - p_2)r^2}{4\mu L} \qquad (A)$$

But the lost head in L feet is $h_L = (p_1 - p_2)/w$; hence

$$v = v_c - \frac{wh_L r^2}{4\mu L} \qquad (B) \text{ and } (6)$$

(b) Since the velocity at the boundary is zero, when $r = r_o$, $v = 0$ in (A), and we have

$$v_c = \frac{(p_1 - p_2)r_o^2}{4\mu L} \quad \text{(at center line)} \qquad (C)$$

Thus, in general,

$$v = \frac{p_1 - p_2}{4\mu L}(r_o^2 - r^2) \qquad (D)$$

7. Develop the expression for the loss of head in a pipe for steady, laminar flow of an incompressible fluid. Refer to Fig. 7-1(d) of Problem 4.

Solution:

$$V_{av} = \frac{Q}{A} = \frac{\int v \, dA}{\int dA} = \frac{\int_0^{r_o} v(2\pi r \, dr)}{\pi r_o^2} = \frac{2\pi(p_1 - p_2)}{\pi r_o^2 (4\mu L)} \int_0^{r_o} (r_o^2 - r^2) r \, dr$$

from which
$$V_{av} = \frac{(p_1 - p_2)r_o^2}{8\mu L} \qquad (A)$$

Thus for laminar flow the average velocity is half the maximum velocity v_c in equation (C) of Problem 6. Rearranging (A), we obtain

$$\frac{p_1 - p_2}{w} = \text{lost head} = \frac{8\mu L V_{av}}{w r_o^2} = \frac{32\mu L V_{av}}{w d^2} \qquad (B)$$

These expressions apply for *laminar flow of all fluids in all pipes and conduits.*

As stated at the beginning of this chapter, the lost head expression for laminar flow in the Darcy form is

$$\text{Lost head} = \frac{64}{R_E} \frac{L}{d} \frac{V^2}{2g} = f \frac{L}{d} \frac{V^2}{2g}$$

8. Determine (a) the shear stress at the walls of a 12″ diameter pipe when water flowing causes a measured lost head of 15 ft in 300 ft of pipe length, (b) the shear stress 2″ from the center line of the pipe, (c) the shear velocity, (d) the average velocity for an f value of .050, (e) the ratio v/v_*.

Solution:

(a) Using equation (B) of Problem 4, when $r = r_o$ the shear stress at the wall is

$$\tau_o = w h_L r_o / 2L = 62.4(15)(\tfrac{1}{2})/600 = 0.78 \text{ psf} = .0053 \text{ psi}$$

(b) Since τ varies linearly from centerline to wall, $\tau = \tfrac{2}{6}(.0053) = .0018$ psi

(c) By equation (5), $v_* = \sqrt{\tau_o/\rho} = \sqrt{0.78/1.94} = 0.634$ ft/sec.

(d) Using $h_L = f \dfrac{L}{d} \dfrac{V^2}{2g}$, we have $15 = .050 \dfrac{300}{1} \dfrac{V^2}{2g}$ and $V = 8.03$ ft/sec.

Otherwise: From equation (3), $\tau_0 = f\rho V^2/8$, $0.78 = .050(1.94)V^2/8$ and $V = 8.03$ ft/sec.

(e) From $\tau_o = \mu(v/y)$ and $\nu = \mu/\rho$ we obtain $\tau_o = \rho\nu(v/y)$ or $\tau_o/\rho = \nu(v/y)$.

Since $\tau_o/\rho = v_*^2$, we have $v_*^2 = \nu(v/y)$, $v/v_*^2 = y/\nu$ and $v/v_* = v_* y/\nu$.

9. If in the preceding problem the water is flowing through a 3 ft by 4 ft rectangular conduit of the same length with the same lost head, what is the shear stress between the water and the pipe wall?

Solution:

For non-circular conduits, the hydraulic radius is the appropriate hydraulic dimension. For a circular pipe,

$$\text{Hydraulic radius } R = \frac{\text{cross sectional area}}{\text{wetted perimeter}} = \frac{\pi d^2/4}{\pi d} = \frac{d}{4} = \frac{r_o}{2}$$

Substituting $r = 2R$ in equation (B) of Problem 4,

$$\tau = \frac{w h_L}{L} R = \frac{62.4(15)}{300} \cdot \frac{(3 \times 4)}{2(3+4)} = 2.67 \text{ psf} = 0.185 \text{ psi}$$

10. Medium lubricating oil, sp gr 0.860, is pumped through 1000 ft of horizontal 2 in. pipe at the rate of .0436 cfs. If the drop in pressure is 30.0 psi, what is the absolute viscosity of the oil?

Solution:

Assuming laminar flow and referring to expression (B) in Problem 7, we obtain

$$(p_1 - p_2) = \frac{32\mu L V_{av}}{d^2} \quad \text{where} \quad V_{av} = \frac{Q}{A} = \frac{.0436}{\frac{1}{4}\pi(2/12)^2} = 2.00 \text{ ft/sec}$$

Then $(30.0 \times 144) = 32\mu(1000)(2.00)/(\tfrac{1}{6})^2$ and $\mu = .00188$ lb sec/ft^2

Checking the original assumption of laminar flow means evaluating the Reynolds number for the conditions of flow. Thus

$$R_E = \frac{Vd}{\nu} = \frac{Vdw}{\mu g} = \frac{2.00 \times \frac{1}{6} \times 0.860 \times 62.4}{.00188 \times 32.2} = 295$$

Since Reynolds Number < 2000, laminar flow exists and the value of μ is correct.

11. Oil of absolute viscosity .00210 lb sec/ft^2 and specific gravity 0.850 flows through 10,000 ft of 12″ cast iron pipe at the rate of 1.57 cfs. What is the lost head in the pipe?

Solution:

$$V = \frac{Q}{A} = \frac{1.57}{\frac{1}{4}\pi(1)^2} = 2.00 \text{ ft/sec}; \quad \text{and} \quad R_E = \frac{Vdw}{\mu g} = \frac{2.00 \times 1.00 \times 0.850 \times 62.4}{0.00210 \times 32.2} = 1570, \text{ which}$$

means laminar flow exists. Hence

$$f = \frac{64}{R_E} = .0407 \quad \text{and} \quad \text{Lost Head} = f\frac{L}{d}\frac{V^2}{2g} = .0407 \times \frac{10,000}{1} \times \frac{(2)^2}{2g} = 25.3 \text{ ft}$$

12. Heavy fuel oil flows from A to B through 3000 ft of horizontal 6″ steel pipe. The pressure at A is 155 psi and at B is 5.0 psi. The kinematic viscosity is .00444 ft^2/sec and the specific gravity is 0.918. What is the flow in cfs?

Solution:

The Bernoulli equation A-B, datum A, gives

$$\left(\frac{155 \times 144}{0.918 \times 62.4} + \frac{V_6^2}{2g} + 0\right) - f\frac{3000}{\frac{1}{2}}\frac{V_6^2}{2g} = \left(\frac{5 \times 144}{0.918 \times 62.4} + \frac{V_6^2}{2g} + 0\right)$$

or $$378 = f(6000)(V_6^2/2g)$$

Both V and f are unknown and are functions of each other. If laminar flow exists, then from equation (B) of Problem 7,

$$V_{av} = \frac{(p_1 - p_2)d^2}{32\mu L} = \frac{(155 - 5)(144) \times (\tfrac{1}{2})^2}{32(.00444 \times 0.918 \times 62.4/32.2)(3000)} = 7.12 \text{ ft/sec}$$

and $R_E = 7.12(\tfrac{1}{2})/.00444 = 800$, hence laminar flow. Thus $Q = A_6 V_6 = \frac{1}{4}\pi(\tfrac{1}{2})^2 \times 7.12 = 1.40$ cfs.

Had the flow been turbulent, equation (B) of Problem 7 would not apply. Another approach will be used in Problem 15 below. Furthermore, had there been a difference in elevation between points A and B, the term $(p_1 - p_2)$ in equation (B) would be replaced by the drop in the hydraulic grade line, in lb/ft^2 units.

13. What size pipe should be installed to carry 0.785 cfs of heavy fuel oil at 60°F if the available lost head in the 1000 ft length of horizontal pipe is 22.0 ft?

Solution:

For the oil, $\nu = .00221$ ft^2/sec and specific gravity $= 0.912$. For such a large value of kinematic viscosity, assume laminar flow. Then

$$\text{Lost Head} = \frac{V_{av} \times 32\mu L}{wd^2} \quad \text{and} \quad V_{av} = \frac{Q}{A} = \frac{0.785}{\frac{1}{4}\pi d^2} = \frac{1.00}{d^2}$$

Substituting, $\quad 22.0 = \dfrac{(1.00/d^2)(32)(.00221 \times 0.912 \times 62.4/32.2)(1000)}{(0.912 \times 62.4)d^2}$, $\quad d = 0.561$ ft.

Check assumption of laminar flow, using $d = 0.561$.

$$R_E = \frac{Vd}{\nu} = \frac{(1.00/d^2)d}{\nu} = \frac{1.00}{0.561 \times .00221} = 807, \quad \text{hence laminar flow.}$$

Use a standard 8″ pipe.

14. Determine the lost head in 1000 ft of new, uncoated 12″ inside diameter cast iron pipe when (a) water at 60°F flows at 5.00 ft/sec, and (b) medium fuel oil at 60°F flows at the same velocity.

Solution:

(a)   When Diagram A-1 is to be used, the relative roughness must be evaluated and then Reynolds Number calculated. From the tabulation on Diagram A-1, the value of ϵ for uncoated cast iron pipe ranges from .0004 ft to .0020 ft. For an inside diameter of 1 ft and the design value of $\epsilon = .0008$ ft, the relative roughness $\epsilon/d = .0008/1 = .0008$.

Using the kinematic viscosity of water from Table 2 in the Appendix,

$$R_E = Vd/\nu = 5.00(1.00)/(1.217 \times 10^{-5}) = 411{,}000 \quad \text{(turbulent flow)}$$

From Diagram A-1, for $\epsilon/d = .0008$ and $R_E = 411{,}000$, $f = .0194$ and

$$\text{Lost Head} = .0194(1000/1)(25/2g) = 7.5 \text{ ft.}$$

Or, using Table 3 in the Appendix (for water only), $f = .0200$ and

$$\text{Lost Head} = f(L/d)(V^2/2g) = .0200(1000/1)(25/2g) = 7.7 \text{ ft}$$

(b)   For the oil, using Table 2, $R_E = 5(1)/(4.75 \times 10^{-5}) = 105{,}000$. For turbulent flow, from Diagram A-1, $f = .0213$ and Lost Head $= .0213(1000/1)(25/2g) = 8.3$ ft.

In general, the degree of roughness of pipes *in service* cannot be estimated with great accuracy and so, in such cases, a precise value of f is not to be anticipated. For this reason, when using Diagrams A-1 and A-2 and Table 3 for other than new surfaces, it is suggested that the third significant figure in f be read or interpolated as either *zero or five*, no greater accuracy being warranted in most practical cases.

For laminar flow, for any pipe and any fluid, use $f = 64/R_E$.

15. Points A and B are 4000 ft apart along a new 6″ I.D. steel pipe. Point B is 50.5 ft higher than A and the pressures at A and B are 123 psi and 48.6 psi respectively. How much medium fuel oil at 70°F will flow from A to B? (From Diagram A-1, $\epsilon = .0002$ ft.)

Solution:

Reynolds Number cannot be calculated immediately. Write the Bernoulli equation A to B, datum A.

$$\left(\frac{123 \times 144}{0.854 \times 62.4} + \frac{V_6^2}{2g} + 0\right) - f\left(\frac{4000}{\frac{1}{2}}\right)\frac{V_6^2}{2g} = \left(\frac{48.6 \times 144}{0.854 \times 62.4} + \frac{V_6^2}{2g} + 50.5\right) \quad \text{and} \quad \frac{V_6^2}{2g} = \frac{151.0}{8000f}$$

Also $R_E = Vd/\nu$. Substituting for V from above,

$$R_E = \frac{d}{\nu}\sqrt{\frac{2g(151.0)}{8000f}} \quad \text{or} \quad R_E\sqrt{f} = \frac{d}{\nu}\sqrt{\frac{2g(151.0)}{8000}} \tag{A}$$

Since the term 151.0 is h_L or the drop in the hydraulic grade line, and the 8000 represents L/d, the general expression to be used *when Q is to be found* is

$$R_E\sqrt{f} = \frac{d}{\nu}\sqrt{\frac{2g(d)(h_L)}{L}} \quad \text{(see Diagram A-2 also)} \tag{B}$$

Then
$$R_E\sqrt{f} \;=\; \frac{0.500}{4.12 \times 10^{-5}} \sqrt{\frac{64.4 \times 151.0}{8000}} \;=\; 13,400$$

Examination of Diagram A-2 will indicate that the flow is turbulent. Then, from Diagram A-2, $f = 0.020$ for $\epsilon/d = .0002/\frac{1}{2} = .0004$. Completing the solution, from the Bernoulli Equation above,

$$\frac{V_6^2}{2g} \;=\; \frac{151.0}{8000(0.020)} \;=\; 0.944, \quad V_6 = 7.79 \text{ ft/sec}, \quad \text{and} \quad Q = A_6 V_6 = \tfrac{1}{4}\pi(\tfrac{1}{2})^2 \times 7.79 = 1.53 \text{ cfs oil}$$

The reader can check the solution by calculating Reynolds Number and finding the value of f from Diagram A-1.

When the flow is laminar, methods illustrated in Problem 12 above should be used.

16. How much water (60°F) would flow under the conditions of Problem 15. Use Table 3.

Solution:

The Bernoulli equation yields $(171.5 - 50.5) = 8000f\dfrac{V_6^2}{2g}, \quad \dfrac{V_6^2}{2g} = \dfrac{121.0}{8000f}.$

The most direct solution in this case is to assume a value of f. Table 3 indicates, for new 6″ pipe, a range in f from .0275 to .0175. Try $f = .0225$. Then

$$V_6^2/2g \;=\; 121.0/(8000 \times .0225) \;=\; 0.672 \text{ ft} \quad \text{and} \quad V_6 = 6.58 \text{ ft/sec}$$

Check on both type of flow and on f in Table 3:

$$R_E \;=\; 6.58(\tfrac{1}{2})/(1.217 \times 10^{-5}) \;=\; 270,000, \quad \text{hence turbulent flow}$$

Now f by interpolation is .0210. Repeating the calculation

$$V_6^2/2g \;=\; 121.0/(8000 \times .0210) \;=\; 0.720 \text{ ft} \quad \text{and} \quad V_6 = 6.81 \text{ ft/sec}$$

From Table 3 to a justifiable accuracy, $f = 0.210$ (check). Then

$$Q \;=\; A_6 V_6 = \tfrac{1}{4}\pi(\tfrac{1}{2})^2 \times 6.81 \;=\; 1.34 \text{ cfs of water}$$

This procedure may also be applied using Diagram A-1, but the method of Problem 15 is preferred.

17. What rate of flow of air at 68°F will be carried by a new horizontal 2″ I.D. steel pipe at an absolute pressure of 3 atmospheres and with a drop of 0.150 psi in 100 ft of pipe? Use $\epsilon = .00025$ ft.

Solution:

From Appendix, for 68°F, $w = .0752$ lb/ft^3 and $\nu = 16.0 \times 10^{-5}$ ft^2/sec at standard atmospheric pressure. At 3 atmospheres, $w = 3 \times .0752 = 0.2256$ lb/ft^3 and $\nu = \frac{1}{3} \times 16.0 \times 10^{-5} = 5.33 \times 10^{-5}$. This kinematic viscosity can also be obtained from

$$\mu \;=\; \frac{w}{g}\nu \;=\; \frac{.0752 \times 16.0 \times 10^{-5}}{32.2} \;=\; 374 \times 10^{-9} \frac{\text{lb sec}}{\text{ft}^2} \quad \text{at 68°F and 14.7 psi absolute}$$

Furthermore, at 3×14.7 psi absolute, $w_{\text{air}} = 0.2256$ lb/ft^3 and

$$\nu \text{ for 3 atmospheres} \;=\; \mu\frac{g}{w} \;=\; 374 \times 10^{-9} \times \frac{32.2}{0.2256} \;=\; 5.33 \times 10^{-5} \text{ ft}^2/\text{sec}$$

To find the flow, the air may be considered as incompressible. Then

$$\frac{p_1 - p_2}{w} = \text{Lost Head} = f\frac{L}{d}\frac{V^2}{2g}, \quad \frac{0.150 \times 144}{0.2256} = 95.7 = f\frac{100}{\frac{2}{12}}\frac{V^2}{2g} \quad \text{and} \quad \frac{V^2}{2g} = \frac{0.160}{f}$$

Also, from Prob. 15, $R_E\sqrt{f} = \dfrac{d}{\nu}\sqrt{\dfrac{2g(d)(h_L)}{L}} = \dfrac{\frac{2}{12}}{5.33 \times 10^{-5}}\sqrt{\dfrac{64.4(\frac{2}{12})(95.7)}{100}} = 10,000$ (turbulent).

From Diagram A-2, $f = .025$ for $\epsilon/d = .00025/(2/12) = .0015$. Then

$$V^2/2g \;=\; 0.160/f \;=\; 6.40 \text{ ft}, \quad V_2 \;=\; 20.3 \text{ ft/sec}, \quad \text{and} \quad Q \;=\; A_2 V_2 \;=\; \tfrac{1}{4}\pi(2/12)^2 \times 20.3 \;=\; 0.443 \text{ cfs air}$$

18. What size of new cast iron pipe, 8000 ft long, will deliver 37.5 cfs of water with a drop in the hydraulic grade line of 215 ft? Use Table 3 for this calculation.

Solution:

The Bernoulli theorem gives $\quad \left(\dfrac{p_A}{w} + \dfrac{V_A^2}{2g} + z_A\right) - f\dfrac{8000}{d}\dfrac{V^2}{2g} \;=\; \left(\dfrac{p_B}{w} + \dfrac{V_B^2}{2g} + z_B\right)$

or $\qquad\qquad \left[\left(\dfrac{p_A}{w} + z_A\right) - \left(\dfrac{p_B}{w} + z_B\right)\right] \;=\; f\dfrac{8000}{d}\dfrac{V^2}{2g}$

The left hand term in the brackets represents the drop in the hydraulic grade line. Expressing V as Q/A and assuming turbulent flow

$$215 \;=\; f\dfrac{8000}{d}\dfrac{(37.5/\tfrac{1}{4}\pi d^2)^2}{2g} \quad \text{which simplifies to} \quad d^5 \;=\; f\dfrac{8000(37.5)^2}{39.7(215)} \;=\; 1317f$$

Assume $f = .020$ (since both d and V are unknown, an assumption is necessary). Then

$$d^5 \;=\; f(1317) \;=\; .020(1317) \;=\; 26.34, \quad d \;=\; 1.92 \text{ ft}$$

From Table 3, for $\quad V = \dfrac{37.5}{\pi(1.92)^2/4} \;=\; 13.0 \text{ ft/sec}, \quad f = .0165$.

For this magnitude of velocity in most pipes, turbulent flow of water exists. Recalculating,

$$d^5 \;=\; .0165(1317) \;=\; 21.70, \quad d \;=\; 1.85 \text{ ft}$$

Checking f, $\quad V = 13.9 \text{ ft/sec}$ and Table 3 gives $f = .0165$ (check).

Use nearest standard size: 2 ft or 24″ pipe. (Check R_E using ν for 70°F water.)

19. Points C and D, at the same elevation, are 500 ft apart in an 8″ pipe and are connected to a differential gage by means of small tubing. When the flow of water is 6.31 cfs, the deflection of the mercury in the gage is 6.43 ft. Determine friction factor f.

Solution:

$$\left(\dfrac{p_C}{w} + \dfrac{V_8^2}{2g} + 0\right) - f\dfrac{500}{\tfrac{2}{3}}\dfrac{V_8^2}{2g} \;=\; \left(\dfrac{p_D}{w} + \dfrac{V_8^2}{2g} + 0\right) \quad \text{or} \quad \left(\dfrac{p_C}{w} - \dfrac{p_D}{w}\right) \;=\; f(750)\dfrac{V_8^2}{2g} \qquad (1)$$

From the differential gage (see Chapter 1), $p_L = p_R$ or

$$p_C/w + 6.43 \;=\; p_D/w + 13.57(6.43), \qquad \text{and} \qquad (p_C/w - p_D/w = 80.9 \text{ ft} \qquad (2)$$

Equating (1) and (2), $\quad 80.9 = f(750)(18.1)^2/2g \quad$ from which $f = .0212$.

20. Medium fuel oil at 50°F is pumped to tank C (see Fig. 7-2) through 6000 ft of new, riveted steel pipe, 16″ inside diameter. The pressure at A is +2.00 psi when the flow is 7.00 cfs. (a) What power must pump AB supply to the oil and (b) what pressure must be maintained at B? Draw the hydraulic grade line.

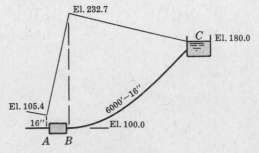

Fig. 7-2

Solution:

$$V_{16} \;=\; \dfrac{Q}{A} \;=\; \dfrac{7.00}{\pi(16/12)^2/4} \;=\; 5.02 \text{ ft/sec} \quad \text{and} \quad R_E \;=\; \dfrac{5.02 \times 16/12}{5.55} \times 10^5 \;=\; 121{,}000.$$

From Diagram A-1, $f = .030$ for $\epsilon/d = .0060/(16/12) = .0045$.

(a) The Bernoulli equation, A to C, datum A, gives

$$\left(\frac{2.00 \times 144}{0.861 \times 62.4} + \frac{(5.02)^2}{2g} + 0\right) + H_p - .030\left(\frac{6000}{16} \times 12\right)\frac{(5.02)^2}{2g} - \frac{(5.02)^2}{2g} = (0 + 0 + 80)$$

Solving, $H_p = 127.3$ ft and H.P. $= \dfrac{wQH_p}{550} = \dfrac{0.861 \times 62.4 \times 7.00 \times 127.3}{550} = 87.0$.

The last term on the left-hand side of the energy equation is the lost head from pipe to tank (see Table 4 in Appendix). In general, when the length to diameter ratio (L/d) is in excess of 2000 to 1, velocity heads and minor losses should be neglected in the Bernoulli equation (here they cancel). Fictitious accuracy results when such minor items are included in the calculation, because f cannot be ascertained to such a degree of accuracy.

(b) The pressure head at B may be calculated by using sections A and B or sections B and C. The former is less work; then

$$\left(5.4 + \frac{V_{16}^2}{2g} + 0\right) + 127.3 = \left(\frac{p_B}{w} + \frac{V_{16}^2}{2g} + 0\right)$$

Thus $p_B/w = 132.7$ ft and $p'_B = wh/144 = (0.861 \times 62.4)(132.7)/144 = 49.5$ psi.

The hydraulic grade line elevations are shown in the figure above.

At A, elevation $(100.0 + 5.4)$ ft $= 105.4$ ft
At B, elevation $(100.0 + 132.7)$ ft $= 232.7$ ft (or $105.4 + 127.3$)
At C, elevation $\pm\ 180.0$ ft

21. At a point A in a horizontal 12″ pipe ($f = .020$) the pressure head is 200 ft. At a distance of 200 ft from A the 12″ pipe reduces suddenly to a 6″ pipe. At a distance of 100 ft from this sudden reduction the 6″ pipe ($f = .015$) suddenly enlarges to a 12″ pipe and point F is 100 ft beyond this change in size. For a velocity of 8.025 ft/sec in the 12″ pipes, draw the energy and hydraulic grade lines. Refer to the figure below.

Solution:

The velocity heads are $V_{12}^2/2g = (8.025)^2/2g = 1.00$ ft and $V_6^2/2g = 16.0$ ft.

The energy line drops in the direction of flow by the amount of the lost head. The hydraulic grade line (gradient) is below the energy line by the amount of the velocity head at any cross section. Note (in Fig. 7-3 below) that the *hydraulic grade line* can rise where a change (enlargement) in size occurs.

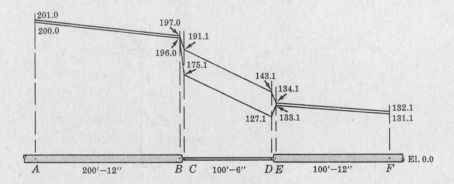

Fig. 7-3

Tabulating the results to the nearest 0.1 ft,

	Lost Head in Feet		Elevation Energy Line	$\dfrac{V^2}{2g}$	Elevation of Hyd. Gradient
At	From	Calculated			
A	(Elevation 0.0)		201.0	1.0	200.0
B	A to B	$.020 \times 200/1 \times 1 = 4.0$	197.0	1.0	196.0
C	B to C	$K_c{}^* \times 16 = .37 \times 16 = 5.9$	191.1	16.0	175.1
D	C to D	$.015 \times 100/\frac{1}{2} \times 16 = 48.0$	143.1	16.0	127.1
E	D to E	$\dfrac{(V_6 - V_{12})^2}{2g} = \dfrac{(32.1 - 8.0)^2}{64.4} = 9.0$	134.1	1.0	133.1
F	E to F	$.020 \times 100/1 \times 1 = 2.0$	132.1	1.0	131.1

*(K_c is from Table 5, Sudden Enlargement term (D to E) from Table 4)

22. Oil flows from tank A through 500 ft of 6″ new asphalt-dipped cast iron pipe to point B at elevation 100.0 in Fig. 7-4. What pressure in psi will be needed at A to cause 0.450 cfs of oil to flow? (Sp gr = 0.840 and $\nu = 2.27 \times 10^{-5}$ ft²/sec.) Use $\epsilon = .0004$ ft.

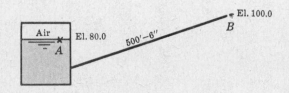

Fig. 7-4

Solution:

$$V_6 = \frac{Q}{A} = \frac{0.450}{0.196} = 2.29 \text{ ft/sec} \quad \text{and} \quad R_E = \frac{Vd}{\nu} = \frac{2.29 \times \frac{1}{2}}{2.27} \times 10^5 = 50,500.$$

From Diagram A-1, $f = .0235$ and the Bernoulli equation, A to B, datum A, gives

$$\left(\frac{p_A}{w} + 0 + 0\right) - 0.50 \frac{(2.29)^2}{2g} - .0235 \frac{500}{\frac{1}{2}} \frac{(2.29)^2}{2g} = \left(0 + \frac{(2.29)^2}{2g} + 20\right)$$

Solving, $p_A/w = 22.0$ ft of oil and $p'_A = wh/144 = (0.840 \times 62.4)(22.0)/144 = 8.00$ psi.

23. The pressure at section A in a new horizontal, wrought iron pipe of 4″ I.D. is 49.5 psi absolute when 0.750 lb/sec of air flow isothermally. Calculate the pressure in the pipe at section B which is 1800 ft from A. (Absolute viscosity $= 390 \times 10^{-9}$ lb sec/ft² and $t = 90°$F.) Use $\epsilon = .0003$ ft.

Solution:

Air has a variable density as the pressure conditions change with flow.

The Bernoulli theorem for compressible fluids was applied in Chapter 6 to conditions involving no loss of head (ideal flow). The basic energy expression with loss of head included for a length of pipe dL and where $z_1 = z_2$ would be

$$\frac{dp}{w} + \frac{V\,dV}{g} + f\frac{dL}{d}\frac{V^2}{2g} = 0$$

Dividing by $\dfrac{V^2}{2g}$,

$$\frac{2g}{V^2}\frac{dp}{w} + \frac{2\,dV}{V} + \frac{f}{d}dL = 0$$

For steady flow, the number of lb/sec flowing is constant; then $W = wQ = wAV$ and W/wA can be substituted for V in the pressure head term, giving

$$\frac{2gw^2A^2}{W^2w}\,dp + \frac{2\,dV}{V} + \frac{f}{d}dL = 0$$

For isothermal conditions, $p_1/w_1 = p_2/w_2 = RT$ or $w = p/RT$. Substituting for w,

$$\frac{2gA^2}{W^2RT}\int_{p_1}^{p_2} p\,dp + 2\int_{V_1}^{V_2}\frac{dV}{V} + \frac{f}{d}\int_0^L dL = 0$$

f being considered a constant, as explained below. Integrating and substituting limits,

$$\frac{gA^2}{W^2RT}(p_2^2 - p_1^2) + 2(\log_e V_2 - \log_e V_1) + f(L/d) = 0 \qquad (A)$$

For comparison with the familiar form of the equation ($z_1 = z_2$), we obtain

$$(Kp_1^2 + 2\log_e V_1) - f(L/d) = (Kp_2^2 + 2\log_e V_2) \qquad (B)$$

where $K = \dfrac{gA^2}{W^2RT}$. Rearranging (A) for ready solution,

$$p_1^2 - p_2^2 = \frac{W^2RT}{gA^2}\left[2\log_e\frac{V_2}{V_1} + f\frac{L}{d}\right] \qquad (C)$$

Now $W^2/A^2 = w_1^2 A_1^2 V_1^2/A_1^2 = w_1^2 V_1^2$ and $RT = p_1/w_1$; hence

$$\frac{W^2RT}{gA^2} = \frac{w_1 V_1^2 p_1}{g} \qquad (D)$$

Then (C) becomes $\quad (p_1 - p_2)(p_1 + p_2) = \dfrac{w_1 p_1 V_1^2}{g}\left[2\log_e\dfrac{V_2}{V_1} + f\dfrac{L}{d}\right]$

$$\frac{(p_1 - p_2)}{w_1} = \frac{2\left[2\log_e\dfrac{V_2}{V_1} + f\dfrac{L}{d}\right]\dfrac{V_1^2}{2g}}{(1 + p_2/p_1)} = \text{Lost Head} \qquad (E)$$

Limiting pressures and velocities will be discussed in Chapter 11.

Before solving this expression, investigation of friction factor f is important, since velocity V is not constant for gases where changes in density may occur.

$$R_E = \frac{Vd}{\mu/\rho} = \frac{Vd\rho}{\mu} = \frac{Wd\rho}{wA\mu}. \quad \text{Since } g = \frac{w}{\rho}, \text{ then } R_E = \frac{Wd}{Ag\mu} \qquad (F)$$

It should be observed that Reynolds Number is constant for steady flow, since μ does not vary if there is no temperature change. Hence friction factor f is constant for the problem even though the velocity will increase as the pressure decreases. Solving (F), using the absolute viscosity given,

$$R_E = \frac{0.750 \times 4/12 \times 10^9}{(\pi/4)(4/12)^2 \times 32.2 \times 390} = 228,000. \quad \text{From Diagram A-1, } f = .0205 \text{ for } \epsilon/d = .0009.$$

Using (C) above, neglecting $2\log_e V_2/V_1$ which is very small compared to the $f(L/d)$ term,

$$(49.5 \times 144)^2 - p_2^2 = \frac{(0.750)^2 \times 53.3(90 + 460)}{32.2[(\pi/4)(4/12)^2]^2}\left[\text{negl.} + (.0205)\frac{1800}{4/12}\right]$$

from which $p_2 = 6590$ psf and $p_2' = 45.7$ psi absolute.

At B: $\quad w_2 = \dfrac{6590}{53.3(90 + 460)} = 0.225 \text{ lb/ft}^3, \quad V_2 = \dfrac{W}{w_2 A} = \dfrac{0.750}{0.225 \times .0873} = 38.2 \text{ ft/sec.}$

At A: $\quad w_1 = \dfrac{(49.5 \times 144)}{53.3(90 + 460)} = 0.243 \text{ lb/ft}^3, \quad V_1 = \dfrac{0.750}{0.243 \times .0873} \doteq 35.4 \text{ ft/sec.}$

Hence $2\log_e V_2/V_1 = 2\log_e 38.2/35.4 = 2 \times 0.077 = 0.154$ which is negligible in terms of the $f(L/d)$ term of 111. Therefore the pressure at Section B is $p_2' = 45.7$ psi.

Had the air been treated as incompressible, then

$$\frac{p_1 - p_2}{w_1} = f\frac{L}{d}\frac{V^2}{2g} = .0205 \times \frac{1800}{4/12} \times \frac{(35.4)^2}{2g} = 2160 \text{ ft}$$

$$\Delta p = w_1 h = 0.243 \times 2160 = 525 \text{ psf} = 3.64 \text{ psi}$$

and $p_2' = 49.5 - 3.6 = 45.9$ psi, an unusually close agreement.

24. A horizontal wrought iron pipe, 6″ inside diameter and somewhat corroded, is transporting 4.50 lb of air per second from A to B. At A the pressure is 70 psi absolute and at B the pressure must be 65 psi absolute. Flow is isothermal at 68°F. What is the length of pipe from A to B? Use $\epsilon = .0013$ ft.

Solution:

Calculating basic values (see Appendix for 68°F at 14.7 psi),

$$w_1 = .0752(70/14.7) = 0.358 \text{ lb/ft}^3, \qquad w_2 = .0752(65/14.7) = 0.333 \text{ lb/ft}^3$$

$$V_1 = \frac{W}{w_1 A} = \frac{4.50}{0.358 \times \frac{1}{4}\pi(\frac{1}{2})^2} = 64.0 \text{ ft/sec}, \qquad V_2 = \frac{4.50}{0.333 \times \frac{1}{4}\pi(\frac{1}{2})^2} = 68.7 \text{ ft/sec}$$

$$R_E = \frac{64.0 \times \frac{1}{2}}{(14.7/70.0)(16.0 \times 10^{-5})} = 952,000. \text{ From Diagram A-1, } f = .025 \text{ for } \epsilon/d = .0026.$$

Using equation (E) of Problem 23,

$$\frac{(70-65)144}{0.358} = \frac{2\,[2\,\log_e 68.7/64.0 + .025(L/\frac{1}{2})]\,(64.0)^2/2g}{(1+65/70)} \quad \text{and} \quad L = 6080 \text{ ft}$$

Note: For the flow of gases in pipelines, if p_2 is not more than 10% smaller than p_1, less than 5% error in pressure drop will result from assuming the fluid to be incompressible and using the Bernoulli equation in its usual form.

25. The elevations of the energy line and hydraulic grade line at point G are 44.0 and 42.0 ft respectively. For the system shown in Fig. 7-5 below, calculate (*a*) the power extracted between G and H if the energy line at H is at Elevation 4.0 and (*b*) the pressure heads at E and F which are at Elevation 20.0. (*c*) Draw, to the nearest 0.1 ft, the energy and hydraulic grade lines, assuming K for valve CD is 0.40 and $f = .010$ for the 6″ pipes.

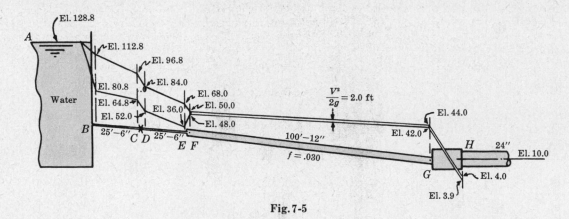

Fig. 7-5

Solution:

The flow must be from the reservoir since the energy line at G is below the reservoir level. GH is a turbine. Before the power extracted can be calculated, flow Q and the head extracted must be obtained.

(*a*) At G, $V_{12}^2/2g = 2.0$ (the difference between the elevations of the energy and hydraulic grade lines).

Also $V_6^2/2g = 16 \times 2.0 = 32.0$ and $V_{24}^2/2g = \frac{1}{16}(2.0) = 0.13$ ft. To obtain Q,

$$V_{12} = 11.34 \text{ ft/sec} \quad \text{and} \quad Q = \frac{1}{4}\pi(1)^2 \times 11.34 = 8.91 \text{ cfs}$$

$$\text{H.P.} = wQH_T/550 = 62.4(8.91)(44.0-4.0)/550 = 40.4 \text{ extracted}$$

(*b*) F to G, datum zero: (Energy at F) $- .030(100/1)(2.0) = $ (Energy at $G = 44.0$)

$$\text{Energy at } F = 44.0 + 6.0 = 50.0 \text{ ft}$$

E to F, datum zero: (Energy at E) $-$ $(45.4 - 11.3)^2/2g$ $=$ (Energy at $F = 50.0$)

$$\text{Energy at } E = 50.0 + 18.0 = 68.0 \text{ ft}$$

$$z + V^2/2g$$

The pressure head at E $=$ $68.0 - (20 + 32)$ $=$ 16.0 ft water.

The pressure head at F $=$ $50.0 - (20 + 2)$ $=$ 28.0 ft water.

(c) Working back from E:

Drop in Energy line D-E $=$ $.010(25/\frac{1}{2})(32.0)$ $=$ 16.0 ft

Drop in Energy line C-D $=$ $0.40(32.0)$ $=$ 12.8 ft

Drop in Energy line B-C $=$ same as D-E $=$ 16.0 ft

Drop in Energy line A-B $=$ $0.50(32.0)$ $=$ 16.0 ft

(Elevation at D $-$ 16.0) $=$ Elevation at E of 68.0, Elev. D $=$ 84.0.

(Elevation at C $-$ 12.8) $=$ Elevation at D of 84.0, Elev. C $=$ 96.8.

(Elevation at B $-$ 16.0) $=$ Elevation at C of 96.8, Elev. B $=$ 112.8.

(Elevation at A $-$ 16.0) $=$ Elevation at B of 112.8, Elev. A $=$ 128.8.

The hydraulic grade line is $V^2/2g$ below the energy line: 32.0 ft in the 6″, 2.0 ft in the 12″, and 0.13 ft in the 24″. The values are shown in the figure above.

26. An old 12″ × 18″ rectangular duct carries air at 15.2 psi absolute and 68°F through 1500 feet with an average velocity of 9.75 ft/sec. Determine the loss of head and the pressure drop, assuming the duct to be horizontal and the size of the surface imperfections is .0018 ft.

Solution:

The lost head term must be revised slightly to apply to non-circular cross sections. The resulting equation will apply to turbulent flow with reasonable accuracy. Substitute for the diameter the *hydraulic radius* which is defined as the cross sectional area divided by the wetted perimeter, or $R = a/p$.

For a circular pipe, $R = \frac{1}{4}\pi d^2/\pi d = d/4$ and the Darcy formula can be rewritten as

$$\text{Lost Head} = \frac{f}{4}\frac{L}{R}\frac{V^2}{2g}$$

For f and its relation to the roughness of the conduit and Reynolds Number, we use

$$R_E = Vd/\nu = V(4R)/\nu$$

For the 12″ × 18″ duct, $R = \dfrac{a}{p} = \dfrac{1 \times 1.5}{2(1 + 1.5)} = 0.30$ ft and

$$R_E = \frac{4VR}{\nu} = \frac{4 \times 9.75 \times 0.30}{(14.7/15.2) \times 16.0} \times 10^5 = 75{,}600$$

From Diagram A-1, $f = .024$ for $\epsilon/d = \epsilon/4R = .0018/(4 \times 0.30) = .0015$. Then

$$\text{Lost Head} = \frac{.024}{4} \times \frac{1500}{0.30} \times \frac{(9.75)^2}{2g} = 44.4 \text{ ft of air}$$

and pressure drop $= wh/144 = (15.2/14.7)(.0752)(44.4)/144 = .024$ psi.

It can be observed that the assumption of constant density of the air is satisfactory.

Supplementary Problems

27. If the shear stress at the wall of a 12″ diameter pipe is 1.00 psf and $f = .040$, what is the average velocity (a) if water at 70°F is flowing, (b) if a fluid with a specific gravity of 0.70 is flowing? *Ans.* 10.1, 12.1 ft/sec

28. What are the shear velocities in the preceding problem?   *Ans.* 0.717, 0.857 ft/sec

29. Water flows through 200 ft of 6″ pipe and the shear stress at the walls is 0.92 psf. Determine the lost head.   *Ans.* 23.6 ft

30. What size pipe will maintain a shear stress at the wall of 0.624 psf when water flows through 300 ft of pipe causing a lost head of 20.0 ft?   *Ans.* $r = 0.30$ ft

31. Compute the critical velocity (lower) for a 4″ pipe carrying water at 80°F.   *Ans.* .0558 ft/sec

32. Compute the critical velocity (lower) for a 4″ pipe carrying heavy fuel oil at 110°F. *Ans.* 2.88 ft/sec

33. What pressure head drop will occur in 300 ft of new, horizontal cast iron pipe, 4″ in diameter, carrying medium fuel oil at 50°F when the velocity is 0.25 ft/sec?   *Ans.* .037 ft

34. What pressure head drop will occur in Prob. 33 if the velocity of the oil is 4.00 ft/sec?   *Ans.* 6.7 ft

35. Considering pipe loss only, how much head is required to deliver 7.85 cfs of heavy fuel oil at 100°F through 3000 ft of new cast iron pipe, 12″ inside diameter? Use $\epsilon = .0008$ ft.   *Ans.* 135 ft

36. In Problem 35, what least value of kinematic viscosity of the oil will produce laminar flow? *Ans.* .00500 ft²/sec

37. Considering pipe loss only, what difference in elevation between two tanks 800 ft apart will deliver 1.10 cfs of medium weight lubricating oil at 50°F through a 6″ diameter pipe?   *Ans.* 49.9 ft

38. Oil of specific gravity 0.802 and kinematic viscosity .00200 ft²/sec flows from Tank A to Tank B through 1000 ft of new pipe at the rate of 3.14 cfs. The available head is 0.527 ft. What size pipe should be used?   *Ans.* 1.97 ft, use 24″ size

39. A pump delivers heavy fuel oil at 60°F through 1000 ft of 2″ brass pipe to a tank 10 ft higher than the supply tank. Neglecting minor losses, for a flow of 0.131 cfs, determine the size of the pump (horsepower) if its efficiency is 80%.   *Ans.* 8.2 hp

40. Water at 100°F flows from A to B through 800 ft of average 12″ inside diameter cast iron pipe ($\epsilon = .0020$ ft). Point B is 30.0 ft above A and the pressure at B must be maintained at 20.0 psi. If 7.85 cfs is to flow through the pipe, what must be the pressure at A in psi?   *Ans.* 45.5 psi

41. An old commercial pipe, horizontal, 36″ inside diameter and 8000 feet long, carries 44.2 cfs of heavy fuel oil with a lost head of 73.5 feet. What pressure in psi must be maintained at A upstream to keep the pressure at B at 20 psi? Use $\epsilon = .045$ ft.   *Ans.* 48.6 psi

42. An old pipe, 24″ inside diameter and 4000 ft long, carries medium fuel oil at 80°F from A to B. The pressures at A and B are 57.0 psi and 20.0 psi respectively, and point B is 60 ft above point A. Calculate the flow in cfs, using $\epsilon = .0016$ ft.   *Ans.* 25.9 cfs

43. Water flows from Tank A whose level is at Elevation 84.0 to Tank B whose level is kept at Elevation 60.0. The tanks are connected by 100 ft of 12″ pipe ($f = .020$) followed by 100 ft of 6″ pipe ($f = .015$). There are two 90° bends in each pipe ($K = 0.50$ each), K for the contraction is 0.75 and the 12″ pipe projects into tank A. If the elevation at the sudden reduction of the pipes is 54.0, find the pressure heads in the 12″ and in the 6″ at this change in size.   *Ans.* 28.8 ft, 22.0 ft

44. In Fig. 7-6 below, point B is 600 ft from the reservoir. When water flows at the rate of 0.500 cfs, calculate (a) the head lost due to the partial obstruction C and (b) the pressure at B in psi absolute. *Ans.* 7.7 ft, 14.2 psia

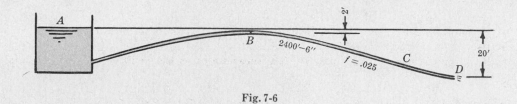

Fig. 7-6

45. A commercial solvent at 70°F flows from tank A to tank B through 500 ft of new 6″ asphalt-dipped, cast iron pipe. The difference in elevation of the liquid levels is 23.0 ft. The pipe projects into tank A and two bends in the line cause a loss of 2 velocity heads. What flow will occur? Use $\epsilon = .00045$ ft. *Ans.* 1.54 cfs

46. A steel conduit, 2″ × 4″, carries 0.64 cfs of water at 60°F average temperature and at a constant pressure, making the hydraulic grade line parallel to the slope of the conduit. How much does the conduit drop in 1000 ft, assuming the size of the surface imperfections is .00085 ft? (Use $\nu = 1.217 \times 10^{-5}$ ft²/sec.)   *Ans.* 260 ft

47. When 1.48 cfs of medium fuel oil flows from A to B through 3500 ft of new 6″ uncoated cast iron pipe the lost head is 143 ft. Sections A and B are at elevations 0.0 and 60.0 respectively and the pressure at B is 50 psi. What pressure in psi must be maintained at A to deliver the flow stated? *Ans.* 125 psi

48. (a) Determine the flow of water through the new cast iron pipes shown in Fig. 7-7 below. (b) What is the pressure head at B, which is 100 ft from the reservoir? (Use Table 3).   *Ans.* 3.49 cfs, 1.91 ft

49. Water at 100°F flows through the system shown in Fig. 7-8 below. Lengths of 3″ and 6″ new, asphalt-dipped cast iron pipe are 180 ft and 100 ft respectively. Loss factors for fittings and valves are: 3″ Bends, $K = 0.40$ each; 6″ Bend, $K = 0.60$; and 6″ Valve, $K = 3.0$. Determine the flow in cfs. *Ans.* 0.457 cfs

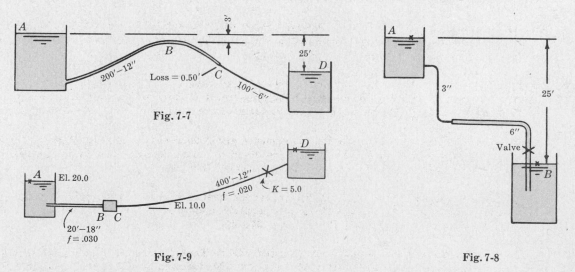

Fig. 7-7

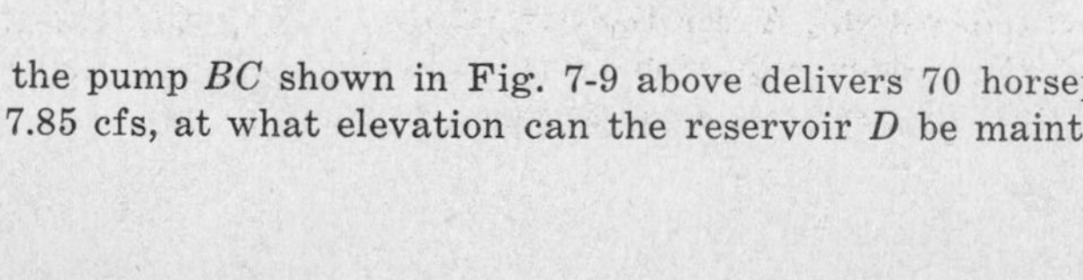

Fig. 7-9

Fig. 7-8

50. If the pump BC shown in Fig. 7-9 above delivers 70 horsepower to the system when the flow of water is 7.85 cfs, at what elevation can the reservoir D be maintained?   *Ans.* 76.6

51. A pump at elevation 10.0 delivers 7.85 cfs of water through a horizontal pipe system to a closed tank whose liquid surface is at elevation 20.0. The pressure head at the 12″ diameter suction side of the pump is −4.0 ft and at the 6″ diameter discharge side is +193.4 ft. The 6″ pipe ($f = .030$) is 100 ft long, suddenly expanding to a 12″ pipe ($f = .020$) which is 600 ft long and which terminates at the tank. A 12″ valve, $K = 1.0$, is located 100 ft from the tank. Determine the pressure in the tank above the water surface. Draw the energy and hydraulic grade lines.   *Ans.* 10.0 psi

52. What size average cast iron pipe should be used to deliver 1.00 cfs of water at 70°F through 4000 ft with a drop in the hydraulic grade line of 70 ft? (Use Table 3.)   *Ans.* $d = 0.53$ ft

53. Pump BC delivers water to reservoir F and the hydraulic grade line is shown in Fig. 7-10 below. Determine (a) the power supplied to the water by pump BC, (b) the power removed by turbine DE and (c) the elevation maintained in reservoir F.   *Ans.* 1010 hp, 71.3 hp, El. 300

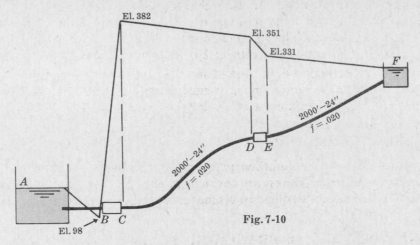

Fig. 7-10

54. Air is blown through a 2″ inside diameter, old wrought iron pipe at a constant temperature of 68°F and at the rate of 0.15 lb/sec. At section A the pressure is 54.7 psi absolute. What will be the pressure 500 ft away from A in the horizontal pipe? Use $\epsilon = .00083$ ft.   *Ans.* 52.9 psia

55. Carbon dioxide at 100°F flows through a horizontal 4″ new wrought iron pipe for 200 ft. The pressure at upstream section A is 120 psi gage and the average velocity is 40.0 ft/sec. Assuming the density change negligible, what is the pressure drop in the 200 ft of pipe? (Absolute viscosity at 100°F = 330×10^{-9} lb sec/ft².)   *Ans.* 1.79 psi

56. Laminar flow occurs through a wide rectangular conduit 9″ high. Assuming that the velocity distribution satisfies the equation $v = 16y(3 - 4y)$, calculate (a) the flow per unit of width, (b) the kinetic energy correction factor, and (c) the ratio of average to maximum velocity.
Ans. 4.50 cfs, $\alpha = 1.52$, 0.67

57. In the laboratory, a 1″ diameter plastic pipe is used to demonstrate laminar flow. If the lower critical velocity is to be 10.0 ft/sec, what should be the kinematic viscosity of the liquid used?   *Ans.* .000417

58. For laminar flow in pipes, $f = 64/R_E$. Using this information, develop the expression for the average velocity in terms of lost head, diameter and other pertinent items.   *Ans.* $V = gd^2 h_L/32\nu L$

59. Determine the flow in a 12″ diameter pipe if the equation of velocity distribution is $v^2 = 400(y - y^2)$, with the origin at the pipe wall.   *Ans.* 5.25 cfs

Chapter 8

Equivalent, Compound, Looping and Branching Pipes

PIPING SYSTEMS

Piping systems which distribute water to a city or to a large industrial plant may become extremely complicated. In this chapter only a few relatively simple conditions will be considered. In most instances water will be the fluid flowing, although the methods of analysis and solution may be applied to other liquids. In general, the length-diameter ratios are large (see Chapter 7, Problem 20) and minor losses may be neglected.

The Hardy Cross method of analyzing the flows in pipe networks is presented in Problems 18, 19 and 20. Flows and pressure drops in the extensive distribution systems of cities can be analyzed by means of analog computers.

EQUIVALENT PIPES

A pipe is equivalent to another pipe or to a piping system when, for a given lost head, the same flow is produced in the equivalent pipe as was produced in the given system. It often proves convenient to replace a complex system by a single equivalent pipe.

COMPOUND, LOOPING, and BRANCHING PIPES

Compound pipes consist of pipes of several sizes in series.

Looping pipes consist of two or more pipes which branch and come together again downstream (in parallel).

Branching pipes consist of two or more pipes which branch and do not come together again downstream.

METHODS of SOLUTION

Methods of solution involve establishing the necessary number of simultaneous equations or employing special modifications of the Darcy formula in which the coefficient depends only on the relative roughness of the pipe. For the flow of water (or of other liquids of approximately the same viscosity), such formulas have been devised by Manning, Schoder, Scobey, Hazen-Williams and others.

HAZEN-WILLIAMS FORMULA

The Hazen-Williams formula will be used in this chapter. Solutions in this book will be accomplished by the use of Diagram B (in the Appendix) and/or by hydraulic slide rule rather than by the more laborious algebraic procedure. Hydraulic slide rules, which may be obtained from the Cast Iron Pipe Research Association or from the Lock Joint Pipe Company, facilitate solutions materially. The formula for velocity is

$$V = 1.318C_1R^{0.63}S^{0.54} \tag{1}$$

where V = velocity in ft/sec, R = hydraulic radius in ft, S = slope of the hydraulic gradient and $C_1 =$ the Hazen-Williams coefficient of relative roughness. Suggested values of C_1 will be found in Table 6 of the Appendix.

The relationship between this empirical formula and the Darcy formula will be shown in Problem 1 below. The greatest advantage of the Hazen-Williams formula is that the coefficient C_1 depends only on the relative roughness.

In Diagram B, flow Q is expressed in million gallons per day (mgd) and not in cfs. The conversion factor is

$$1 \text{ mgd} = 1.547 \text{ cfs}$$

Solved Problems

1. Convert the Hazen-Williams formula to the Darcy-type formula.

   Solution:

   $$V = 1.318 C_1 R^{0.63} S^{0.54}$$

   The value of $S = h/L$ and $R = d/4$ (see Chapter 7, Problem 26). Solving for h,

   $$h^{0.54} = \frac{4^{0.63}}{1.318} \frac{L^{0.54}}{d^{0.63}} \frac{V}{C_1}$$

   or

   $$h = \frac{2g(4)^{1.165}}{(1.318)^{1.850}} \left(\frac{L}{d}\right) \frac{V^2}{2g} \left[\frac{d^{-0.015}}{V^{0.150} d^{0.150} C_1^{1.850}}\right] = \frac{194 d^{-0.015}}{C_1^{1.850}} \left(\frac{L}{d}\right) \frac{V^2}{2g} \left[\frac{1}{d^{0.150} V^{0.150}}\right]$$

   To include Reynolds Number in the equation, multiply by $(\nu/\nu)^{0.150}$, obtaining

   $$h = \frac{194 d^{-0.015}}{C_1^{1.850} \nu^{0.150}} \left(\frac{L}{d}\right) \frac{V^2}{2g} \left[\frac{\nu^{0.150}}{V^{0.150} d^{0.150}}\right] = \frac{194 d^{-0.015}}{C_1^{1.850} \nu^{0.150} R_E^{0.150}} \left(\frac{L}{d}\right) \frac{V^2}{2g} = f_1 \left(\frac{L}{d}\right) \frac{V^2}{2g}.$$

   It will be observed that, if the small term $d^{-0.015}$ is omitted, the frictional factor f_1 is a function of Reynolds Number and the roughness factor C_1 for any liquid whose viscosity does not change greatly (percentagewise) with temperature changes. In such cases, an average (or representative) value of viscosity would be used as a constant in the Darcy-type formula.

2. Compare results obtained by algebraic solution with the values determined by Diagram B for (a) the flow produced in a new 12″ pipe with a drop in the hydraulic grade line of 14.4 ft in 5000 ft of pipe and (b) the lost head occurring in 6000 ft of old 24″ cast iron pipe when the flow is 6 mgd.

   Solution:

   (a) **Algebraically.** $S = 14.4/5000 = .00288$ and $R = \frac{1}{4}d = \frac{1}{4}$ ft.
   From Table 6 in the Appendix, $C_1 = 130$. Then

   $$Q = AV = \tfrac{1}{4}\pi(1)^2 [1.318 \times 130 (\tfrac{1}{4})^{0.63} (.00288)^{0.54}] = 2.38 \text{ cfs} = 1.54 \text{ mgd}$$

   By Diagram. Diagram B is drawn for $C_1 = 100$.
   $D = 12″$ and $S = .00288$ or 2.88 ft/1000 ft.

   Using these values, $Q_{100} = 1.17$ mgd (by reading diagram according to instructions thereon).

   Examination of the Hazen-Williams formula indicates that V and Q vary directly with coefficient C_1. Thus the flow for C_1 of 130 is

   $$Q_{130} = (130/100)(1.17) \text{ mgd} = 1.53 \text{ mgd}$$

(b) **Algebraically** ($C_1 = 100$). $Q = 6$ mgd $= 9.28$ cfs.

$$9.28 = \tfrac{1}{4}\pi(2)^2[1.318 \times 100(2/4)^{0.63}S^{0.54}] \quad \text{and} \quad S = .0020$$

By Diagram. $Q = 6$ mgd, $D = 24''$.

$$S = 2.0 \text{ ft/1000 ft} = .002 \text{ (from diagram)}.$$

3. An average 12″ cast iron pipe carries 3.50 cfs of water. What head is lost in 4000 ft of pipe (a) using the Darcy formula and (b) using the Hazen-Williams formula?
Solution:

(a) $V_{12} = 3.50/[\tfrac{1}{4}\pi(1)^2] = 4.46$ ft/sec. From Table 3 in the Appendix, $f = .0260$.

$$\text{Lost head} = f\frac{L}{d}\frac{V^2}{2g} = .0260\frac{4000}{1}\frac{(4.46)^2}{2g} = 32.2 \text{ ft}$$

(b) $Q = 3.50/1.547 = 2.26$ mgd and $C_1 = 110$. Q for C_1 of $100 = (100/110)(2.26) = 2.06$ mgd.

From Diagram B, $S = 8.2$ ft/1000 ft and the Lost Head $= 8.2 \times 4 = 32.8$ ft.

The agreement is unusually close in this case. Hydraulic slide rule gives $S = 8.0$ ft/1000 ft and a lost head of 32.0 ft.

Experience and judgment in the selection of coefficient C_1 will produce satisfactory results for the flow of water and other liquids of comparable viscosity.

4. For a lost head of 5.0 ft/1000 ft, and using $C_1 = 100$ for all pipes, how many 8″ pipes are equivalent to a 16″ pipe? to a 24″ pipe?
Solution:

Using Diagram B, for $S = 5.0$ ft/1000 ft: Q for 8″ pipe $= 0.55$ mgd
Q for 16″ pipe $= 3.40$ mgd
Q for 24″ pipe $= 10.0$ mgd.

Thus it would take 3.40/0.55 or 6.2 8″ pipes to be hydraulically equivalent to a 16″ pipe of the same relative roughness. Likewise, 10.0/0.55 or 18.2 8″ pipes are equivalent to a 24″ pipe for a lost head of 5.0 ft/1000 ft or for any other lost head condition.

5. A compound piping system consists of 6000 ft of 20″, 4000 ft of 16″, and 2000 ft of 12″ new cast iron pipe. Convert the system to (a) an equivalent length of 16″ pipe and (b) an equivalent size pipe 12,000 ft long.
Solution:

Use $C_1 = 130$ for new cast iron pipe.

(a) Because the common hydraulic quantity for a compound piping system is the flow, assume a value of 2.6 mgd (any convenient value will serve). In order to use Diagram B, change Q_{130} to Q_{100}, that is,

$$Q_{100} = (100/130)(2.6) = 2.0 \text{ mgd}$$

$S_{20} = 0.64$ ft/1000 ft   and the Lost Head $= 0.64 \times 6 = $ 3.8 ft, (13.9%)
$S_{16} = 1.87$ ft/1000 ft   Lost Head $= 1.87 \times 4 = $ 7.5 ft, (27.5%)
$S_{12} = 8.0$ ft/1000 ft   Lost Head $= 8.0 \ \times 2 = \underline{16.0 \text{ ft}}$, (58.6%)
For $Q = 2.6$ mgd: Total Lost Head $= 27.3$ ft, (100.0%)

The equivalent 16″ pipe must carry 2.6 mgd with a lost head of 27.3 ft ($C_1 = 130$).

$$S_{16} = 1.87 \text{ ft/1000 ft} = \frac{\text{lost head in ft}}{\text{equivalent length in 1000 ft units}} = \frac{27.3}{L_E}$$

and $L_E = 14.6$ in 1000 ft units, or 14,600 ft.

(b) The 12,000 ft of pipe, $C_1 = 130$, must carry the 2.6 mgd with a lost head of 27.3 ft.

$$S_E = \frac{\text{lost head in ft}}{\text{length in 1000 ft units}} = \frac{27.3}{12} = 2.28 \text{ ft/1000 ft}$$

From Diagram B, using $Q_{100} = 2$ mgd, $D = 15.5''$ (approximately).

6. Convert the piping system shown in Fig. 8-1 below to an equivalent length of 6″ pipe.

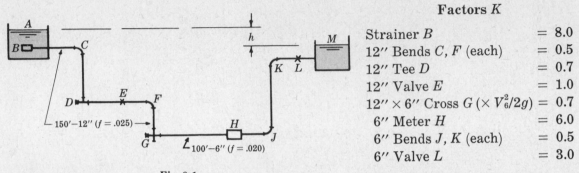

Factors K

Strainer B	$= 8.0$
12″ Bends C, F (each)	$= 0.5$
12″ Tee D	$= 0.7$
12″ Valve E	$= 1.0$
12″ × 6″ Cross G ($\times V_6^2/2g$)	$= 0.7$
6″ Meter H	$= 6.0$
6″ Bends J, K (each)	$= 0.5$
6″ Valve L	$= 3.0$

Fig. 8-1

Solution:

This problem will be solved by use of the Bernoulli equation, A to M, datum M, as follows.

$$(0 + 0 + h) - \overset{\text{Bends}}{\left(8.0 + 2 \times 0.5 + 0.7 + 1.0 + .025 \times \frac{150}{1}\right)\frac{V_{12}^2}{2g}}$$

$$- \overset{\text{Bends} \quad\quad \text{Exit}}{\left(0.7 + 6.0 + 2 \times 0.5 + 3.0 + 1.0 + .020 \times \frac{100}{\frac{1}{2}}\right)\frac{V_6^2}{2g}} = (0 + 0 + 0)$$

Then $h = 14.5\dfrac{V_{12}^2}{2g} + 15.7\dfrac{V_6^2}{2g} = \left(14.5 \times \dfrac{1}{16} + 15.7\right)\dfrac{V_6^2}{2g} = 16.6\dfrac{V_6^2}{2g}.$

For any available head h, the lost head is $16.6(V_6^2/2g)$. The lost head in L_E ft of 6″ pipe is $f(L_E/d)(V_6^2/2g)$. Equating the two values,

$$16.6\frac{V_6^2}{2g} = .020\frac{L_E}{\frac{1}{2}}\frac{V_6^2}{2g} \quad\quad \text{and} \quad\quad L_E = 415 \text{ ft}$$

The velocity heads in this equality cancelled each other. It should be remembered that exact hydraulic equivalence depends upon f which is not constant over wide ranges of velocity.

7. For the compound piping system in Problem 5, what flow will be produced for a total lost head of 70.0 ft, (a) using the equivalent pipe method and (b) using the percentage method?

Solution:

(a) From Prob. 5, 14,600 ft of 16″ pipe is equivalent to the compound system. For a lost head of 70.0 ft,

$$S_{16} = 70.0/14.6 = 4.8 \text{ ft/1000 ft} \quad \text{and, from Diagram } B, \quad Q_{100} = 3.4 \text{ mgd}$$

Hence $Q_{130} = (130/100)(3.4) = 4.4$ mgd

(b) The percentage method requires the calculation of lost head values for an assumed flow Q. Although values are available from Problem 5, an additional calculation will be made to serve as a check on the solution. Assuming $Q_{130} = 3.9$ mgd, then $Q_{100} = 100/130 \times 3.9 = 3.0$ mgd, and, from Diagram B,

$$S_{20} = 1.35 \text{ ft/1000 ft} \quad \text{and} \quad \text{Lost Head} = 1.35 \times 6 = 8.1 \text{ ft,} \quad (14.0\%)$$
$$S_{16} = 4.0 \text{ ft/1000 ft} \quad \text{and} \quad \text{Lost Head} = 4.0 \times 4 = 16.0 \text{ ft,} \quad (27.5\%)$$
$$S_{12} = 17.0 \text{ ft/1000 ft} \quad \text{and} \quad \text{Lost Head} = 17.0 \times 2 = 34.0 \text{ ft,} \quad (58.5\%)$$
$$\text{For} \quad Q = 3.9 \text{ mgd: Total Lost Head} = 58.1 \text{ ft,} \quad (100.0\%).$$

The same percentages appear here as are shown in Problem 5. Apply these percentages to the given total lost head of 70 ft, i.e.,

$$LH_{20} = 70 \times 14.0\% = 9.8 \text{ ft,} \quad S = 9.8/6 = 1.63 \text{ ft/1000 ft,} \quad Q = 130/100 \times 3.3 = 4.3 \text{ mgd}$$
$$LH_{16} = 70 \times 27.5\% = 19.2 \text{ ft,} \quad S = 19.2/4 = 4.80 \text{ ft/1000 ft,} \quad Q = 130/100 \times 3.4 = 4.4 \text{ mgd}$$
$$LH_{12} = 70 \times 58.5\% = 41.0 \text{ ft,} \quad S = 41.0/2 = 20.5 \text{ ft/1000 ft,} \quad Q = 130/100 \times 3.4 = 4.4 \text{ mgd}$$

The calculation for one size is sufficient to obtain flow Q, but the check calculations give assurance that no mistakes have been made.

8. For the system shown in the adjacent figure, when the flow from reservoir A to main D is 3.25 mgd, the pressure at D is 20.0 psi. The flow to D must be increased to 4.25 mgd with the pressure at 40.0 psi. What size pipe, 5000 ft long, should be laid from B to C (shown dotted) parallel to the existing 12″ to accomplish this result?

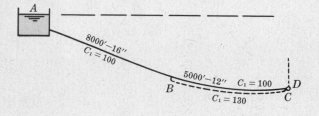

Fig. 8-2

Solution:

The elevation of reservoir A can be determined by using the data in the first sentence of the problem. From Diagram B,

$$\text{for} \quad Q = 3.25 \text{ mgd,} \quad S_{16} = 4.6 \text{ ft/1000 ft,} \quad \text{Lost Head} = 4.6 \times 8 = 36.8 \text{ ft}$$
$$S_{12} = 19.0 \text{ ft/1000 ft,} \quad \text{Lost Head} = 19.0 \times 5 = \underline{95.0 \text{ ft}}$$
$$\text{Total Lost Head} = 131.8 \text{ ft}$$

The hydraulic grade line drops 131.8 ft to an elevation of 46.2 ft above D (equivalent to 20.0 psi). Thus reservoir A is $(131.8 + 46.2) = 178.0$ ft above point D.

For a pressure of 40.0 psi, the elevation of the hydraulic grade line at D will be 92.4 ft above D, or the available head for the flow of 4.25 mgd will be $(178.0 - 92.4) = 85.6$ ft.

In the 16″, $Q = 4.25$ mgd, $S = 7.5$ ft/1000 ft, Lost Head $= 7.5 \times 8 = 60.0$ ft. Hence

$$\text{Lost Head from } B \text{ to } C = 85.6 - 60.0 = 25.6 \text{ ft}$$

For the existing 12″, $S = 25.6/5 = 5.1$ ft/1000 ft, $Q = 1.6$ mgd and the flow in the new pipe must be $(4.25 - 1.6) = 2.65$ mgd with an available head (drop in the hydraulic grade line) of 25.6 ft from B to C.

$$S = 25.6/5 = 5.1 \text{ ft/1000 ft} \quad \text{and} \quad Q_{100} = 100/130 \times 2.65 = 2.04 \text{ mgd}$$

Diagram B gives $D = 13″$ approximately (use stock size, 14″).

9. For the looping pipe system in Fig. 8-3, the pressure head at A is 120.0 ft of water and the pressure head at E is 72.0 ft of water. Assuming the pipes are in a horizontal plane, what are the flows in each branch of the loop?

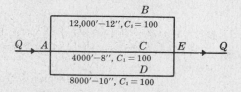

Fig. 8-3

Solution:

The drop in the hydraulic grade line A to E is $(120 - 72) = 48$ ft, neglecting the minor values of velocity head differences. The flows can be calculated inasmuch as the slopes of the grade lines are known. Thus, using Diagram B,

$$S_{12} = 48/12 = 4.0 \text{ ft/1000 ft}, \quad Q_{12} = 1.4 \text{ mgd}, \quad (41.1\%)$$
$$S_8 = 48/4 = 12.0 \text{ ft/1000 ft}, \quad Q_8 = 0.9 \text{ mgd}, \quad (26.5\%)$$
$$S_{10} = 48/8 = 6.0 \text{ ft/1000 ft}, \quad Q_{10} = \underline{1.1 \text{ mgd}}, \quad (32.4\%)$$
$$\text{Total } Q = 3.4 \text{ mgd}, \quad (100.0\%)$$

10. In Problem 9, had the total flow Q been 6.50 mgd, how much lost head occurs between A and E and how does Q divide in the loop? Use two solutions, the percentage method and the equivalent pipe method.

Solution:

In a looping system, the common hydraulic quantity is the lost head across the loop (AE). The solution will proceed as if Problem 9 had not been solved.

Assuming a lost head, A to E, of 24 ft, the values of flow for the assumed lost head may be obtained from Diagram B.

$$S_{12} = 24/12 = 2.0 \text{ ft/1000 ft}, \quad Q_{12} = 0.95 \text{ mgd}, \quad (41.3\%)$$
$$S_8 = 24/4 = 6.0 \text{ ft/1000 ft}, \quad Q_8 = 0.60 \text{ mgd}, \quad (26.2\%)$$
$$S_{10} = 24/8 = 3.0 \text{ ft/1000 ft}, \quad Q_{10} = \underline{0.75 \text{ mgd}}, \quad (32.5\%)$$
$$\text{Total } Q = 2.30 \text{ mgd}, \quad (100.0\%)$$

(a) **Percentage Method.**

The flow in each branch of the loop will be a constant percentage of the total flow through the loop for any reasonable range of lost heads across the loop. The percentages shown above agree favorably with the percentages tabulated in Problem 9 (within the accuracy of the Diagram or of the slide rule). Apply these percentages to the given flow of 6.50 mgd.

$$Q_{12} = 41.3\% \times 6.50 = 2.69 \text{ mgd}, \quad S_{12} = 13.5 \text{ ft/1000 ft}, \quad LH_{A-E} = 162 \text{ ft}$$
$$Q_8 = 26.2\% \times 6.50 = 1.70 \text{ mgd}, \quad S_8 = 40.0 \text{ ft/1000 ft}, \quad LH_{A-E} = 160 \text{ ft}$$
$$Q_{10} = 32.5\% \times 6.50 = \underline{2.11 \text{ mgd}}, \quad S_{10} = 20.0 \text{ ft/1000 ft}, \quad LH_{A-E} = 160 \text{ ft}$$
$$Q = 6.50 \text{ mgd}$$

This method gives a check on the calculations, as indicated above for the three lost head values. It is the preferred method.

(b) **Equivalent Pipe Method** (use a 12″ diameter).

The calculation of flows for an assumed lost head must be made, as for the first method. Using the values above, for a lost head of 24 ft a flow of 2.30 mgd is produced in the given looping system. An equivalent pipe must produce the same flow for the lost head of 24 ft, i.e.,

$$Q = 2.30 \text{ mgd}, \quad \text{Lost Head} = 24 \text{ ft} \quad \text{and} \quad S_{12} = 10.1 \text{ ft/1000 ft from Diagram } B$$

From $S = \dfrac{h}{L}$, $\quad 10.1 = \dfrac{24 \text{ ft}}{L_E \text{ in 1000 ft}}$ and $L_E = 2380$ ft (of 12″ pipe, $C_1 = 100$).

For the given flow of 6.50 mgd, $S_{12} = 68$ ft/1000 ft and total lost head A-$E = 68 \times 2.38 = 162$ ft. With this lost head, the three values of flow can be obtained.

11. For the system shown in Fig. 8-4 below, (a) what flow will occur when the drop in the hydraulic grade line from A to B is 200 ft? (b) What length of 20″ pipe ($C_1 = 120$) is equivalent to Section AB?

Solution:

(a) The most direct solution can be obtained by assuming a drop in the hydraulic grade line (lost head) from W to Z and following that assumption to a logical conclusion.

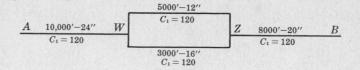

Fig. 8-4

For example, assume a lost head of 30 ft from W to Z. Then, from Diagram B,

$$S_{12} = 30/5 = 6.0 \text{ ft/1000 ft} \quad \text{and} \quad Q_{12} = 120/100 \times 1.74 = 2.1 \text{ mgd}, \quad (26.0\%)$$
$$S_{16} = 30/3 = 10.0 \text{ ft/1000 ft} \quad \text{and} \quad Q_{16} = 120/100 \times 5.00 = 6.0 \text{ mgd}, \quad (74.0\%)$$
$$\text{Total } Q = \underline{8.1 \text{ mgd}, \ (100.0\%)}$$

The lost head from A to B can be computed for this total flow of 8.1 mgd. To employ Diagram B, use $Q_{100} = 100/120 \times 8.1 = 6.75$ mgd.

A to W, $S_{24} = 2.45$ ft/1000 ft, Lost Head $= 2.45 \times 10 =$	24.5 ft,	(23.9%)
W to Z, (as assumed above) $=$	30.0 ft,	(29.3%)
Z to B, $S_{20} = 6.0$ ft/1000 ft, Lost Head $= 6.0 \times 8 =$	48.0 ft,	(46.8%)
Total Lost Head for Q of 8.1 mgd $=$	102.5 ft,	(100.0%)

Applying these percentage values to the given lost head of 200 ft produces:

actual Lost Head$_{A-w}$ $= 200 \times 23.9\% = 47.8$ ft, $S_{24} = 47.8/10 = 4.78$ ft/1000 ft;
actual Lost Head$_{w-z}$ $= 200 \times 29.3\% = 58.6$ ft;
actual Lost Head$_{z-B}$ $= 200 \times 46.8\% = 93.6$ ft, $S_{20} = 93.6/8 = 11.7$ ft/1000 ft.

From Diagram B, the flow in the 24″ is $(120/100)(9.75) = 11.7$ mgd.

As a check, in the 20″ pipe $Q = (120/100)(9.8) = 11.7$ mgd.

This flow divides in the loop WZ at the percentages calculated above, namely 26.0% and 74.0%.

(b) Using the information above for the system from A to B, a flow of 8.1 mgd is produced with a drop in the hydraulic grade line of 102.5 ft. For 8.1 mgd in a 20″ pipe, $C_1 = 120$,

$$S_{20} = 6.0 \text{ ft/1000 ft} = 102.5/L_E \quad \text{or} \quad L_E = 17,100 \text{ ft}$$

12. In Fig. 8-5, when pump YA delivers 5.00 cfs, find the pressure heads at A and B. Draw the hydraulic grade lines.

Solution:

Reduce the section (loop) BC to an equivalent pipe, 16″ in diameter, $C_1 = 100$. By so doing, a single-size pipe of the same relative roughness is readily handled for all conditions of flow. Assuming a drop in the grade line of 22 ft from B to C, the following values are obtained by hydraulic slide rule (the reader may check the values using Diagram B).

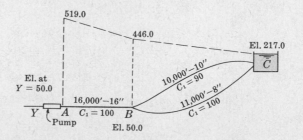

Fig. 8-5

$$S_{10} = 22/10 = 2.2 \text{ ft/1000 ft}, \quad Q_{10} = 0.57 \text{ mgd}$$
$$S_8 = 22/11 = 2.0 \text{ ft/1000 ft}, \quad Q_8 = \underline{0.34 \text{ mgd}}$$
$$\text{Total } Q = 0.91 \text{ mgd}$$

For $Q = 0.91$ mgd and $D = 16''$ $(C_1 = 100)$, $S_{16} = 0.435$ ft/1000 ft $= 22.0/L_E$ and $L_E = 50,500$ ft.

The flow from pump to reservoir is 5.00 cfs or 3.23 mgd. For $(50,500 + 16,000) = 66,500$ ft of equivalent 16'' pipe, the lost head from A to C will be

$$S_{16} = 4.55 \text{ ft/1000 ft}, \quad \text{Lost Head} = 4.55 \times 66.5 = 302 \text{ ft}$$

Thus the elevation of the hydraulic grade line at A is $(217 + 302) = 519$, as shown in the figure. The drop from A to $B = 4.55 \times 16 = 73$ ft and the elevation at B becomes $(519 - 73) = 446$.

$$\text{Pressure head at } A = 519 - 50 = 469 \text{ ft}$$
$$\text{Pressure head at } B = 446 - 50 = 396 \text{ ft}$$

13. In Fig. 8-6 below, which system has the greater capacity, $ABCD$ or $EFGH$?   ($C_1 = 120$ for all pipes.)

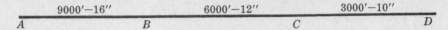

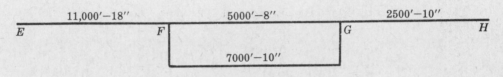

Fig. 8-6

Solution:

Assume $= 2$ mgd in $ABCD$. Then, using the hydraulic slide rule,

$$S_{16} = 1.33 \text{ ft/1000 ft}, \quad \text{Lost Head} = 1.33 \times 9 = 12.0 \text{ ft}$$
$$S_{12} = 5.35 \text{ ft/1000 ft}, \quad \text{Lost Head} = 5.35 \times 6 = 32.1 \text{ ft}$$
$$S_{10} = 13.0 \text{ ft/1000 ft}, \quad \text{Lost Head} = 13.0 \times 3 = \underline{39.0 \text{ ft}}$$
$$\text{For } Q = 2 \text{ mgd, total Lost Head} = 83.1 \text{ ft}$$

For the loop FG in $EFGH$, find the percentage of any flow Q in each branch. Assume the lost head from F to G to be 24 ft. Then

$$S_8 = 24/5 = 4.80 \text{ ft/1000 ft} \quad \text{and} \quad Q_8 = 0.65 \text{ mgd,} \quad (40.1\%)$$
$$S_{10} = 24/7 = 3.43 \text{ ft/1000 ft} \quad \text{and} \quad Q_{10} = \underline{0.97 \text{ mgd,}} \quad (59.9\%)$$
$$\text{Total } Q = 1.62 \text{ mgd,} \quad (100.0\%)$$

To compare capacities, several alternatives present themselves. Rather than using equivalent pipes, we might calculate the lost head caused by a flow of 2 mgd through each system. The system with the smaller lost head would have the greater capacity. Or we might find flow Q caused by the identical drop in the hydraulic grade line in each system. The pipe system with the larger flow would have the greater capacity. In this problem, compare the lost head of 83.1 ft in $ABCD$ when $Q = 2$ mgd with the value of lost head obtained for $EFGH$ for the same flow.

(*a*) For $Q_{18} = 2$ mgd, $S_{18} = 0.75$ ft/1000 ft,   Lost Head$_{EF} = 8.3$ ft.

(*b*) For $Q_8 = 40.1\% \times 2$ mgd
$$= 0.80 \text{ mgd}, \quad S_8 = 7.1 \text{ ft/1000 ft}, \quad \text{Lost Head}_{FG} = 35.5 \text{ ft,}$$

   or   for $Q_{10} = 59.9\% \times 2$ mgd (as check)
$$= 1.20 \text{ mgd}, \quad S_{10} = 5.1 \text{ ft/1000 ft}, \quad (\text{L.H.} = 5.1 \times 7 = 35.7 \text{ ft}).$$

(*c*) For $Q_{10} = 2$ mgd, $S_{10} = 13.0$ ft/1000 ft,   Lost Head $= 32.5$ ft.
For $Q = 2$ mgd, total Lost Head from E to $H = 76.3$ ft.

Hence the system $EFGH$ has the greater capacity.

14. In Fig. 8-7 below, the flow from reservoir A is 10 mgd.   Determine the power extracted by turbine DE if the pressure head at E is -10.0 ft.   Indicate grade lines.

Solution:

The analysis of a branching system of pipes should concentrate itself on the junction point C. First, the sum of the flows toward C must equal the sum of the flows away from C.   Second, the elevation of the hydraulic grade line at C is often the key to the solution.

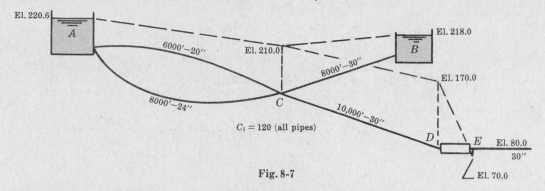

Fig. 8-7

To calculate the elevation of the hydraulic grade line at C, assume the lost head, A to C, to be 24 ft.   Then

$$S_{20} = 24/6 = 4.0 \text{ ft/1000 ft}, \quad Q_{20} = 6.5 \text{ mgd}, \quad (42.0\%)$$
$$S_{24} = 24/8 = 3.0 \text{ ft/1000 ft}, \quad Q_{24} = 9.0 \text{ mgd}, \quad (58.0\%)$$
$$\text{Total } Q \quad = 15.5 \text{ mgd}, \ (100.0\%)$$

Apply these percentages to the given 10 mgd flow from A to C.

$$Q_{20} = 4.2 \text{ mgd}, \quad S_{20} = 1.77 \text{ ft/1000 ft}, \quad \text{Lost Head} = 10.6 \text{ ft}$$
$$Q_{24} = 5.8 \text{ mgd}, \quad S_{24} = 1.33 \text{ ft/1000 ft}, \quad \text{Lost Head} = 10.6 \text{ ft (check)}$$

Thus the elevation of the hydraulic grade line at $C = 220.6 - 10.6 = 210.0$. From this information, the grade line drops 8 ft from B to C and the flow must be from B to C.   Then

$$S_{30} = 8/8 = 1.0 \text{ ft/1000 ft}, \quad Q = 8.95 \text{ mgd}$$

Also,   flow from C = flow to C
$$Q_{C-D} = 10.0 + 8.95 = 18.95 \text{ mgd}$$

Thus $S_{30} = 4.0$ ft/1000 ft, $LH_{C-D} = 40.0$ ft, and elevation of grade line at $D = 210 - 40 = 170.0$.

$$\text{Horsepower extracted} = \frac{wQH_T}{550} = \frac{62.4(18.95 \times 1.547)(170 - 70)}{550} = 333 \text{ hp}.$$

15. In Fig. 8-8 below, valve F is partly closed, creating a lost head of 3.60 ft when the flow through the valve is 0.646 mgd.   What is the length of 10″ pipe to reservoir A?

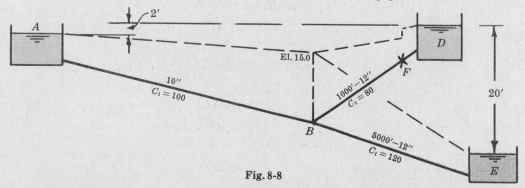

Fig. 8-8

Solution:

For DB, flow $Q = 0.646$ mgd and $S_{12} = 1.40$ ft/1000 ft.

The total lost head from D to B = $1.40 \times 1 + 3.60 = 5.00$ ft, giving the grade line elevation at B of 15.0 (calling Elevation $E = 0$).

For BE, $S_{12} = (15.0 - 0.0)/5 = 3$ ft/1000 ft and $Q = 1.46$ mgd.

For AB, flow $Q = 1.46 - 0.646 = 0.81$ mgd and $S_{10} = 3.45$ ft/1000 ft (by hydraulic slide rule).

Then, from $S = h/L$, $L = h/S = 3.00/3.45 = 0.870$ in 1000 ft units, or $L = 870$ ft.

16. Water is to be pumped at the rate of 2.00 cfs through 4000 ft of new cast iron pipe to a reservoir whose surface is 120 ft above the lower water level. The annual cost of pumping the 2.00 cfs is \$5.00 per ft pumped against, and the annual cost of the pipe itself is 10% of its initial cost. Assume cast iron pipe in place costs \$140.00 per ton, with class B (200 ft head) pipe having the following weights per foot: 6″, 33.3 lb; 8″, 47.5 lb; 10″, 63.8 lb; 12″, 82.1 lb; and 16″, 125 lb. Determine the economical pipe size for this installation.

Solution:

Calculations for the 12″ pipe will be shown in detail with a summary of all results given in the table below. The lost head in the 12″, by hydraulic slide rule using $C_1 = 130$ for new cast iron pipe, is 2.06 ft/1000 ft.

Hence, total head pumped against = $120 + 4 \times 2.06 = 128.2$ ft.

$$\text{Cost of pumping} = 128.2 \times \$5 = \$641 \text{ per year}$$
$$\text{Cost of pipe in place} = \$140 \times 4000 \times 82.1/2000 = \$23,000$$
$$\text{Annual cost of pipe} = 10\% \times \$23,000 = \$2300$$

Tabulating these values for comparison with the costs of the other sizes under consideration gives the following.

D inches	S ft/1000 ft	Lost Head in ft	Total Pumping Head = 120 + LH	Annual Costs for 2 cfs Pumping	+ Pipe Cost	= Total
6″	59.0	236.	356. ft	\$1780	\$ 933	\$2713
8″	14.8	59.2	179.2 ft	896	1330	2226
10″	5.00	20.0	140.0 ft	700	1790	2490
12″	2.06	8.2	128.2 ft	641	2300	2941
16″	0.515	2.1	122.1 ft	611	3500	4111

The economical size is 8″.

17. For the constant elevations of the water surfaces shown in Fig. 8-9(a) below, what flows will occur?

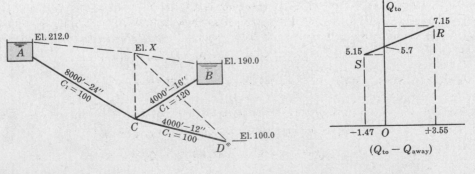

Fig. 8-9(a)   Fig. 8-9(b)

Solution:

Because the elevation of the hydraulic grade line at C cannot be computed (all flows unknown), the problem will be solved by successive trials. A convenient assumption is to choose the elevation of the hydraulic grade line at C at 190.0. By so assuming, flow to or from reservoir B will be zero, reducing the number of calculations thereby.

For elevation of hydraulic grade line at $C = 190.0$,

$$S_{24} = (212 - 190)/8 = 2.75 \text{ ft/1000 ft} \quad \text{and} \quad Q = 7.15 \text{ mgd to } C$$
$$S_{12} = (190 - 100)/4 = 22.5 \text{ ft/1000 ft} \quad \text{and} \quad Q = 3.60 \text{ mgd away from } C$$

Examination of these values of flow indicates that the grade line at C must be higher, thereby reducing the flow from A and increasing the flow to D as well as adding flow to B. In an endeavor to "straddle" the correct elevation at C, assume a value of 200.0. Thus, for elevation at $C = 200.0$,

$$S_{24} = (212 - 200)/8 = 1.50 \text{ ft/1000 ft} \quad \text{and} \quad Q = 5.15 \text{ mgd to } C$$
$$S_{16} = (200 - 190)/4 = 2.50 \text{ ft/1000 ft} \quad \text{and} \quad Q = 2.82 \text{ mgd away from } C$$
$$S_{12} = (200 - 100)/4 = 25.0 \text{ ft/1000 ft} \quad \text{and} \quad Q = 3.80 \text{ mgd away from } C$$

The flow away from $C = 6.62$ mgd against the flow to C of 5.15 mgd. Using Fig. (b) above to obtain a guide regarding a reasonable third assumption, connect plotted points R and S. The line so drawn intersects the $(Q_{\text{to}} - Q_{\text{away}})$ zero abscissa at $Q_{\text{to}} = 5.7$ mgd (scaled). Inasmuch as the values plotted do not vary linearly, use a flow to C slightly larger, say 6.0 mgd.

For $Q = 6.0$ mgd to C, $S_{24} = 1.98$ ft/1000 ft, Lost Head$_{A-c} = 1.98 \times 8 = 15.8$ ft, and the hydraulic grade line at C is at elevation $(212 - 15.8) = 196.2$. Then

$$S_{16} = 6.2/4 = 1.55 \text{ ft/1000 ft}, \quad Q = 2.17 \text{ mgd from } C$$
$$S_{12} = 96.2/4 = 24.05 \text{ ft/1000 ft}, \quad Q = \underline{3.73} \text{ mgd from } C$$
$$\text{Total } Q \text{ from } C = \overline{5.90} \text{ mgd}$$

These resulting flows agree sufficiently well as not to require further calculations. (An elevation of the hydraulic grade line at C of about 196.5 would probably prove in, with flows to C and from C of about 5.95 mgd.)

18. Develop the expression utilized to study flows in a pipe network.

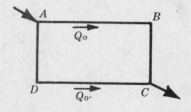

Fig. 8-10

Solution:

The method of attack, developed by Professor Hardy Cross, consists of assuming flows throughout the network, and then balancing the calculated head losses. In the simple looping pipe system shown, for the correct flow in each branch of the loop,

$$\text{LH}_{ABC} = \text{LH}_{ADC} \quad \text{or} \quad \text{LH}_{ABC} - \text{LH}_{ADC} = 0 \quad (1)$$

In order to utilize this relationship, the flow formula to be used must be written in the form $\text{LH} = kQ^n$. For the Hazen-Williams formula, this expression is $\text{LH} = kQ^{1.85}$.

But, since we are assuming flows Q_o, the correct flow Q in any pipe of a network can be expressed as $Q = Q_o + \Delta$, where Δ is the correction to be applied to Q_o. Then, using the binomial theorem,

$$kQ^{1.85} = k(Q_o + \Delta)^{1.85} = k(Q_o^{1.85} + 1.85 \, Q_o^{1.85-1} \Delta + \cdots)$$

Terms beyond the second can be neglected since Δ is small compared with Q_o.

For the loop above, substituting in expression (1) we obtain

$$k(Q_o^{1.85} + 1.85 \, Q_o^{0.85} \Delta) - k(Q_{o'}^{1.85} + 1.85 \, Q_{o'}^{0.85} \Delta) = 0$$
$$k(Q_o^{1.85} - Q_{o'}^{1.85}) + 1.85k(Q_o^{0.85} - Q_{o'}^{0.85})\Delta = 0$$

Solving for Δ,
$$\Delta = -\frac{k(Q_o^{1.85} - Q_{o'}^{1.85})}{1.85k(Q_o^{0.85} - Q_{o'}^{0.85})} \qquad (2)$$

In general, we may write for a more complicated loop,

$$\Delta = -\frac{\Sigma \, kQ_o^{1.85}}{1.85 \, \Sigma \, kQ_o^{0.85}} \tag{3}$$

But $kQ_o^{1.85} = \text{LH}$ and $kQ_o^{0.85} = (\text{LH})/Q_o$.   Therefore

$$\Delta = -\frac{\Sigma \, (\text{LH})}{1.85 \, \Sigma \, (\text{LH}/Q_o)} \qquad \text{for each loop of a network} \tag{4}$$

In utilizing expression (4), care must be exercised regarding the sign of the numerator. Expression (1) indicates that clockwise flows may be considered as producing clockwise losses, and counterclockwise flows, counter-clockwise losses. This means that the minus sign is assigned to all counterclockwise conditions in a loop, namely flow Q and lost head LH. Hence to avoid mistakes, this sign notation must be observed in carrying out a solution. On the other hand, the denominator of (4) is always positive.

The next two problems will illustrate how equation (4) is utilized.

19. The looping system shown in Fig. 8-11 is the same system as appears as part of Problem 11.   For $Q = 11.7$ mgd total, how much flow occurs in each branch of the loop, using the Hardy Cross procedure?

Solution:

Values of Q_{12} and Q_{16} are assumed to be 4.0 mgd and 7.7 mgd respectively. The tabulation below is prepared, (note the -7.70 mgd), the values of S calculated by means of the hydraulic slide rule (or chart), then $\text{LH} = S \times L$, and LH/Q_o can be calculated. Note that the large ΣLH indicates that the Q's are not well-balanced. (The values were assumed deliberately to produce this large ΣLH, to illustrate the procedure.)

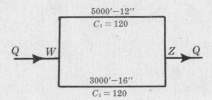

Fig. 8-11

D	L	Assumed Q_o mgd	S ft/1000'	LH, ft	LH/Q_o	Δ	Q_1
12"	5000'	4.00	19.5	97.5	24.4	−0.85	3.15
16"	3000'	−7.70	−16.3	−48.9	6.4	−0.85	−8.55
		$\Sigma = 11.70$		$\Sigma = +48.6$	30.8		11.70

$$\Delta = -\frac{\Sigma\text{LH}}{1.85 \, \Sigma \, (\text{LH}/Q)} = -\frac{+48.6}{1.85(30.8)} = -0.85 \text{ mgd}$$

Then the Q_1 values become $(4.00 - 0.85) = 3.15$ mgd and $(-7.70 - 0.85) = -8.55$ mgd. Repeating the calculation produces

S	LH	LH/Q_1	Δ	Q_2
12.5	62.5	19.80	−0.06	3.09
−19.8	−59.4	6.95	−0.06	−8.61
	$\Sigma = +3.1$	26.75		11.70

No further calculation is necessary, since the slide rule or chart cannot be read to the accuracy of .06 mgd. Ideally, ΣLH should equal zero, but this goal is seldom attained.

It will be noted that in Problem 11 the quantity flowing in the 12" was 26% of 11.7 mgd, or 3.05 mgd, a satisfactory check.

20. Water flows through the piping system shown, with certain measured flows indicated on the sketch. At point A, the elevation is 200.0 ft and the pressure head is 150.0 ft. The elevation at I is 100.0 ft. Determine (a) the flows throughout the network and (b) the pressure head at I. (Use $C_1 = 120$.)

Solution:

(a)   The method of attack may be outlined as follows:

(1) Assume any distribution of flow, proceeding loop by loop — in this case loops I, II, III and IV. Examine carefully each junction point so that the flow *to* the point equals the flow *away from* the point (continuity principle).

(2) Compute for each loop the head loss in each pipe of the loop (by equation, diagram or hydraulic slide rule).

(3) Sum up the lost heads around each loop, with due regard to sign (should the sum of the lost heads for the loop be zero, flows Q_1 are correct).

(4) Sum up the LH/Q_1 values, and calculate correction term Δ for each loop.

(5) Apply the Δ value to each pipeline, thereby increasing or decreasing the assumed Q's. For cases where a pipe is in two loops, the difference between the two Δ's must be applied as the proper correction to the assumed flow Q_1 (see application below).

(6) Proceed until the Δ values are negligible.

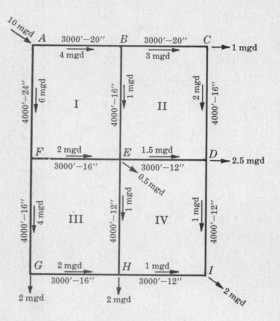

Fig. 8-12

The steps listed above have been carried out in tabular form, using the hydraulic slide rule to obtain the lost head terms in feet per thousand feet (S). Values of LH are obtained by multiplying S by the number of thousand feet in the pipe being considered. Values of LH divided by Q are also tabulated.

Line	D, in.	L, ft	Assumed Q_1 mgd	S ft/1000'	LH, ft	$\dfrac{\text{LH}}{Q_1}$	Δ	Q_2
AB	20	3000	4.0	1.62	4.86	1.22	$+0.31$	4.31
BE	16	4000	1.0	0.37	1.48	1.48	$+0.31 - (0.13) = +0.18$	1.18
EF	16	3000	-2.0	-1.33	-3.99	2.00	$+0.31 - (0.50) = -0.19$	-2.19
FA	24	4000	-6.0	-1.41	-5.64	0.94	$+0.31$	-5.69
					$\Sigma = -3.29$	5.64		
BC	20	3000	3.0	0.95	2.85	0.95	$+0.13$	3.13
CD	16	4000	2.0	1.33	5.32	2.66	$+0.13$	2.13
DE	12	3000	-1.5	-3.15	-9.45	6.30	$+0.13 - (-0.12) = +0.25$	-1.25
EB	16	4000	-1.0	-0.37	-1.48	1.48	$+0.13 - (0.31) = -0.18$	-1.18
					$\Sigma = -2.76$	11.39		
FE	16	3000	2.0	1.33	3.99	2.00	$+0.50 - (+0.31) = +0.19$	2.19
EH	12	4000	1.0	1.48	5.92	5.92	$+0.50 - (-0.12) = +0.62$	1.62
HG	16	3000	-2.0	-1.33	-3.99	2.00	$+0.50$	-1.50
GF	16	4000	-4.0	-4.85	-19.40	4.85	$+0.50$	-3.50
					$\Sigma = -13.48$	14.77		
ED	12	3000	1.5	3.15	9.45	6.30	$-0.12 - (0.13) = -0.25$	1.25
DI	12	4000	1.0	1.48	5.92	5.92	-0.12	0.88
IH	12	3000	-1.0	-1.48	-4.44	4.44	-0.12	-1.12
HE	12	4000	-1.0	-1.48	-5.92	5.92	$-0.12 - (0.50) = -0.62$	-1.62
					$\Sigma = +5.01$	22.58		

The Δ terms are calculated [expression (4), Problem 18], as follows:

$$\Delta_{\mathrm{I}} = \frac{-(-3.29)}{1.85(5.64)} = +0.31 \qquad\qquad \Delta_{\mathrm{III}} = \frac{-(-13.48)}{1.85(14.77)} = +0.50$$

$$\Delta_{\mathrm{II}} = \frac{-(-2.76)}{1.85(11.39)} = +0.13 \qquad\qquad \Delta_{\mathrm{IV}} = \frac{-(+5.01)}{1.85(22.58)} = -0.12$$

For line EF in loop I, its net Δ term is $(\Delta_{\mathrm{I}} - \Delta_{\mathrm{III}})$ or $[+0.31 - (+0.50)] = -0.19$. It will be observed that the Δ for loop I is combined with that of the Δ for loop III since the line EF occurs in each loop. In similar fashion, for line FE in loop III, the net Δ term is $(\Delta_{\mathrm{III}} - \Delta_{\mathrm{I}})$ or $[+0.50 - (+0.31)] = +0.19$. Note that the net Δ's have the same magnitude but *opposite signs*. This can readily be understood since flow in EF is counter-clockwise for loop I, whereas flow in FE in loop III is clockwise.

In determining the values of Q_2 for the second calculation,

$$Q_{AB} = (4.00 + 0.31) = 4.31 \text{ mgd}$$

whereas

$$Q_{EF} = (-2.00 - 0.19) = -2.19 \text{ mgd} \qquad \text{and} \qquad Q_{FA} = (-6.00 + 0.31) = -5.69 \text{ mgd}$$

The procedure is carried along until the Δ terms are insignificant with regard to the accuracy expected, keeping in mind the accuracy of the values of C_1 and of the slide rule itself. Reference to the last column of the table on the next page will indicate the final values of Q in the several pipelines.

Since the sums of the lost head values are small for all the loops, we may consider the flow values listed in the last column in the table on the next page as correct within the accuracy expected. The reader may practice by calculating the next values of Δ, then Q_5, etc.

(b)   The elevation of the hydraulic grade line at A is $(200.0 + 150.0) = 350.0$. The lost head to I can be calculated by any route from A to I, adding the losses in the usual manner, i.e., in the direction of flow. Using $ABEHI$ we obtain $\mathrm{LH}_{A \text{ to } I} = (6.06 + 2.72 + 10.72 + 3.93) = 23.43$ ft. As a check, using $ABEDI$, $\mathrm{LH} = (6.06 + 2.72 + 8.91 + 6.72) = 24.41$ ft. Using a value of 24 ft, the elevation of the hydraulic grade line at $I = (350.0 - 24.0) = 326.0$ ft. Hence the pressure head at $I = (326.0 - 100.0) = 226.0$ ft.

Line	Q_2	S	LH	LH/Q	Δ
AB	4.31	1.86	5.58	1.29	+0.20
BE	1.18	0.51	2.04	1.72	+0.20 + negl = +0.20
EF	−2.19	−1.57	−4.71	2.15	+0.20 − (−.06) = +0.26
FA	−5.69	−1.28	−5.12	0.90	+0.20
			$\Sigma = -2.21$	6.06	
BC	3.13	1.02	3.06	0.98	negl
CD	2.13	1.48	5.92	2.79	negl
DE	−1.25	−2.28	−6.84	5.50	negl − 0.19 = −0.19
EB	−1.18	−0.51	−2.04	1.72	negl − 0.20 = −0.20
			$\Sigma = +0.10$	10.99	
FE	2.19	1.57	4.71	2.15	−0.06 − 0.20 = −0.26
EH	1.62	3.65	14.60	9.02	−0.06 − 0.19 = −0.25
HG	−1.50	−0.79	−2.37	1.58	−0.06
GF	−3.50	−3.75	−15.00	4.28	−0.06
			$\Sigma = +1.94$	17.03	
ED	1.25	2.28	6.84	5.42	+0.19 + negl = +0.19
DI	0.88	1.18	4.72	5.38	+0.19
IH	−1.12	−1.83	−5.49	4.90	+0.19
HE	−1.62	−3.65	−14.60	9.02	+0.19 − (−0.06) = +0.25
			$\Sigma = -8.53$	24.72	

Line	Q_3	S	LH	LH/Q	Δ	Q_4
AB	4.51	2.02	6.06	1.34	-0.02	4.49
BE	1.39	0.68	2.72	1.95	$-0.02 - 0.12 = -0.14$	1.25
EF	-1.93	-1.25	-3.75	1.94	$-0.02 - 0.12 = -0.14$	-2.07
FA	-5.49	-1.20	-4.80	0.88	-0.02	-5.51
			$\Sigma = +0.23$	6.11		
BC	3.12	1.02	3.06	0.98	$+0.12$	3.24
CD	2.12	1.49	5.96	2.81	$+0.12$	2.24
DE	-1.45	-2.97	-8.91	6.15	$+0.12 + 0.02 = +0.14$	-1.31
EB	-1.39	-0.68	-2.72	1.95	$+0.12 + 0.02 = +0.14$	-1.25
			$\Sigma = -2.61$	11.89		
FE	1.93	1.25	3.75	1.94	$+0.12 + 0.02 = +0.14$	2.07
EH	1.37	2.68	10.72	7.83	$+0.12 + 0.02 = +0.14$	1.51
HG	-1.56	-0.84	-2.52	1.62	$+0.12$	-1.44
GF	-3.56	-3.90	-15.60	4.38	$+0.12$	-3.44
			$\Sigma = -3.65$	15.77		
ED	1.45	2.97	8.91	6.15	$-0.02 - 0.12 = -0.14$	1.31
DI	1.07	1.68	6.72	6.28	-0.02	1.05
IH	-0.93	-1.31	-3.93	4.23	-0.02	-0.95
HE	-1.37	-2.68	-10.72	7.83	$-0.02 - 0.12 = -0.14$	-1.51
			$\Sigma = +0.98$	24.49		

Supplementary Problems

(Problems 23-49 were solved using the hydraulic slide rule.)

21. Using Diagram B, calculate the expected flow in a 16″ pipe with a drop in the hydraulic grade line of 6 ft in a mile. (Use $C_1 = 100$.)   *Ans.* 1.5 mgd

22. Had the pipe in Problem 21 been new cast iron pipe, what flow could be expected?   *Ans.* 2.0 mgd

23. In a test of a 20″ cast iron pipe the flow was steady at 4.0 mgd and the grade line dropped 4.0 ft in a length of 2000 ft of pipe. What is the indicated value of C_1?   *Ans.* 107

24. What size new cast iron pipe will deliver 12.6 mgd through 6000 ft with a lost head of 30 ft?
Ans. 24″

25. A flow of 12.0 mgd is required through old cast iron pipe ($C_1 = 100$) with a slope of the hydraulic grade line of 1.0 ft/1000 ft. Theoretically, how many 16″ pipes would be required? 20″ pipes? 24″ pipes? 36″ pipes?   *Ans.* 8.46, 4.68, 2.90, 1

26. Check the ratios in Problem 25 by using a flow of 12 mgd with any assumed slope of the grade line.

27. What lost head in a 16″ new cast iron pipe will create the same flow as occurs in a new 20″ pipe with a drop in the hydraulic grade line of 1.0 ft/1000 ft?   *Ans.* 2.97 ft/1000 ft

28. Compound Pipe $ABCD$ consists of 20,000 ft of 16″, 10,000 ft of 12″ and 5000 ft of 8″ ($C_1 = 100$). (a) Find the flow when the lost head is 200 ft from A to D. (b) What size pipe 5000 ft long laid parallel to the existing 8″ and joining it at C and D will make the new C-D section equivalent to the ABC section (use $C_1 = 100$)? (c) If 8000 ft of 12″ pipe were laid from C to D, paralleling the 8″ pipe CD, what would be the total loss of head from A to D when $Q = 3$ cfs?
    Ans. 1.40 mgd, 6.6″, 135 ft

29. Compound Pipe $ABCD$ consists of 10,000 ft of 20″, 8000 ft of 16″ and L ft of 12″ ($C_1 = 120$). What length L will make the $ABCD$ pipes equivalent to a 15″ pipe, 16,500 ft long, $C_1 = 100$? If the length of the 12″ pipe from C to D were 3000 ft, what flow would occur for a lost head of 135 ft from A to D?
    Ans. 5000 ft, 4.4 mgd

30. Convert 3000 ft of 10″, 1500 ft of 8″ and 500 ft of 6″ in series to an equivalent length of 8″ pipe (all values of $C_1 = 120$).   *Ans.* 4550 ft

31. Reservoirs A and D are connected by 8000 ft of 20″ pipe (A-B), 6000 ft of 16″ pipe (B-C), and 2000 ft of unknown size (C-D). The difference in levels in the reservoirs is 85 ft. (a) Find the size of pipe CD so that 4.50 mgd will flow from A to D, using $C_1 = 120$ for all pipes. (b) How much flow will be produced if pipe CD is 14″ in diameter and if a 12″ pipe is connected at B, paralleling BCD and 9000 ft long?   *Ans.* 13″, 6.45 mgd

32. A piping system ($C_1 = 120$) consists of 10,000 ft of 30″ (AB), 8000 ft of 24″ (BC) and, from C to D, two 16″ pipes in parallel, each 6000 ft long. (a) For a flow of 9 mgd from A to D, what is the lost head? (b) If a valve in one of the 16″ pipes were closed, what change in lost head would occur for the same flow?   *Ans.* 70 ft, change = 94 ft

33. In Fig. 8-13 below, for a pressure head at D of 100 ft (a) compute the power supplied the turbine DE. (b) If the dotted pipe were installed (3000 ft of 24″), what power would be available to the turbine if the flow were 13.0 mgd? ($C_1 = 120$).   *Ans.* 154 hp, 208 hp

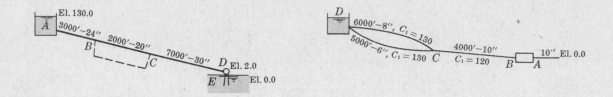

Fig. 8-13   Fig. 8-14

34. In Fig. 8-14 above, when the pressure heads at A and B are 10.0 ft and 295 ft respectively, the pump AB adds 100 hp to the system shown. What elevation can be maintained at reservoir D?
    Ans. El. 150.6

35. It is necessary to deliver 13.7 mgd to point D, in Fig. 8-15 below, at a pressure of 40 psi. Determine the pressure at A in psi.   *Ans.* 45.8 psi

Fig. 8-15   Fig. 8-16

36. (a) In Fig. 8-16 above, the pressure in supply main D is 30.0 psi when the flow from A is 6.00 mgd. Valves B and C are closed. Find the elevation of reservoir A. (b) The flow and pressure in (a) are unchanged but valve C is open fully and valve B is only partly open. If the new elevation of reservoir A is 211.3, what is the loss through valve B?   *Ans.* El. 224, 20 ft

37. Determine the flow through each pipe of the system shown in Fig. 8-17 below.
    Ans. 4.67 mgd, 3.52 and 1.15 mgd

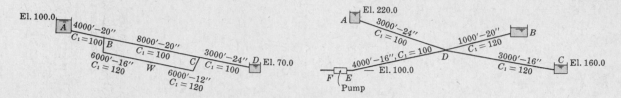

Fig. 8-17   Fig. 8-18

38. Pump XY at elevation 20.0 delivers 3.0 mgd through 6000 ft of 16″ new cast iron pipe YW. The discharge pressure at Y is 38.7 psi. At W two pipes connect to the 16″, 2500 ft of 12″ ($C_1 = 100$) going to reservoir A whose flow line is at elevation 100.0 and 2000 ft of 10″ ($C_1 = 130$) going to reservoir B. Determine the elevation of B and the flows to or from the reservoirs.
    Ans. El. 14.0, 1.0 mgd and 4.0 mgd

39. When $Q_{ED} = Q_{DC} = 6.50$ mgd, determine the gage reading at E in psi and the elevation of reservoir B in Fig. 8-18 above.   *Ans.* 70.0 psi, El. 181.2

40. In the Fig. 8-19 below, water flows through the 36″ pipe at the rate of 22.3 mgd. Determine the horsepower of the pump XA (78.5% efficiency) which will produce the flows and elevations for the system if the pressure head at X is zero ft. (Draw grade lines.)   *Ans.* 250 hp

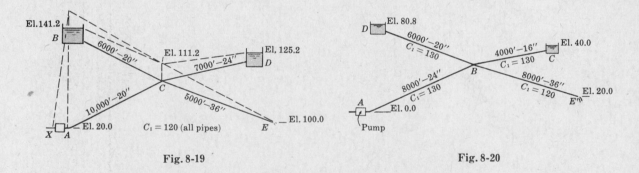

Fig. 8-19   Fig. 8-20

41. How much water must the pump supply when the flow through the 36″ pipe is 30.0 mgd and what is the pressure head at A in Fig. 8-20 above?   *Ans.* 25.2 mgd, 190 ft

42. The pressure head at A in pump AB is 120 ft when the energy change in the system (see Fig. 8-21 below) due to the pump is 153 horsepower. The lost head through valve Z is 10 ft. Find all the flows and the elevation of reservoir T. Sketch the hydraulic grade lines.   *Ans.* 87.4 ft

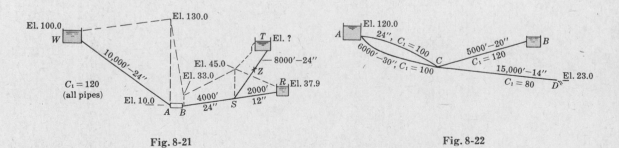

Fig. 8-21   Fig. 8-22

43. The total flow from A, in Fig. 8-22 above, is 9.00 mgd and the flow to B is 6.84 mgd. Find (a) the elevation of B and (b) the length of the 24″ pipe.   *Ans.* El. 94.2, 7280 ft

44. What are the rates of flow to or from each reservoir in Fig. 8-23 below?
     Ans. 3.42, 0, 1.87, 1.53 mgd

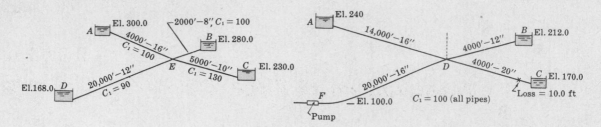

Fig. 8-23   Fig. 8-24

45. If the pressure head at F is 150.0 ft, find the flows through the system in Fig. 8-24 above.
     Ans. 2.35, 2.51, 1.22, 6.1 mgd

46. Using the piping system in Problem 9, for $Q = 5.0$ mgd what is the flow in each branch and what is
     the lost head? Use the Hardy Cross method.   *Ans.* 98 ft, $Q_{12} = 2.1$, $Q_{10} = 1.6$, $Q_8 = 1.3$ mgd

47. Solve Problem 35, using the Hardy Cross method.

48. Three piping systems A, B and C are being studied. Which system has the greatest capacity? Use
     $C_1 = 120$ for all pipes in the sketch.   *Ans.* B

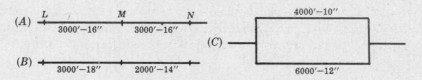

Fig. 8-25

49. In the preceding problem, what size pipe, 3000 ft long and laid parallel to MN in system A (thus
     forming a loop from M to N) would make new system A have 50% more capacity than system C?
     Ans. $d = 14.8$ in.

Chapter 9

Measurement of Flow of Fluids

INTRODUCTION

Numerous devices are used in engineering practice to measure the flow of fluids. Velocity measurements are made with Pitot tubes, current meters, and rotating and hot-wire anemometers. In model studies, photographic methods are often used. Quantity measurements are accomplished by means of orifices, tubes, nozzles, Venturi meters and flumes, elbow meters, weirs, numerous modifications of the foregoing and various patented meters. In order to apply the hydraulic devices intelligently, use of the Bernoulli equation and additional knowledge of the characteristics and coefficients of each device are imperative. In the absence of reliable values of coefficients, a device should be calibrated for the expected operating conditions.

Formulas developed for incompressible fluids may be used for compressible fluids where the pressure differential is small relative to the total pressure. In many practical cases such small differentials occur. However, where compressibility must be considered, special formulas will be developed and used (see Problems 5-8 and 23-28).

PITOT TUBE

The pitot tube measures the velocity at a point by virtue of the fact that the tube measures the stagnation pressure, which exceeds the local static pressure by $w(V^2/2g)$ psf. In an open stream of fluid, since the local pressure is zero gage, the height to which the liquid rises in the tube measures the velocity head. Problems 1 and 5 develop expressions for the flow of incompressible and compressible fluids respectively.

COEFFICIENT of DISCHARGE

The coefficient of discharge (c) is the ratio of the actual discharge through the device to the ideal discharge. This coefficient may be expressed as

$$c = \frac{\text{actual flow } Q \text{ in cfs}}{\text{ideal flow } Q \text{ in cfs}} = \frac{Q}{A\sqrt{2gH}} \tag{1}$$

More practically, when the coefficient of discharge c has been determined experimentally,

$$Q = cA\sqrt{2gH} \quad \text{in cfs} \tag{2}$$

where A = cross sectional area of device in square feet

H = total head causing flow, in feet of the fluid.

The coefficient of discharge may also be written in terms of the coefficient of velocity and the coefficient of contraction, i.e.,

$$c = c_v \times c_c \tag{3}$$

133

The coefficient of discharge is not constant. For a given device, it varies with Reynolds number.   In the Appendix the following information will be found:

(1) Table 7 contains coefficients of discharge for circular orifices discharging water at about 60°F into the atmosphere. Little authoritative data are available for all fluids throughout wide ranges of Reynolds number.

(2) Diagram C indicates the variation of c' with Reynolds number for three Pipe Orifice ratios. No authoritative data are available below Reynolds number of about 10,000.

(3) Diagram D shows the variation of c with Reynolds number for three long-radius Flow Nozzle ratios (pipe line nozzles).

(4) Diagram E indicates the variation of c with Reynolds number for five sizes of Venturi Meters of diameter ratios of 0.500.

COEFFICIENT of VELOCITY

The coefficient of velocity (c_v) is the ratio of the actual mean velocity in the cross section of the stream (jet) to the ideal mean velocity which would occur without friction. Thus

$$c_v = \frac{\text{actual mean velocity in ft/sec}}{\text{ideal mean velocity in ft/sec}} = \frac{V}{\sqrt{2gH}} \qquad (4)$$

COEFFICIENT of CONTRACTION

The coefficient of contraction (c_c) is the ratio of the area of the contracted section of a stream (jet) to the area of the opening through which the fluid flows.   Thus

$$c_c = \frac{\text{area of stream (jet)}}{\text{area of opening}} = \frac{A_{\text{jet}}}{A_o} \qquad (5)$$

LOST HEAD

The lost head in orifices, tubes, nozzles and Venturi meters is expressed as

$$\text{Lost Head in feet of the fluid} = \left(\frac{1}{c_v^2} - 1\right)\frac{V_{\text{jet}}^2}{2g} \qquad (6)$$

When this expression is applied to a Venturi Meter, V_{jet} = throat velocity and $c_v = c$.

WEIRS

Weirs measure the flow of liquids in open channels, usually water. A number of empirical formulas are available in engineering literature, each with its limitations. Only a few will be listed below. Most weirs are rectangular: the *suppressed* weir with no end contractions and generally used for larger flows, and the *contracted* weir for smaller flows. Other weirs are triangular, trapezoidal, parabolic and proportional flow. For accurate results, a weir should be calibrated in place under the conditions for which it is to be used.

THEORETICAL WEIR FORMULA

The theoretical weir formula for rectangular weirs, developed in Problem 29, is

$$Q = \frac{2}{3}cb\sqrt{2g}\left[\left(H + \frac{V^2}{2g}\right)^{3/2} - \left(\frac{V^2}{2g}\right)^{3/2}\right] \tag{7}$$

where　Q = flow in cfs

　　　c = coefficient (to be determined experimentally)

　　　b = length of weir crest in feet

　　　H = head on weir in feet (height of level liquid surface above crest)

　　　V = average velocity of approach in ft/sec.

FRANCIS FORMULA

The Francis formula, based upon experiments on rectangular weirs from 3.5 ft to 17 ft long under heads from 0.6 ft to 1.6 ft, is

$$Q = 3.33\left(b - \frac{nH}{10}\right)\left[\left(H + \frac{V^2}{2g}\right)^{3/2} - \left(\frac{V^2}{2g}\right)^{3/2}\right] \tag{8}$$

where the notation is the same as above and

　　　$n = 0$ for a suppressed weir

　　　$n = 1$ for a weir with one contraction

　　　$n = 2$ for a fully contracted weir.

BAZIN FORMULA

The Bazin formula (lengths from 1.64 ft to 6.56 ft under heads from 0.164 ft to 1.969 ft) is

$$Q = \left(3.25 + \frac{0.0789}{H}\right)\left[1 + 0.55\left(\frac{H}{H+Z}\right)^2\right]bH^{3/2} \tag{9}$$

where　Z = height of the weir crest above the channel bottom.

The bracketed term becomes negligible for low velocities of approach.

FTELEY and STEARNS FORMULA

The Fteley and Stearns formula (lengths 5 ft and 19 ft under heads from 0.07 ft to 1.63 ft) for suppressed weirs is

$$Q = 3.31b\left(H + \alpha\frac{V^2}{2g}\right)^{3/2} + 0.007b \tag{10}$$

where　α = factor dependent upon crest height Z (table of values required).

TRIANGULAR WEIR FORMULA (developed in Problem 30) is

$$Q = \frac{8}{15}c\tan\frac{\theta}{2}\sqrt{2g}\,H^{5/2} \tag{11}$$

or, for a given weir,　　　　　$Q = mH^{5/2}$ 　　　　　(12)

TRAPEZOIDAL WEIR FORMULA (of Cipolletti) is

$$Q = 3.367bH^{3/2} \tag{13}$$

This weir has side (end) slopes of 1 horizontal to 4 vertical.

For **DAMS USED as WEIRS** the expression for approximate flow is

$$Q = mbH^{3/2} \qquad (14)$$

where m = experimental factor, usually from model studies.

Non-uniform flow over broad-crested weirs is discussed in Chapter 10, Problem 52.

TIME to EMPTY TANKS by means of an orifice is (see Problem 38)

$$t = \frac{2A_T}{cA_o\sqrt{2g}}(h_1^{1/2} - h_2^{1/2}) \qquad \text{(constant cross section, no inflow)} \qquad (15)$$

$$t = \int_{h_1}^{h_2} \frac{-A_T\,dh}{Q_{out} - Q_{in}} \qquad \text{(inflow < outflow, constant cross section)} \qquad (16)$$

For a tank whose cross section is not constant, see Problem 41.

TIME to EMPTY TANKS by means of weirs is calculated by using (see Problem 43)

$$t = \frac{2A_T}{mL}(H_2^{-1/2} - H_1^{-1/2}) \qquad (17)$$

TIME to ESTABLISH FLOW in a pipeline is (see Problem 45)

$$t = \frac{LV_f}{2gH}\ln\left(\frac{V_f + V}{V_f - V}\right) \qquad (18)$$

Solved Problems

1. A Pitot tube having a coefficient of 0.98 is used to measure the velocity of water at the center of a pipe. The stagnation pressure head is 18.6 ft and the static pressure head in the pipe is 15.5 ft. What is the velocity?

Solution:

If the tube is shaped and positioned properly, a point of zero velocity (stagnation point) is developed at B in front of the open end of the tube (see Fig. 9-1). Applying the Bernoulli theorem from A in the undisturbed liquid to B yields

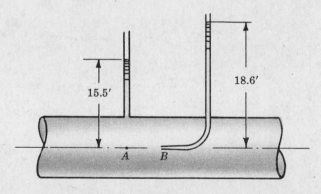

Fig. 9-1

$$\left(\frac{p_A}{w} + \frac{V_A^2}{2g} + 0\right) - \underset{\text{(assumed)}}{\text{no loss}} = \left(\frac{p_B}{w} + 0 + 0\right) \qquad (1)$$

Then, for an ideal "frictionless" fluid,

$$\frac{V_A^2}{2g} = \frac{p_B}{w} - \frac{p_A}{w} \qquad \text{or} \qquad V_A = \sqrt{2g\left(\frac{p_B}{w} - \frac{p_A}{w}\right)} \qquad (2)$$

For the actual tube, a coefficient c, which depends upon the design of the tube, must be introduced. The actual velocity for the problem above would be

$$V_A = c\sqrt{2g(p_B/w - p_A/w)} = 0.98\sqrt{2g(18.6 - 15.5)} = 13.8 \text{ ft/sec}$$

The above equation will apply to all incompressible fluids. The value of c may be taken as unity in most engineering problems. Solving (1) above for the stagnation pressure at B gives

$$p_B = p_A + \tfrac{1}{2}\rho V^2 \qquad \text{where } \rho = w/g \tag{3}$$

2. Air flows through a duct, and the Pitot-static tube measuring the velocity is attached to a differential gage containing water. If the deflection of the gage is 4 in., calculate the air velocity, assuming the specific weight of air is constant at 0.0761 lb/ft³ and that the coefficient of the tube is 0.98.
   Solution:

   For the differential gage,

   $$(p_B - p_A)/w = (4/12)(62.4)/0.0761 = 273 \text{ ft air.} \qquad \text{Then } V = 0.98\sqrt{64.4(273)} = 130 \text{ ft/sec.}$$

   (See Problems 26-28 and Chapter 11 for acoustic velocity considerations.)

3. Carbon tetrachloride (sp gr 1.60) flows through a pipe. The differential gage attached to the Pitot-static tube shows a 3 in. deflection of mercury. Assuming $c = 1.00$, find the velocity.
   Solution:

   $$p_B - p_A = (3/12)(13.6 - 1.60)62.4 = 187.2 \text{ psf}, \qquad V = \sqrt{64.4[187.2/(1.60 \times 62.4)]} = 11.0 \text{ ft/sec}$$

4. Water flows at a velocity of 4.65 ft/sec. A differential gage which contains a liquid of specific gravity 1.25 is attached to the Pitot-static tube. What is the deflection of the gage fluid?
   Solution:

   $$V = c\sqrt{2g(\Delta p/w)}, \qquad 4.65 = 1.00\sqrt{64.4(\Delta p/w)} \qquad \text{and} \qquad \Delta p/w = 0.336 \text{ ft water}$$

   Applying differential gage principles, $0.336 = (1.25 - 1)h$ and $h = 1.34 \text{ ft deflection.}$

5. Develop the expression for measuring the flow of a gas with a Pitot tube.
   Solution:

   The flow from A to B in the figure of Problem 1 above may be considered adiabatic and with negligible loss. Using the Bernoulli equation D in Problem 20 of Chapter 6, A to B, we obtain

   $$\left[\left(\frac{k}{k-1}\right)\frac{p_A}{w_A} + \frac{V_A^2}{2g} + 0\right] - \text{negligible loss} = \left[\left(\frac{k}{k-1}\right)\left(\frac{p_A}{w_A}\right)\left(\frac{p_B}{p_A}\right)^{(k-1)/k} + 0 + 0\right]$$

   or

   $$\frac{V_A^2}{2g} = \left(\frac{k}{k-1}\right)\left(\frac{p_A}{w_A}\right)\left[\left(\frac{p_B}{p_A}\right)^{(k-1)/k} - 1\right] \tag{1}$$

   The term p_B is the stagnation pressure. This expression (1) is usually rearranged, introducing the ratio of the velocity at A to the acoustic velocity c of the undisturbed fluid.

   From Chapter 1, the acoustic velocity $c = \sqrt{E/\rho} = \sqrt{kp/\rho} = \sqrt{kpg/w}$. Combining with equation (1) above,

   $$\frac{V_A^2}{2} = \left(\frac{c^2}{k-1}\right)\left[\left(\frac{p_B}{p_A}\right)^{(k-1)/k} - 1\right] \qquad \text{or,} \qquad \frac{p_B}{p_A} = \left[1 + \left(\frac{k-1}{2}\right)\left(\frac{V_A}{c}\right)^2\right]^{k/(k-1)} \tag{2}$$

   Expanding by the binomial theorem,

   $$\frac{p_B}{p_A} = 1 + \frac{k}{2}\left(\frac{V_A}{c}\right)^2\left[1 + \frac{1}{4}\left(\frac{V_A}{c}\right)^2 - \frac{k-2}{24}\left(\frac{V_A}{c}\right)^4 + \cdots\right] \tag{3}$$

In order to compare this expression with formula (3) of Problem 1, multiply through by p_A and replace kp_A/c^2 by ρ_A, obtaining

$$p_B \;=\; p_A + \frac{1}{2}\rho_A V_A^2\left[1 + \frac{1}{4}\left(\frac{V_A}{c}\right)^2 - \frac{k-2}{24}\left(\frac{V_A}{c}\right)^4 + \cdots\right] \qquad (4)$$

The above expressions apply to all compressible fluids for ratios of V/c less than unity. For ratios over unity, shockwave and other phenomena occur, the adiabatic assumption is not sufficiently accurate and the derivation no longer applies. The ratio V/c is called the *Mach number*.

The bracketed term in (4) is greater than unity and the first two terms provide sufficient accuracy. The effect of compressibility is to increase the stagnation-point pressure over that of an incompressible fluid (see expression (3) of Problem 1).

Acoustic velocities will be discussed in Problems 26-28 and in Chapter 11.

6. Air flowing under atmospheric conditions ($w = 0.0763$ at $60°F$) at a velocity of 300 ft/sec is measured by a Pitot tube. Calculate the error in the stagnation pressure by assuming the air to be incompressible.

Solution:

Using formula (3) of Problem 1 above,

$$p_B \;=\; p_A + \tfrac{1}{2}\rho V^2 \;=\; 14.7(144) + \tfrac{1}{2}(0.0763/32.2)(300)^2 \;=\; 2223 \text{ lb/ft}^2 \text{ absolute}$$

Using formula (4) of Problem 5 above and $c = \sqrt{kgRT} = \sqrt{1.4(32.2)(53.3)(520)} = 1117$ ft/sec,

$$p_B \;=\; 14.7(144) + \tfrac{1}{2}(0.0763/32.2)(300)^2[1 + \tfrac{1}{4}(300/1117)^2 \;\cdots]$$
$$= 2117 + 107[1 + 0.0181] \;=\; 2225 \text{ lb/ft}^2 \text{ absolute}$$

The error in the stagnation pressure is less than 0.1% and the error in $(p_B - p_A)$ is about 1.8%.

7. The difference between the stagnation pressure and the static pressure measured by a Pitot-static device is 412 lb/ft². The static pressure is 14.5 psi absolute and the temperature in the air stream is $60°F$. What is the velocity of the air, (a) assuming air is compressible and (b) assuming air is incompressible?

Solution:

(a) $p_A = 14.5(144) = 2090$ psf absolute　and　$c = \sqrt{kgRT} = \sqrt{1.4(32.2)(53.3)(520)} = 1117$ ft/sec.

From equation (2) of Prob. 5,　$\dfrac{p_B}{p_A} = \left[1 + \left(\dfrac{k-1}{2}\right)\left(\dfrac{V_A}{c}\right)^2\right]^{k/(k-1)}$

$$\frac{2090 + 412}{2090} \;=\; \left[1 + \left(\frac{1.4-1}{2}\right)\left(\frac{V_A}{1117}\right)^2\right]^{1.4/0.4}, \qquad V_A = 576 \text{ ft/sec.}$$

(b) $w = \dfrac{14.5(144)}{53.3(520)} = 0.0752$ lb/ft³　and　$V = \sqrt{2g(p_B/w - p_A/w)} = \sqrt{2g(412/0.0752)} = 593$ ft/sec.

8. Air flows at 800 ft/sec through a duct. At standard barometer, the stagnation pressure is -5.70 ft of water, gage. The stagnation temperature is $145°F$. What is the static pressure in the duct?

Solution:

With two unknowns in equation (2) of Problem 5, assume a V/c ratio (Mach number) of 0.72. Then

$$(-5.70 + 34.0)62.4 \;=\; p_A[1 + \tfrac{1}{2}(1.4 - 1)(0.72)^2]^{1.4/0.4}$$

and　$p_A = 62.4(28.3)/1.402 = 1260$ lb/ft² absolute.

Checking the assumption, using the adiabatic relation

$$\frac{T_B}{T_A} = \left(\frac{p_B}{p_A}\right)^{(k-1)/k}, \qquad \frac{460 + 145}{T_A} = \left(\frac{28.3 \times 62.4}{1260}\right)^{0.4/1.4}, \qquad T_A = 548° \text{ Rankine}$$

Also $c = \sqrt{kgRT} = \sqrt{1.4(32.2)(53.3)(548)} = 1145$ ft/sec.

Then $V/c = 800/1145 = 0.699$ and $p_A = \dfrac{62.4 \times 28.3}{[1 + 0.2(0.699)^2]^{1.4/0.4}} = 1270$ lb/ft² absolute.

No further refinement is necessary.

9. A 4″ diameter standard orifice discharges water under a 20.0 ft
head. What is the flow in cfs?
   Solution:
   Applying the Bernoulli equation, A to B in the adjoining figure,
   datum B,

$$(0 + 0 + 20) - \left(\frac{1}{c_v^2} - 1\right)\frac{V_{jet}^2}{2g} = \left(\frac{V_{jet}^2}{2g} + \frac{p_B}{w} + 0\right)$$

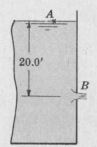

But the pressure head at B is zero (as discussed in Chap. 4, Prob. 6).
Then
$$V_{jet} = c_v\sqrt{2g \times 20}$$

Also $Q = A_{jet}V_{jet}$ which, using the definitions of the coefficients, becomes

Fig. 9-2

$$Q = (c_c A_o)c_v\sqrt{2g \times 20} = cA_o\sqrt{2g \times 20}$$

From Table 7, $c = 0.594$ for $D = 4″$ and $h = 20$ ft. Hence $Q = 0.594[\frac{1}{4}\pi(\frac{1}{3})^2]\sqrt{2g \times 20} = 1.86$ cfs.

10. The actual velocity in the contracted section of a jet of liquid flowing from a 2″ diameter
orifice is 28.0 ft/sec under a head of 15 ft. (a) What is the value of the coefficient of
velocity? (b) If the measured discharge is 0.403 cfs, determine the coefficients of
contraction and discharge.
   Solution:

(a) Actual velocity $= c_v\sqrt{2gH}$, $28.0 = c_v\sqrt{64.4 \times 15}$, $c_v = 0.900$.

(b) Actual $Q = cA\sqrt{2gH}$, $0.403 = c[\frac{1}{4}\pi(\frac{2}{12})^2]\sqrt{64.4 \times 15}$, $c = 0.594$.

   From $c = c_v \times c_c$, $c_c = 0.594/0.900 = 0.660$.

11. Oil flows through a standard 1″ diameter orifice under an 18.0 ft head at the rate of
0.111 cfs. The jet strikes a wall 5.00 ft away and 0.390 ft vertically below the centerline of the contracted section of the jet. Compute the coefficients.
   Solution:

(a) $Q = cA\sqrt{2gH}$, $0.111 = c[\frac{1}{4}\pi(\frac{1}{12})^2]\sqrt{2g(18.0)}$, $c = 0.598$.

(b) From kinematic mechanics, $x = Vt$ and $y = \frac{1}{2}gt^2$. Here x and y represent the coordinates of the
jet, as measured.

   Eliminate t and obtain $x^2 = (2V^2/g)y$.

   Substituting, $(5.00)^2 = (2V^2/32.2)(0.390)$ and   actual $V = 32.1$ ft/sec in jet.

   Then $32.1 = c_v\sqrt{2g(18.0)}$ and $c_v = 0.946$.   Finally, $c_c = c/c_v = 0.598/0.946 = 0.632$.

12. The tank in Problem 9 is closed and the air space above the water is under pressure, causing the flow to increase to 2.65 cfs. Find the pressure in the air space in psi.

Solution:

$$Q = cA_o\sqrt{2gH} \qquad \text{or} \qquad 2.65 = c[\tfrac{1}{4}\pi(\tfrac{1}{3})^2]\sqrt{2g(20 + p/w)}$$

Table 7 indicates that c does not change appreciably at the range of head under consideration. Using $c = 0.593$ and solving, $p/w = 20.7$ ft water (the assumed c checks for total head H). Then

$$p' = wh/144 = 62.4(20.7)/144 = 8.95 \text{ psi}.$$

13. Oil of specific gravity 0.720 flows through a 3″ diameter orifice whose coefficients of velocity and contraction are 0.950 and 0.650 respectively. What must be the reading of gage A in Fig. 9-3 in order that the power in the jet C be 8.00 hp?

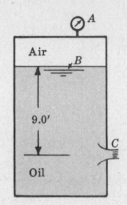

Solution:

The velocity in the jet may be calculated from the value of the power in the jet:

$$\text{horsepower in jet} = \frac{wQH_\text{jet}}{550} = \frac{w(c_cA_oV_\text{jet})(0 + V_\text{jet}^2/2g + 0)}{550}$$

$$8.00 = \frac{(0.720 \times 62.4)(0.650)[\tfrac{1}{4}\pi(\tfrac{1}{4})^2]V_\text{jet}^3/2g}{550}.$$

Solving, $V_\text{jet}^3 = 197{,}500$ and $V_\text{jet} = 58.2$ ft/sec.

Applying the Bernoulli equation, B to C, datum C,

Fig. 9-3

$$\left(\frac{p_A}{w} + \text{negl.} + 9.0\right) - \left[\frac{1}{(0.95)^2} - 1\right]\frac{(58.2)^2}{2g} = \left(0 + \frac{(58.2)^2}{2g} + 0\right)$$

and $p_A/w = 49.5$ ft of oil. Then $p_A' = wh/144 = (0.720 \times 62.4)(49.5)/144 = 15.4$ psi.

Note: The reader should not confuse the total head H causing flow with the value of H_jet in the horsepower expression. They are not the same.

14. For the 4″ diameter short tube shown in Fig. 9-4, (a) what flow of water at 75°F will occur under a head of 30 ft? (b) What is the pressure head at Section B? (c) What maximum head can be used if the tube is to flow full at exit? (Use $c_v = 0.82$.)

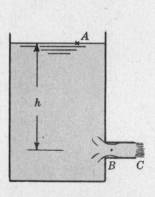

Solution:

For a standard short tube, the stream contracts at B to about 0.62 of the area of the tube. The lost head from A to B has been measured at about 0.042 times the velocity head at B.

(a) Applying the Bernoulli equation, A to C, datum C,

Fig. 9-4

$$(0 + \text{negl.} + 30) - \left[\frac{1}{(0.82)^2} - 1\right]\frac{V_\text{jet}^2}{2g} = \left(0 + \frac{V_\text{jet}^2}{2g} + 0\right)$$

and $V_\text{jet} = 36.0$ ft/sec. Then $Q = A_\text{jet}V_\text{jet} = [1.00 \times \tfrac{1}{4}\pi(\tfrac{1}{3})^2](36.0) = 3.14$ cfs.

(b) Now the Bernoulli equation, A to B, datum B, gives

$$(0 + \text{negl.} + 30) - 0.042\frac{V_B^2}{2g} = \left(\frac{p_B}{w} + \frac{V_B^2}{2g} + 0\right) \qquad (A)$$

Also, $Q = A_B V_B = A_C V_C$ or $c_c A V_B = A V_C$ or $V_B = V_{jet}/c_c = 36.0/0.62 = 58.1$ ft/sec.

Substituting in equation (A), $30 = \dfrac{p_B}{w} + 1.042\dfrac{(58.1)^2}{2g}$ and $\dfrac{p_B}{w} = -24.4$ ft of water.

(c)   As the head causing flow through the short tube is increased, the pressure head at B will become less and less. For steady flow (and with the tube full at exit), the pressure head at B must not be less than the vapor pressure head for the liquid at the particular temperature. From Table 1 in the Appendix, for water at 75°F this value is 0.43 psia or about 1.0 ft absolute at sea level (-33.0 ft gage).

From (A) above, $h = \dfrac{p_B}{w} + 1.042\dfrac{V_B^2}{2g} = -33.0 + 1.042\dfrac{V_B^2}{2g}$ (B)

Also, $c_c A V_B = A V_C = A c_v\sqrt{2gh}$

Thus $V_B = \dfrac{c_v}{c_c}\sqrt{2gh}$ or $\dfrac{V_B^2}{2g} = \left(\dfrac{c_v}{c_c}\right)^2 h = \left(\dfrac{0.82}{0.62}\right)^2 h = 1.75h$

Substituting in (B), $h = -33.0 + 1.042(1.75\,h)$ and $h = 40.1$ ft of water (75°F).

Any head over 40 ft will cause the stream to spring free of the sides of the tube. The tube will then function as an orifice.

Cavitation may result at vapor pressure conditions (see Chapter 12).

15. Water flows through a 4″ pipe at the rate of 0.952 cfs and thence through a nozzle attached to the end of the pipe. The nozzle tip is 2″ in diameter and the coefficients of velocity and contraction for the nozzle are 0.950 and 0.930 respectively. What pressure head must be maintained at the base of the nozzle if atmospheric pressure surrounds the jet?

Solution:

Apply the Bernoulli equation, base of nozzle to jet.

$$\left(\frac{p}{w} + \frac{V_4^2}{2g} + 0\right) - \left[\frac{1}{(0.950)^2} - 1\right]\frac{V_{jet}^2}{2g} = \left(0 + \frac{V_{jet}^2}{2g} + 0\right)$$

and the velocities are computed from $Q = AV$: $0.952 = A_4 V_4 = A_{jet} V_{jet} = (c_c A_2) V_{jet}$.   Thus

$$V_4 = \frac{0.952}{\frac{1}{4}\pi(4/12)^2} = 10.9 \text{ ft/sec} \quad \text{and} \quad V_{jet} = \frac{0.952}{0.930[\frac{1}{4}\pi(2/12)^2]} = 46.9 \text{ ft/sec}.$$

Substituting above and solving, $p/w = 37.9 - 1.85 = 36.0$ ft of water.

Had the formula $V_{jet} = c_v\sqrt{2gH}$ been used, H would be $(p/w + V_4^2/2g)$ or

$$46.9 = 0.950\sqrt{2g[p/w + (10.9)^2/2g]}$$

from which $\sqrt{p/w + 1.85} = 6.16$ and $p/w = 36.0$ ft of water, as before.

16. A 4″ base diameter by 2″ tip diameter nozzle points downward and the pressure head at the base of the nozzle is 26.0 ft of water. The base of the nozzle is 3.0 ft above the tip and the coefficient of velocity is 0.962. Determine the horsepower in the jet of water.

Solution:

For a nozzle, unless c_c is given, it may be taken as unity. Therefore $V_{jet} = V_{2''}$.

Before the horsepower may be calculated, both V and Q must be found. Using the Bernoulli equation, base to tip, datum tip, gives

$$\left(26.0 + \frac{V_4^2}{2g} + 3.0\right) - \left[\frac{1}{(0.962)^2} - 1\right]\frac{V_2^2}{2g} = \left(0 + \frac{V_2^2}{2g} + 0\right)$$

and $A_4 V_4 = A_2 V_2$ or $V_4^2 = (2/4)^4 V_2^2 = \frac{1}{16} V_2^2$.   Solving, $V_2 = 42.8$ ft/sec.

$$\text{Horsepower in jet} = \frac{wQH_{jet}}{550} = \frac{62.4[\frac{1}{4}\pi(2/12)^2(42.8)][0 + (42.8)^2/2g + 0]}{550} = 3.01 \text{ hp.}$$

17. Water flows through a $12'' \times 6''$ Venturi meter at the rate of 1.49 cfs and the differential gage is deflected 3.50 ft, as shown in the figure below. The specific gravity of the gage liquid is 1.25. Determine the coefficient of the meter.

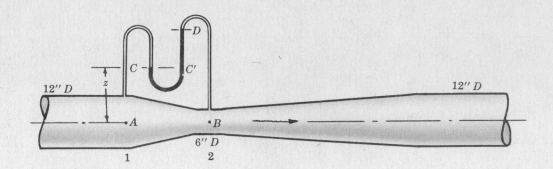

Fig. 9-5

Solution:

The coefficient of a Venturi meter is the same as the coefficient of discharge ($c_c = 1.00$ and thus $c = c_v$). Flow coefficient K should not be confused with meter coefficient c. Clarification will be made at the end of this problem.

Applying the Bernoulli equation, A to B, ideal case, yields

$$\left(\frac{p_A}{w} + \frac{V_{12}^2}{2g} + 0\right) - \text{no lost head} = \left(\frac{p_B}{w} + \frac{V_6^2}{2g} + 0\right)$$

and $V_{12}^2 = (A_6/A_{12})^2 V_6^2$. Solving, $V_6 = \sqrt{\dfrac{2g(p_A/w - p_B/w)}{1 - (A_6/A_{12})^2}}$ (no lost head).

The true velocity (and hence the true value of flow Q) will be obtained by multiplying the ideal value by the coefficient c of the meter. Thus

$$Q = A_6 V_6 = A_6 c \sqrt{\frac{2g(p_A/w - p_B/w)}{1 - (A_6/A_{12})^2}} \tag{1}$$

To obtain the differential pressure head indicated above, the principles of the differential gage must be used.

$$p_C = p_{C'}$$
$$(p_A/w - z) = p_B/w - (z + 3.50) + 1.25(3.50) \quad \text{or} \quad (p_A/w - p_B/w) = 0.875 \text{ ft}$$

Substituting in (1), $\quad 1.49 = \frac{1}{4}\pi(\frac{1}{2})^2 c\sqrt{2g(0.875)/(1 - 1/16)}$ and $\quad c = 0.978$ (use 0.98).

Note: Equation (1) is sometimes written $Q = KA_2\sqrt{2g(\Delta p/w)}$ where K is called the flow coefficient. It is apparent that

$$K = \frac{c}{\sqrt{1 - (A_2/A_1)^2}} \quad \text{or} \quad \frac{c}{\sqrt{1 - (D_2/D_1)^4}}$$

Tables or charts which give K can readily be used to obtain c if it is so desired. Diagrams in this book give values of c. The conversion factors to obtain the values of K for certain diameter-ratio devices are indicated on the several diagrams in the Appendix.

18. Water flows upward through a vertical $12'' \times 6''$ Venturi meter whose coefficient is 0.980. The differential gage deflection is 3.88 ft of liquid of specific gravity 1.25, as shown in the adjoining figure. Determine the flow in cfs.

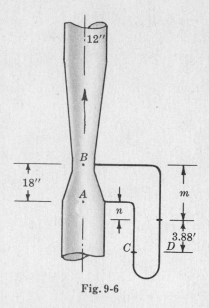

Solution:

Reference to the Bernoulli equation in Problem 17 indicates that, for this problem, $z_A = 0$ and $z_B = 1.50$ ft. Then

$$Q = cA_6\sqrt{\frac{2g[(p_A/w - p_B/w) - 1.50]}{1 - (1/2)^4}}$$

Using the principles of the differential gage to obtain $\Delta p/w$,

$$p_C/w = p_D/w \quad \text{(ft of water units)}$$
$$p_A/w + (n + 3.88) = p_B/w + m + 1.25(3.88)$$

$$[(p_A/w - p_B/w) - (m - n)] = 3.88(1.25 - 1.00)$$
$$[(p_A/w - p_B/w) - 1.50] = 0.97 \text{ ft of water}$$

Fig. 9-6

Substituting in the equation for flow, $Q = 0.980(\tfrac{1}{4}\pi)(\tfrac{1}{2})^2\sqrt{2g(0.97)/(1 - 1/16)} = 1.57$ cfs.

19. Water at $100°$F flows at the rate of 0.525 cfs through a $4''$ diameter orifice used in an $8''$ pipe. What is the difference in pressure head between the upstream section and the contracted section (vena contracta section)?

Solution:

In Diagram C of the Appendix, it is observed that c' varies with Reynolds number. Note that Reynolds number must be calculated for the orifice cross section, not for the contracted section of the jet nor for the pipe section. This value is

$$R_E = \frac{V_o D_o}{\nu} = \frac{(4Q/\pi D_o^2)D_o}{\nu} = \frac{4Q}{\nu\pi D_o} = \frac{4(0.525)}{\pi(.00000739)(4/12)} = 272{,}000$$

Diagram C for $\beta = 0.500$ gives $c' = 0.604$.

Applying the Bernoulli theorem, pipe section to jet section, produces the general equation for incompressible fluids, as follows:

$$\left(\frac{p_8}{w} + \frac{V_8^2}{2g} + 0\right) - \left[\frac{1}{c_v^2} - 1\right]\frac{V_{\text{jet}}^2}{2g} = \left(\frac{p_{\text{jet}}}{w} + \frac{V_{\text{jet}}^2}{2g} + 0\right)$$

and

$$Q = \dot{A}_8 V_8 = (c_c A_4)V_{\text{jet}}$$

Substituting for V_8 in terms of V_{jet} and solving,

$$\frac{V_{\text{jet}}^2}{2g} = c_v^2\left(\frac{p_8/w - p_{\text{jet}}/w}{1 - c^2(A_4/A_8)^2}\right) \quad \text{or} \quad V_{\text{jet}} = c_v\sqrt{\frac{2g(p_8/w - p_{\text{jet}}/w)}{1 - c^2(D_4/D_8)^4}}$$

Then $Q = A_{\text{jet}} V_{\text{jet}} = (c_c A_4) \times c_v\sqrt{\dfrac{2g(p_8/w - p_{\text{jet}}/w)}{1 - c^2(D_4/D_8)^4}} = cA_4\sqrt{\dfrac{2g(p_8/w - p_{\text{jet}}/w)}{1 - c^2(D_4/D_3)^4}}.$

More conveniently, for an orifice with velocity of approach and a contracted jet, the equation can be written

$$Q = \frac{c'A_4}{\sqrt{1 - (D_4/D_8)^4}}\sqrt{2g(\Delta p/w)} \tag{1}$$

or

$$Q = KA_4\sqrt{2g(\Delta p/w)} \tag{2}$$

where K is called the flow coefficient. The meter coefficient c' may be determined experimentally for a given ratio of diameter of orifice to diameter of pipe, or the flow coefficient K may be preferred.

Proceeding with the solution by substituting in the above expression (1),

$$0.525 = \frac{0.604 \times \frac{1}{4}\pi(4/12)^2}{\sqrt{1 - (1/2)^4}} \sqrt{2g(\Delta p/w)} \quad \text{and} \quad \Delta p/w = (p_8/w - p_{\text{jet}}/w) = 1.44 \text{ ft water.}$$

20. For the pipe orifice in Problem 19, what pressure difference in psi would cause the same quantity of turpentine at 68°F to flow? (See Appendix for sp gr and ν.)

Solution:

$$R_E = \frac{4Q}{\pi\nu D_o} = \frac{4(0.525)}{\pi(.0000186)(4/12)} = 108,000. \quad \text{From Diagram } C, \text{ for } \beta = 0.500, \ c' = 0.607.$$

Then

$$0.525 = \frac{0.607 \times \frac{1}{4}\pi(4/12)^2}{\sqrt{1 - (1/2)^4}} \sqrt{2g(\Delta p/w)} \qquad \text{from which}$$

$$\Delta\frac{p}{w} = \left(\frac{p_8}{w} - \frac{p_{\text{jet}}}{w}\right) = 1.42 \text{ ft turpentine} \quad \text{and} \quad \Delta p' = \frac{wh}{144} = \frac{(0.862 \times 62.4)(1.42)}{144} = 0.530 \text{ psi.}$$

21. Determine the flow of water at 70°F through a 6″ orifice installed in a 10″ pipeline if the pressure head differential for vena-contracta taps is 3.62 ft of water.

Solution:

This type of problem was met in the flow of fluids in pipes. The value of c' cannot be found inasmuch as Reynolds number cannot be computed. Referring to Diagram C, for $\beta = 0.600$, a value of c' will be assumed at 0.610. Using this assumed value,

$$Q = \frac{0.610 \times \frac{1}{4}\pi(6/12)^2}{\sqrt{1 - (0.60)^4}} \sqrt{64.4(3.62)} = 1.96 \text{ cfs}$$

Then

$$R_E = \frac{4(1.96)}{\pi(.00001059)(6/12)} = 472,000 \text{ (trial value)}$$

From Diagram C, $\beta = 0.600$, $c' = 0.609$. Recalculation of the flow using $c' = 0.609$ gives $Q = 1.95$ cfs. (Reynolds number is unaffected.)

Special Note: Professor R. C. Binder of Purdue University suggests on pages 132-3 of his Fluid Mechanics text (Second Edition) that this type problem need not be a "cut and try" proposition. He proposes that special lines be drawn on the coefficient — Reynolds number chart. In the case of the pipe orifice, equation (1) of Problem 19 can be written

$$\frac{Q}{A_4} = \frac{c'\sqrt{2g(\Delta p/w)}}{\sqrt{1 - (D_4/D_8)^4}} = V_4 \qquad \text{since } Q = AV$$

But

$$R_E = \frac{V_4 D_4}{\nu} = \frac{c'\sqrt{2g(\Delta p/w)} \times D_4}{\nu\sqrt{1 - (4/8)^4}} \quad \text{or} \quad \frac{R_E}{c'} = \frac{D_4\sqrt{2g(\Delta p/w)}}{\nu\sqrt{1 - (4/8)^4}}$$

or, in general,

$$\frac{R_E}{c'} = \frac{D_o\sqrt{2g(\Delta p/w)}}{\nu\sqrt{1 - (D_o/D_p)^4}}$$

Two straight lines called T-lines have been drawn on Diagram C, one for $R_E/c' = 700,000$ and one for $R_E/c' = 800,000$. For Problem 21, the calculated

$$\frac{R_E}{c'} = \frac{(6/12)\sqrt{64.4(3.62)}}{.00001059\sqrt{1 - (0.60)^4}} = 775,000$$

As near as can be read, the 775,000 line cuts the $\beta = 0.600$ curve at $c' = 0.609$. The flow Q is then calculated readily.

22. A nozzle with a 4″ diameter tip is installed in a 10″ pipe. Medium fuel oil at 80°F flows through the nozzle at the rate of 3.49 cfs. Assume the calibration of the nozzle is represented by curve $\beta = 0.40$ on Diagram D. Calculate the differential gage reading if a liquid of specific gravity 13.6 is the gage liquid.

Solution:

The Bernoulli equation, pipe section to jet, yields the same equation as was obtained in Problem 17 for the Venturi meter, since the nozzle is designed for a coefficient of contraction of unity.

$$Q = A_4 V_4 = A_4 c \sqrt{\frac{2g(p_A/w - p_B/w)}{1 - (4/10)^4}} \tag{1}$$

Diagram D indicates that c varies with Reynolds number.

$$V_4 = \frac{Q}{A_4} = \frac{3.49}{\frac{1}{4}\pi(4/12)^2} = 40.0 \text{ ft/sec} \quad \text{and} \quad R_E = \frac{40.0 \times 4/12}{3.65 \times 10^{-5}} = 365,000$$

Curve for $\beta = 0.40$ gives $c = 0.993$. Thus

$$3.49 = \frac{1}{4}\pi(4/12)^2 \times 0.993 \sqrt{\frac{2g(p_A/w - p_B/w)}{1 - (4/10)^4}}$$

and $(p_A/w - p_B/w) = 24.6$ ft of fuel oil.

Differential gage principles produce, using sp gr of the oil $= 0.851$ from the Appendix,

$$24.6 = h(13.6/0.851 - 1) \quad \text{and} \quad h = 1.64 \text{ ft (gage reading)}$$

Had the gage differential reading been given, the procedure used in the preceding problem would be utilized, e.g., a value of c assumed, Q calculated, Reynolds number obtained and c read from the appropriate curve on Diagram D. If c differs from the assumed value, the calculation is repeated until the coefficient checks in.

23. Derive an expression for the flow of a compressible fluid through a nozzle flow-meter and a Venturi meter.

Solution:

Since the change in velocity takes place in a very short period of time, little heat can escape and adiabatic conditions will be assumed. The Bernoulli theorem for compressible flow was shown in Chapter 6, equation (D) of Problem 20, to give

$$\left[\left(\frac{k}{k-1}\right)\frac{p_1}{w_1} + \frac{V_1^2}{2g} + z_1\right] - H_L = \left[\left(\frac{k}{k-1}\right)\frac{p_1}{w_1}\left(\frac{p_2}{p_1}\right)^{(k-1)/k} + \frac{V_2^2}{2g} + z_2\right]$$

For a nozzle meter and for a horizontal Venturi meter, $z_1 = z_2$ and the lost head will be taken care of by means of the coefficient of discharge. Also, since $c_c = 1.00$,

$$W = w_1 A_1 V_1 = w_2 A_2 V_2 \quad \text{(lb/sec)}$$

Then upstream $V_1 = W/w_1 A_1$, downstream $V_2 = W/w_2 A_2$. Substituting and solving for W,

$$\frac{W^2}{w_2^2 A_2^2} - \frac{W^2}{w_1^2 A_1^2} = 2g\left(\frac{k}{k-1}\right)\left(\frac{p_1}{w_1}\right)\left[1 - \left(\frac{p_2}{p_1}\right)^{(k-1)/k}\right]$$

or　　(ideal)　$W = \dfrac{w_2 A_2}{\sqrt{1 - (w_2/w_1)^2 (A_2/A_1)^2}} \sqrt{\dfrac{2gk}{k-1}(p_1/w_1) \times [1 - (p_2/p_1)^{(k-1)/k}]}$

It may be more practical to eliminate w_2 under the radical. Since $w_2/w_1 = (p_2/p_1)^{1/k}$,

$$(\text{ideal}) \quad W = w_2 A_2 \sqrt{\frac{\dfrac{2gk}{k-1}(p_1/w_1) \times [1 - (p_2/p_1)^{(k-1)/k}]}{1 - (A_2/A_1)^2(p_2/p_1)^{2/k}}} \tag{1}$$

The true value of W in lb/sec is obtained by multiplying the right hand side of the equation by coefficient c.

For comparison, equation (1) of Problem 17 and equation (1) of Problem 22 (for incompressible fluids) may be written

$$W = wQ = \frac{wA_2 c}{\sqrt{1 - (A_2/A_1)^2}} \sqrt{2g(\Delta p/w)}$$

or

$$W = wKA_2\sqrt{2g(\Delta p/w)}$$

The above equation can be expressed more generally so that it will apply both to compressible and incompressible fluids. An expansion (adiabatic) factor Y is introduced and the value of w_1 at inlet is specified. The fundamental relation is then

$$W = w_1 KA_2 Y\sqrt{2g(\Delta p/w_1)} \qquad\qquad (2)$$

For incompressible fluids, $Y = 1$. For compressible fluids, equate expressions (1) and (2) and solve for Y. By so doing,

$$Y = \sqrt{\frac{1 - (A_2/A_1)^2}{1 - (A_2/A_1)^2 (p_2/p_1)^{2/k}} \times \frac{[k/(k-1)][1 - (p_2/p_1)^{(k-1)/k}](p_2/p_1)^{2/k}}{1 - p_2/p_1}}$$

This expansion factor Y is a function of three dimensionless ratios. Table 8 lists some typical values for nozzle flow meters and for Venturi meters.

Note: Values of Y' for orifices and for orifice meters should be determined experimentally. The values differ from the above value of Y since the coefficient of contraction is not unity nor is it a constant. Knowing Y', solutions are identical to those which follow for flow nozzles and Venturi meters. The reader is referred to experiments by H. B. Reynolds and J. A. Perry as two sources of material.

24. Air at a temperature of 80°F flows through a 4″ pipe and through a 2″ flow nozzle. The pressure differential is 0.522 ft of oil, sp gr 0.910. The pressure upstream from the nozzle is 28.3 psi gage. How many pounds per second are flowing for a barometric reading of 14.7 psi, (a) assuming the air has constant density and (b) assuming adiabatic conditions?

Solution:

(a)
$$w_1 = \frac{(28.3 + 14.7)144}{53.3(460 + 80)} = 0.215 \text{ lb/ft}^3$$

From differential gage principles, using pressure heads in feet of air,

$$\frac{\Delta p}{w_1} = 0.522\left(\frac{w_{oil}}{w_{air}} - 1\right) = 0.522\left(\frac{0.910 \times 62.4}{0.215} - 1\right) = 137 \text{ ft of air}$$

Assuming $c = 0.980$ and using equation (1) of Problem 22 after multiplying by w_1, we have

$$W = w_1 Q = 0.215 \times \tfrac{1}{4}\pi(2/12)^2(0.980)\sqrt{\frac{2g(137)}{1 - (2/4)^4}} = 0.445 \text{ lb/sec}$$

To check the value of c, find Reynolds number and use the appropriate curve on Diagram D. (Here $w_1 = w_2$ and $\nu = 16.9 \times 10^{-5}$ at standard atmosphere from Table 1B.)

$$V_2 = \frac{W}{A_2 w_2} = \frac{W}{(\pi d_2^2/4)w_2}$$

Then
$$R_E = \frac{V_2 d_2}{\nu} = \frac{4W}{\pi d_2 \nu w_2} = \frac{4(0.445)}{\pi(2/12)(16.9 \times 14.7/43.0)10^{-5}(0.215)} = 274,000$$

From Diagram D, $c = 0.986$. Recalculating, $W = 0.447$ lb/sec.

Further refinement in calculation is not warranted inasmuch as Reynolds number will not be changed materially, nor will the value of c read from Diagram D.

(b)   Calculate pressures and specific weights first.

$$p_1 = (28.3 + 14.7)144 = 6200 \text{ lb/ft}^2, \quad p_2 = (6200 - 137 \times 0.215) = 6170 \text{ lb/ft}^2$$

$$\frac{p_2}{p_1} = \frac{6170}{6200} = 0.994 \quad \text{and} \quad \left(\frac{w_2}{w_1}\right)^k = 0.994 \text{ (see Chapter 1).} \quad \text{Then} \quad w_2 = 0.214 \text{ lb/ft}^3$$

Table 8 gives some values of expansion factor Y referred to in Problem 23. Interpolation may be used, in this case between the pressure ratio of 0.95 and 1.00 to obtain Y for $p_2/p_1 = 0.994$. For $k = 1.40$ and $d_2/d_1 = 0.50$, we obtain $Y = 0.997$.

Assuming $c = 0.980$, from examination of Diagram D and noting that $K = 1.032c$, equation (2) of Problem 23 becomes

$$W = w_1 K A_2 Y \sqrt{2g(\Delta p/w_1)}$$
$$= (0.215)(1.032 \times 0.980) \times \tfrac{1}{4}\pi(2/12)^2 \times 0.997\sqrt{64.4(137)} = 0.444 \text{ lb/sec}$$

Checking c, $\quad R_E = \dfrac{4W}{\pi d_2 \nu w_2} = \dfrac{4(0.444)}{\pi(2/12)(16.9 \times 14.7/43.0)10^{-5}(0.214)} = 275{,}000$

and $c = 0.986$ (Diagram D, curve $\beta = 0.50$).

Recalculating, $W = 0.447$ lb/sec. Further refinement is not essential. Note that no error is introduced in part (a) by assuming a constant density of air.

25. An $8'' \times 4''$ Venturi meter is used to measure the flow of carbon dioxide at $68°F$. The deflection of the water column in the differential gage is 71.8 in. and the barometer reads 30.0 in. of mercury. For a pressure at entrance of 18.0 psi absolute, calculate the weight flow.

Solution:

The absolute pressure at entrance $= p_1 = 18.0 \times 144 = 2590$ psf absolute and the specific weight w_1 of the carbon dioxide is

$$w_1 = \frac{2590}{34.9(460 + 68)} = 0.1405 \text{ lb/ft}^3$$

The pressure difference $= (71.8/12)(62.4 - 0.141) = 372$ psf and hence the absolute pressure at throat $= p_2 = 2590 - 372 = 2220$ psf absolute.

To obtain the specific weight w_2, we use $\dfrac{p_2}{p_1} = \dfrac{2220}{2590} = 0.858$ and $\dfrac{w_2}{w_1} = (0.858)^{1/k}$ (see Chap. 1).

Thus $\quad w_2 = (0.1405)(0.858)^{1/1.30} = 0.1250 \text{ lb/ft}^3$.

$$W = w_1 K A_2 Y \sqrt{2g(\Delta p/w_1)} \quad \text{in lb/sec}$$

Using $k = 1.30$, $d_2/d_1 = 0.50$ and $p_2/p_1 = 0.858$, Y (Table 8) $= 0.909$ by interpolation. Assuming $c = 0.985$, from Diagram E, and noting that $K = 1.032c$, we have

$$W = (0.1405)(1.032 \times 0.985) \times \tfrac{1}{4}\pi(4/12)^2 \times 0.909\sqrt{2g(372/0.1405)} = 4.67 \text{ lb/sec}$$

To check the assumed value of c, determine Reynolds number and use the appropriate curve on Diagram E. From Problem 24,

$$R_E = \frac{4W}{\pi d_2 \nu w_2} = \frac{4(4.67)}{\pi(4/12)(9.1 \times 14.7/18.0 \times 10^{-5})(0.1250)} = 1.92 \times 10^6$$

From Diagram E, $c = 0.984$. Recalculating, $W = 4.66$ lb/sec.

26. Establish the relationship which limits the velocity of a compressible fluid in convergent passages (acoustic velocity).

Solution:

Neglecting the velocity of approach in the Bernoulli Equation (D) of Problem 20, Chapter 6, for an ideal fluid we obtain

$$\frac{V_2^2}{2g} = \frac{k}{k-1}\left(\frac{p_1}{w_1}\right)\left[1 - \left(\frac{p_2}{p_1}\right)^{(k-1)/k}\right] \tag{1}$$

Also, had $(p_2/w_2)^{1/k}$ been substituted for $(p_1/w_1)^{1/k}$ before the integration which produced equation (D), the velocity head would have been

$$\frac{V_2^2}{2g} = \frac{k}{k-1}\left(\frac{p_2}{w_2}\right)\left[\left(\frac{p_1}{p_2}\right)^{(k-1)/k} - 1\right] \qquad (2)$$

If the fluid attains the acoustic velocity c_2 at Section 2, then $V_2 = c_2$ and $V_2^2 = c_2^2 = kp_2g/w_2$, (see Chapter 1). Substituting in equation (2),

$$\frac{kp_2g}{2gw_2} = \frac{k}{k-1}\left(\frac{p_2}{w_2}\right)\left[\left(\frac{p_1}{p_2}\right)^{(k-1)/k} - 1\right]$$

which simplifies to $\qquad\qquad \frac{p_2}{p_1} = \left(\frac{2}{k+1}\right)^{k/(k-1)} \qquad\qquad (3)$

This ratio p_2/p_1 is called the *critical pressure ratio* and depends upon the fluid flowing. For values of p_2/p_1 equal to or less than the critical pressure ratio, a gas will flow at the acoustic velocity. The pressure in a free jet flowing at the acoustic velocity will *equal or exceed* the pressure which surrounds it.

27. Carbon dioxide discharges through a $\frac{1}{2}''$ hole in the wall of a tank in which the pressure is 110 psi gage and the temperature is 68°F. What is the velocity in the jet (standard barometer)?

Solution:

From Table 1A, $R = 34.9$ and $k = 1.30$.

$$w_1 = \frac{p_1}{RT_1} = \frac{(110+14.7)144}{34.9(460+68)} = 0.0973 \text{ lb/ft}^3$$

$$\text{Critical } \left(\frac{p_2}{p_1}\right) = \left(\frac{2}{k+1}\right)^{k/(k-1)} = \left(\frac{2.00}{2.30}\right)^{1.30/0.30} = 0.542$$

$$\text{Ratio } \left(\frac{\text{atmosphere}}{\text{tank pressure}}\right) = \frac{14.7}{124.7} = 0.118$$

Since this latter ratio is less than the critical pressure ratio, the pressure of the escaping gas $= 0.542 \times p_1$. Hence $p_2 = 0.542 \times 124.7 = 67.5$ psi absolute.

$$V_2 = c_2 = \sqrt{1.3 \times 32.2 \times 34.9 \times T_2} = \sqrt{1460 T_2}$$

where $T_2/T_1 = (p_2/p_1)^{(k-1)/k} = (0.542)^{0.30/1.30} = 0.868$, $T_2 = 458°$. Then $V_2 = \sqrt{1460 \times 458} = 817$ fps.

28. Nitrogen flows through a duct in which changes in cross section occur. At a particular cross section the velocity is 1200 ft/sec, the pressure is 12.0 psi absolute and the temperature is 90°F. Assuming no friction losses and adiabatic conditions, (a) what is the velocity at a section where the pressure is 18.0 psi absolute and (b) what is the Mach number at this section?

Solution:

For nitrogen, $R = 55.1$ and $k = 1.40$, from Table 1A of the Appendix.

(a) From Problem 20 of Chapter 6, equation (D) for adiabatic conditions may be written

$$\frac{V_2^2}{2g} - \frac{V_1^2}{2g} = \frac{k}{k-1}\left(\frac{p_1}{w_1}\right)\left[1 - \left(\frac{p_2}{p_1}\right)^{(k-1)/k}\right]$$

where no lost head is considered and $z_1 = z_2$.

Calculate the specific weight of nitrogen at cross section 1.

$$w_1 = \frac{p_1}{RT_1} = \frac{12.0 \times 144}{55.1(460 + 90)} = 0.0569 \text{ lb/ft}^3 \quad \text{(or use } p_1/w_1 = RT_1)$$

Then $\dfrac{V_2^2}{2g} - \dfrac{(1200)^2}{2g} = \dfrac{1.40}{0.40}\Big(\dfrac{12.0 \times 144}{0.0569}\Big)\Big[1 - \Big(\dfrac{18 \times 144}{12 \times 144}\Big)^{0.40/1.40}\Big]$ from which $V_2 = 773$ ft/sec.

(b) Mach number $= \dfrac{V_2}{c_2} = \dfrac{773}{\sqrt{kgRT_2}}$ where $\dfrac{T_2}{T_1} = \Big(\dfrac{p_2}{p_1}\Big)^{(k-1)/k}$ or $\dfrac{T_2}{550} = \Big(\dfrac{18 \times 144}{12 \times 144}\Big)^{0.286} = 1.123.$

Then $T_2 = 618°$ Rankine and Mach number $= \dfrac{773}{\sqrt{1.40 \times 32.2 \times 55.1 \times 618}} = 0.623.$

29. Develop the theoretical formula for flow over a rectangular weir.

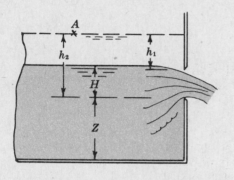 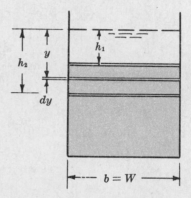

Fig. 9-7

Solution:

Consider the rectangular opening in Fig. 9-7 above to extend the full width W of the channel ($b = W$). With the liquid surface in the dotted position, application of the Bernoulli theorem between A and an elemental strip of height dy in the jet produces, for ideal conditions,

$$(0 + V_A^2/2g + y) - \text{no losses} = (0 + V_{\text{jet}}^2/2g + 0)$$

where V_A represents the average velocity of the particles approaching the opening.

Thus the ideal $V_{\text{jet}} = \sqrt{2g(y + V_A^2/2g)}$

and ideal $dQ = dA\,V_{\text{jet}} = (b\,dy)V_{\text{jet}} = b\sqrt{2g}\,(y + V_A^2/2g)^{1/2}\,dy$

$$\text{ideal } Q = b\sqrt{2g}\int_{h_1}^{h_2}(y + V_A^2/2g)^{1/2}\,dy$$

A weir exists when $h_1 = 0$. Let H replace h_2 and introduce a coefficient of discharge c to obtain the actual flow. Then

$$Q = cb\sqrt{2g}\int_0^H (y + V_A^2/2g)^{1/2}\,dy$$

$$= \tfrac{2}{3}cb\sqrt{2g}\,[(H + V_A^2/2g)^{3/2} - (V_A^2/2g)^{3/2}]$$

$$= mb[(H + V_A^2/2g)^{3/2} - (V_A^2/2g)^{3/2}] \tag{1}$$

Notes:

(1) For a fully contracted rectangular weir the end contractions cause a reduction in flow. Length b is corrected to recognize this condition, and the formula becomes

$$Q = m(b - \tfrac{2}{10}H)[(H + V_A^2/2g)^{3/2} - (V_A^2/2g)^{3/2}] \tag{2}$$

(2) For high weirs and most contracted weirs the velocity head of approach is negligible and

$$Q = m(b - \tfrac{2}{10}H)H^{3/2} \quad \text{for contracted weirs} \tag{3}$$

or $$Q = mbH^{3/2} \quad \text{for suppressed weirs} \tag{4}$$

(3) Coefficient of discharge c is not constant. It embraces the many complexities not included in the derivation, such as surface tension, viscosity, density, non-uniform velocity distribution, secondary flows and possibly others.

30. Derive the theoretical formula for flow through a triangular-notched weir. Refer to the adjacent figure.

Solution:

From Problem 29 above,

$$V_{\text{jet}} = \sqrt{2g(y + \text{negligible } V^2/2g)}$$

and $$\text{ideal } dQ = dA\, V_{\text{jet}} = x\, dy\sqrt{2gy}$$

By similar triangles,

$$\frac{x}{b} = \frac{H-y}{H} \quad \text{and} \quad b = 2H \tan\frac{\theta}{2}$$

Then actual $Q = (b/H)c\sqrt{2g}\int_o^H (H-y)y^{1/2}\, dy$.

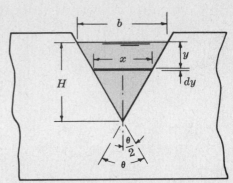

Fig. 9-8

Integrating and substituting, $$Q = \tfrac{8}{15} c\sqrt{2g}\, H^{5/2}\tan\tfrac{1}{2}\theta \tag{1}$$

A common V-notch is the 90° opening. Expression (1) then becomes $Q = 4.28cH^{5/2}$ where an *average* value of c is about 0.60 for heads above 1.0 ft.

31. During a test on an 8 ft suppressed weir which was 3 ft high, the head was maintained constant at 1.000 ft. In 38.0 seconds 7600 gallons of water were collected. Find weir factor m in equations (1) and (4) of Problem 29.

Solution:

(a) Change the measured flow to cfs. $Q = 7600/(7.48 \times 38) = 26.7$ cfs.

(b) Check the velocity of approach. $V = Q/A = 26.7/(8 \times 4) = 0.834$ ft/sec. Then
$$V^2/2g = (0.834)^2/2g = 0.0108 \text{ ft}$$

(c) Using (1), $$Q = mb[(H + V^2/2g)^{3/2} - (V^2/2g)^{3/2}]$$
or $$26.7 = m \times 8[(1.000 + .011)^{3/2} - (.0108)^{3/2}]$$
and $m = 3.28$ (slide rule accuracy).

Using (4), $$Q = 26.7 = mbH^{3/2} = m \times 8 \times (1.000)^{3/2}$$
and $m = 3.34$ (about 1.8% higher neglecting the velocity of approach terms).

32. Determine the flow over a suppressed weir 10.0 ft long and 4.00 ft high under a head of 3.000 ft. The value of m is 3.46.

Solution:

Since the velocity head term cannot be calculated, an approximate flow is
$$Q = mbH^{3/2} = 3.46(10)(3.000)^{3/2} = 178.8 \text{ cfs}$$

For this flow, $V = 178.8/(10 \times 7) = 2.554$ ft/sec and $V^2/2g = 0.101$ ft.   Using equation (1) of Prob. 29,

$$Q = 3.46(10)[(3.000 + 0.101)^{3/2} - (0.101)^{3/2}] = 188 \text{ cfs}$$

This second calculation shows an increase of 9 cfs or about 5.0% over the first calculation.   Further calculation will generally produce an unwarranted refinement, i.e., beyond the accuracy of the formula itself.   However, to illustrate, the revised velocity of approach would be

$$V = 188/(10 \times 7) = 2.69 \text{ ft/sec} \quad \text{and} \quad V^2/2g = 0.112 \text{ ft}$$

and
$$Q = 3.46(10)[(3.000 + 0.112)^{3/2} - (0.112)^{3/2}] = 189 \text{ cfs}$$

33. A suppressed weir, 25.0 ft long, is to discharge 375.0 cfs into a channel.   The weir factor $m = 3.42$.   To what height Z (nearest 1/100 ft) may the weir be built, if the water behind the weir must not exceed 6 ft in depth?

Solution:

Velocity of approach $V = Q/A = 375.0/(25 \times 6) = 2.50$ ft/sec.

Then $375.0 = 3.42 \times 25.0\left[\left(H + \dfrac{(2.50)^2}{2g}\right)^{3/2} - \left(\dfrac{(2.50)^2}{2g}\right)^{3/2}\right]$ and $H = 2.59$ ft.

Height of weir is $Z = 6.00 - 2.59 = 3.41$ ft.

34. A contracted weir, 4.00 ft high, is to be installed in a channel 8 ft wide.   The maximum flow over the weir is 60.0 cfs when the total depth back of the weir is 7.00 ft.   What length of weir should be installed if $m = 3.40$?

Solution:

Velocity of approach $V = Q/A = 60.0/(8 \times 7) = 1.071$ ft/sec.   It appears that the velocity head is negligible in this case.

$$Q = m(b - \tfrac{2}{10}H)(H)^{3/2}, \qquad 60.0 = 3.40(b - \tfrac{2}{10} \times 3.00)(3.00)^{3/2}, \qquad b = 4.00 \text{ ft long.}$$

35. The discharge from a 6″ diameter orifice, under a 10.0 ft head, $c = 0.600$, flows into a rectangular weir channel and over a contracted weir.   The channel is 6 ft wide and, for the weir, $Z = 5.00$ ft and $b = 1.00$ ft.   Determine the depth of water in the channel if $m = 3.35$.

Solution:

The discharge through the orifice is

$$Q = cA\sqrt{2gh} = 0.600 \times \tfrac{1}{4}\pi(\tfrac{1}{2})^2\sqrt{2g(10.0)} = 2.99 \text{ cfs}$$

For the weir, $Q = m(b - \tfrac{2}{10}H)H^{3/2}$ (velocity head neglected)

or $2.99 = 3.35(1.00 - 0.20H)H^{3/2}$ and $H^{3/2} - 0.20H^{5/2} = 0.893$

By successive trials, $H = 1.09$ ft; and the depth $= Z + H = 5.00 + 1.09 = 6.09$ ft.

36. The discharge of water over a 45° triangular weir is 0.750 cfs.   For $c = 0.580$, determine the head on the weir.

Solution:

$$Q = \tfrac{8}{15}c\sqrt{2g}\,(\tan \tfrac{1}{2}\theta)H^{5/2}, \qquad 0.750 = \tfrac{8}{15}(0.580)\sqrt{2g}\,(\tan 22\tfrac{1}{2}°)H^{5/2}, \qquad H = 0.885 \text{ ft}$$

37. What length of trapezoidal (Cipolletti) weir should be constructed so that the head will be 1.54 ft when the discharge is 122 cfs?

Solution:

$$Q = 3.367bH^{3/2}, \qquad 122 = 3.367b(1.54)^{3/2}, \qquad b = 19.0 \text{ ft}$$

38. Establish the formula to determine the time to lower the liquid level in a tank of constant cross section by means of an orifice. Refer to the adjacent figure.

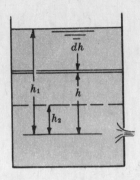

Solution:

Inasmuch as the head is changing with time, we know that $\partial V/\partial t \neq 0$, i.e., we do not have steady flow. This means that the energy equation should be amended to include an acceleration term, which complicates the solution materially. As long as the head does not change too rapidly, no appreciable error will be introduced by assuming steady flow, thus neglecting the acceleration-head term. An approximate check on the error introduced is given in Problem 39.

Case A.

With *no inflow* taking place, the instantaneous flow will be

$$Q = cA_o\sqrt{2gh} \quad \text{cfs}$$

Fig. 9-9

In time interval dt, the small volume dV discharged will be $Q\,dt$. In the same time interval, the head will decrease dh ft and the volume discharged will be the area of the tank A_T times dh. Equating these values,

$$(c\,A_o\,\sqrt{2gh})\,dt = -A_T\,dh$$

where the negative sign signifies that h decreases as t increases. Solving for t yields

$$t = \int_{t_1}^{t_2} dt = \frac{-A_T}{cA_o\sqrt{2g}} \int_{h_1}^{h_2} h^{-1/2}\,dh$$

or

$$t = t_2 - t_1 = \frac{2A_T}{cA_o\sqrt{2g}}(h_1^{1/2} - h_2^{1/2}) \qquad (1)$$

In using this expression, an average value of coefficient of discharge c may be used without producing significant error in the result. As h_2 approaches zero, a vortex will form and the orifice will cease to flow full. However, using $h_2 = 0$ will not produce serious error in most cases.

Equation (1) can be rewritten by multiplying and dividing by $(h_1^{1/2} + h_2^{1/2})$. There results

$$t = t_2 - t_1 = \frac{A_T(h_1 - h_2)}{\frac{1}{2}(cA_o\sqrt{2gh_1} + cA_o\sqrt{2gh_2})} \qquad (2)$$

Noting that the volume discharged in time $(t_2 - t_1)$ is $A_T(h_1 - h_2)$, this equation simplifies to

$$t = t_2 - t_1 = \frac{\text{volume discharged}}{\frac{1}{2}(Q_1 + Q_2)} = \frac{\text{volume discharged in ft}^3}{\text{average flow } Q \text{ in ft}^3/\text{sec}} \qquad (3)$$

Problem 41 will illustrate a case where tank cross section is not constant but can be expressed as a function of h. Other cases, such as reservoirs emptying, are beyond the scope of this book (see Water Supply Engineering texts).

Case B.

With a constant rate of inflow less than the flow through the orifice taking place,

$$-A_T\,dh = (Q_{out} - Q_{in})dt \qquad \text{and} \qquad t = t_2 - t_1 = \int_{h_1}^{h_2} \frac{-A_T\,dh}{Q_{out} - Q_{in}}$$

Should Q_{in} exceed Q_{out}, the head would increase, as would be expected.

39. A 4 ft diameter tank contains oil of specific gravity 0.75. A 3″ diameter short tube is installed near the bottom of the tank ($c = 0.85$). How long will it take to lower the level of the oil from 6 ft above the tube to 4 ft above the tube?

Solution:

$$t = t_2 - t_1 = \frac{2A_T}{cA_o\sqrt{2g}}(h_1^{1/2} - h_2^{1/2}) = \frac{2 \times \frac{1}{4}\pi(4)^2}{0.85 \times \frac{1}{4}\pi(3/12)^2\sqrt{2g}}(6^{1/2} - 4^{1/2}) = 33.8 \text{ sec}$$

In order to evaluate the approximate effect of assuming steady flow, the change of velocity with time t is estimated as

$$\frac{\partial V}{\partial t} \cong \frac{\Delta V}{\Delta t} = \frac{\sqrt{2g(6)} - \sqrt{2g(4)}}{33.8} = \frac{8.025(2.45 - 2)}{33.8} = 0.107 \text{ ft/sec}^2$$

This is about $\frac{1}{3}\%$ of acceleration g, a negligible addition to the acceleration g. Such an accuracy is not warranted in these illustrations of unsteady flow, particularly as orifice coefficients are not known to such an accuracy.

40. The initial head on an orifice was 9 ft and when the flow was terminated the head was measured at 4 ft. Under what constant head H would the same orifice discharge the same volume of water in the same time interval? Assume coefficient c constant.

Solution:

Volume under falling head = volume under constant head

$$\tfrac{1}{2}cA_o\sqrt{2g}(h_1^{1/2} + h_2^{1/2}) \times t = cA_o\sqrt{2gH} \times t$$

Substituting and solving, $\tfrac{1}{2}(\sqrt{9} + \sqrt{4}) = \sqrt{H}$ and $H = 6.25$ ft.

41. A tank has the form of a frustum of a cone, with the 8 ft diameter uppermost and the 4 ft diameter at the bottom. The bottom contains an orifice whose average coefficient of discharge may be taken as 0.60. What size orifice will empty the tank in 6 minutes if the depth full is 10.0 ft? Refer to the adjoining figure.

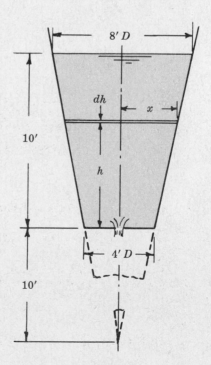

Fig. 9-10

Solution:

From Problem 38,

$$Q\,dt = -A_T\,dh$$

$$cA_o\sqrt{2gh}\,dt = -\pi x^2\,dh$$

and, by similar triangles, $x/4 = (10 + h)/20$.   Then

$$(0.60 \times \tfrac{1}{4}\pi d_o^2\sqrt{2g})dt = -\pi\frac{(10 + h)^2}{25}h^{-1/2}\,dh$$

$$d_o^2\int dt = \frac{-4\pi}{25\pi \times 0.60\sqrt{2g}}\int_{10}^{0}(10 + h)^2 h^{-1/2}\,dh$$

Since $\int dt = 360$ seconds,

$$d_o^2 = \frac{+4}{360 \times 25 \times 0.60\sqrt{2g}}\int_0^{10}(100h^{-1/2} + 20h^{1/2} + h^{3/2})dh$$

Integrating and solving, we obtain $d^2 = 0.109$ and $d = 0.33$ ft.   Use $d = 4''$ orifice.

42. Two square tanks have a common wall in which an orifice, area = 0.25 ft² and coefficient = 0.80, is located. Tank A is 8 ft on a side and the initial depth above the orifice is 10.0 ft. Tank B is 4 ft on a side and the initial depth above the orifice is 3.0 ft. How long will it take for the water surfaces to be at the same elevation?

Solution:

At any instant the difference in level of the surfaces may be taken as head h. Then

$$Q = 0.80 \times 0.25\sqrt{2gh}$$

and the change in volume $dv = Q\,dt = 1.605\sqrt{h}\,dt$.

In this interval of time dt the change in head is dh. Consider the level in Tank A to have fallen dy; then the corresponding rise in level in Tank B will be the ratio of the areas times dy, or $(64/16)dy$. The change in head is thus

$$dh = dy + (64/16)dy = 5dy$$

The change in volume $dv = 8 \times 8 \times dy$ $[= 4 \times 4 \times (64/16)dy$ also]

or, using dh, $dv = (64/5)dh = 12.8dh$

Equating values of dv, $1.605\sqrt{h}\,dt = -12.8dh$, $dt = \dfrac{-12.8}{1.605}\displaystyle\int_{7}^{0} h^{-1/2}\,dh$, $t = 42.1$ sec.

The problem may be solved using the average rate of discharge expressed in (3) of Problem 38.

$$Q_{av} = \tfrac{1}{2}[\,0.80 \times 0.25\sqrt{2g(7)}\,] = 2.13 \text{ cfs}$$

Tank A lowers y ft while Tank B rises $(64/16)y$ ft with the total change in level of 7 ft; then $y + 4y = 7$ and $y = 1.40$ ft. Thus, change in volume $= 8 \times 8 \times 1.40 = 89.6$ ft³ and

$$t = \frac{\text{change in volume}}{\text{average } Q} = \frac{89.6}{2.13} = 42.1 \text{ sec}$$

43. Develop the expression for the time to lower the liquid level in a tank, lock or canal by means of a suppressed weir.

Solution:

$Q\,dt = -A_T\,dH$ (as before)   or $(mLH^{3/2})dt = -A_T\,dH$.

Then $t = \displaystyle\int_{t_1}^{t_2} dt = \frac{-A_T}{mL}\int_{H_1}^{H_2} H^{-3/2}\,dH$ or $t = t_2 - t_1 = \dfrac{2A_T}{mL}(H_2^{-1/2} - H_1^{-1/2})$.

44. A rectangular flume, 50 ft long and 10 ft wide, feeds a suppressed weir under a head of 1.000 ft. If the supply to the flume is cut off, how long will it take for the head on the weir to decrease to 4 in? Use $m = 3.33$.

Solution:

From Problem 43, $t = \dfrac{2(50 \times 10)}{3.33 \times 10}\left[\dfrac{1}{\sqrt{0.333}} - \dfrac{1}{\sqrt{1.000}}\right] = 22.0$ sec.

45. Determine the time required to establish flow in a pipeline of length L under a constant head H discharging into the atmosphere, assuming an inelastic pipe, an incompressible fluid, and constant friction factor f.

Solution:

The final velocity V_f can be determined from the Bernoulli Equation, as follows:

$$H - f\frac{L}{d}\frac{V_f^2}{2g} - k\frac{V_f^2}{2g} = \left(0 + \frac{V_f^2}{2g} + 0\right)$$

In this equation, the minor losses are represented by the $kV_f^2/2g$ term, and the energy in the jet at the end of the pipe is kinetic energy represented by $V_f^2/2g$. This equation may be written in the form

$$\left[H - f\frac{L_E}{d}\frac{V_f^2}{2g}\right] = 0 \tag{1}$$

where L_E is the equivalent length of pipe for the system (see Problem 6, Chapter 8).

From Newton's equation of motion, at any instant

$$w(AH_e) = M\frac{dV}{dt} = \frac{w}{g}(AL)\frac{dV}{dt}$$

where H_e is the effective head at any instant and V is a function of time and not length. Rearranging the equation,

$$dt = \left(\frac{wAL}{gwAH_e}\right)dV \qquad \text{or} \qquad dt = \frac{L\,dV}{gH_e} \tag{2}$$

In equation (1), for all intermediate values of V the term in the brackets is not zero, but is the effective head available to cause the acceleration of the liquid. Hence expression (2) may be written

$$\int dt = \int\frac{L\,dV}{g\left(H - f\dfrac{L_E}{d}\dfrac{V^2}{2g}\right)} = \int\frac{L\,dV}{g\left(f\dfrac{L_E}{d}\dfrac{V_f^2}{2g} - f\dfrac{L_E}{d}\dfrac{V^2}{2g}\right)} \tag{3A}$$

Since from (1), $\dfrac{fL_E}{2gd} = \dfrac{H}{V_f^2}$, $\quad\displaystyle\int dt = \int\frac{L\,dV}{g(H - HV^2/V_f^2)}$ $\tag{3B}$

or

$$\int_0^t dt = \frac{L}{gH}\int_0^{V_f}\frac{V_f^2}{V_f^2 - V^2}dV$$

Integrating, $\qquad\qquad t = \dfrac{LV_f}{2gH}\ln\left(\dfrac{V_f + V}{V_f - V}\right)$ $\tag{4}$

It will be noted that, as V approaches final velocity V_f, $(V_f - V)$ approaches zero. Thus mathematically, time t approaches infinity.

Equation (3B) may be rearranged, using symbol ϕ for the ratio V/V_f. Then

$$\frac{dV}{dt} = \frac{gH}{L}(1 - V^2/V_f^2) = \frac{gH}{L}(1 - \phi^2) \tag{5}$$

Using $\quad V = V_f\phi\quad$ and $\quad\dfrac{dV}{dt} = V_f(d\phi/dt)$, we obtain

$$\frac{d\phi}{1 - \phi^2} = \frac{gH\,dt}{V_f L}$$

Integrating, $\qquad\qquad \frac{1}{2}\ln\left(\dfrac{1 + \phi}{1 - \phi}\right) = \dfrac{gHt}{V_f L} + C$

and when $t = 0$, $C = 0$. Then

$$\frac{1 + \phi}{1 - \phi} = e^{2gHt/V_f L}$$

Using hyperbolic functions, $\quad \phi = \tanh(gHt/V_f L)$, and since $\phi = V/V_f$,

$$V = V_f\tanh\frac{gHt}{V_f L} \tag{6}$$

The advantage of expression (6) is that the value of velocity V in terms of the final velocity V_f can be calculated for any chosen time.

46. Simplify equation (4) in the preceding problem to give the time to establish flow such that velocity V equals (a) 0.75, (b) 0.90 and (c) 0.99 times final velocity V_f.

Solution:

(a) $t = \dfrac{LV_f}{2gH} \ln \left[\dfrac{V_f + 0.75V_f}{V_f - 0.75V_f} \right] = \left(\dfrac{LV_f}{2gH} \right)(2.3026) \log \dfrac{1.75}{0.25} = 0.973 \dfrac{LV_f}{gH}$

(b) $t = \dfrac{LV_f}{2gH} \ln \dfrac{1.90}{0.10} = \left(\dfrac{LV_f}{2gH} \right)(2.3026) \log \dfrac{1.90}{0.10} = 1.472 \dfrac{LV_f}{gH}$

(c) $t = \dfrac{LV_f}{2gH} \ln \dfrac{1.99}{0.01} = \left(\dfrac{LV_f}{2gH} \right)(2.3026) \log \dfrac{1.99}{0.01} = 2.647 \dfrac{LV_f}{gH}$

47. Water is discharged from a tank through 2000 ft of 12″ pipe ($f = .020$). The head is constant at 20 ft. Valves and fittings in the line produce losses of $21(V^2/2g)$. After a valve is opened, how long will it take to attain a velocity of 0.90 of the final velocity?

Solution:

Bernoulli's Equation, tank surface to end of pipe, will yield

$$(0 + 0 + H) - [f(L/d) + 21.0]V^2/2g = (0 + V^2/2g + 0)$$

or $H = [.020(2000/1) + 22.0]V^2/2g = 62.0(V^2/2g)$. Then using the procedure in Prob. 6 of Chap. 8,

$$62.0(V^2/2g) = .020(L_E/1)(V^2/2g) \quad \text{or} \quad L_E = 3100 \text{ ft}$$

Inasmuch as equation (4) in Problem 45 does not contain L_E, the final velocity must be calculated, as follows:

$$H = f \frac{L_E}{d} \frac{V_f^2}{2g} \quad \text{or} \quad V_f = \sqrt{\frac{2gdH}{fL_E}} = \sqrt{\frac{64.4(1)(20.0)}{.020(3100)}} = 4.55 \text{ ft/sec}$$

Substituting in (b) of Problem 46 yields $\quad t = 1.472 \dfrac{(2000)(4.55)}{(32.2)(20.0)} = 20.8$ seconds.

48. In Problem 47, what velocity will be attained in 10 seconds and in 15 seconds?

Solution:

In equation (6) of Problem 45, evaluate gHt/V_fL.

For 10 sec, $\dfrac{32.2 \times 20 \times 10}{4.55 \times 2000} = 0.707$.   For 15 sec, $\dfrac{32.2 \times 20 \times 15}{4.55 \times 2000} = 1.062$.

Using a table of hyperbolic functions and equation (6), $V = V_f \tanh gHt/V_fL$, we obtain

for 10 sec, $V = 4.55 \tanh 0.707 = 4.55 \times .6151 = 2.80$ ft/sec

for 15 sec, $V = 4.55 \tanh 1.062 = 4.55 \times .7864 = 3.57$ ft/sec

It will be noted that the value V/V_f is represented by the value of the hyperbolic tangent. In the solution above, 61.5% and 78.6% of the final velocity are attained in the 10 and 15 seconds respectively.

Supplementary Problems

49. Turpentine at 68°F flows through a pipe in which a Pitot-static tube having a coefficient of 0.97 is centered. The differential gage containing mercury shows a deflection of 4 inches. What is the center velocity?   *Ans.* 17.3 ft/sec

50. Air at 120°F flows by a Pitot-static at a velocity of 60.0 ft/sec. If the coefficient of the tube is 0.95, what water differential reading is expected, assuming constant specific weight of the air at atmospheric pressure.   *Ans.* 0.816 in.

51. The lost head through a 2″ diameter orifice under a certain head is 0.540 ft and the velocity of the water in the jet is 22.5 ft/sec. If the coefficient of discharge is 0.61, determine the head causing flow, the diameter of the jet and the coefficient of velocity.   *Ans.* 8.39 ft, 1.59″, 0.97

52. What size standard orifice is required to discharge 0.565 cfs of water under a head of 28.5 ft?   *Ans.* 2″

53. A sharp-edged orifice has a diameter of 1″ and coefficients of velocity and contraction of 0.98 and 0.62 respectively. If the jet drops 3.08 ft in a horizontal distance of 8.19 ft determine the flow in cfs and the head on the orifice.   *Ans.* .0632 cfs, 5.66 ft

54. Oil of specific gravity 0.800 flows from a closed tank through a 3″ diameter orifice at the rate of 0.914 cfs. The diameter of the jet is 2.305 in. The level of the oil is 24.5 ft above the orifice and the air pressure is equivalent to −6.0 in. of mercury. Determine the three coefficients of the orifice.   *Ans.* 0.580, 0.590, 0.982

55. Refer to Figure 9-11 below. The 3″ diameter orifice has coefficients of velocity and contraction of 0.950 and 0.632 respectively. Determine (*a*) the flow for the deflection of the mercury indicated and (*b*) the power in the jet.   *Ans.* 1.04 cfs, 2.06 hp

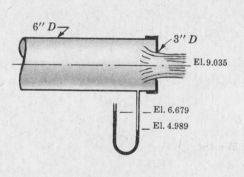

Fig. 9-11

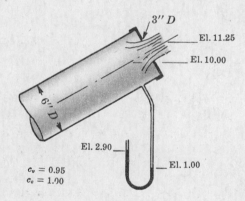

Fig. 9-12

56. Refer to Figure 9-12 above. Heavy fuel oil at 60°F flows through the 3″ diameter end-of-pipe orifice causing the deflection of the mercury in the U-tube gage. Determine the power in the jet.   *Ans.* 2.90 hp

57. Steam locomotives sometimes take on water by means of a scoop which dips into a long, narrow tank between the rails. If the lift into the tank is 9 ft, at what velocity must the train travel, in mph (neglecting friction)?   *Ans.* 16.4 mph

58. Air at 60°F flows through a large duct and thence through a 3″ diameter hole in the thin metal (*c* = 0.62). A U-tube gage containing water registers 1.25 in. Considering the specific weight of air constant, what is the flow through the opening in lb/min?   *Ans.* 10.3 lb/min

59. An oil having specific gravity 0.926 and viscosity 350 Saybolt-seconds flows through a 3″ diameter pipe orifice placed in a 5″ diameter pipe. The differential gage registers a pressure drop of 21.5 psi. Determine flow *Q*.   *Ans.* 1.93 cfs

60. A nozzle with a 2″ diameter tip is attached at the end of an 8″ horizontal pipeline. The coefficients of velocity and contraction are respectively 0.976 and 0.909. A pressure gage attached at the base of the nozzle and located 7.1 ft above its center line reads 32.0 psi. Determine the flow of water in cfs. *Ans.* 1.40 cfs

61. When the flow of water through a horizontal 12″ × 6″ Venturi meter ($c = 0.95$) is 3.93 cfs, find the deflection of the mercury in the differential gage attached to the meter.   *Ans.* 6.18 in.

62. When 4.20 cfs of water flows through a 12″ × 6″ Venturi meter the differential gage records a difference of pressure heads of 7.20 ft. What is the indicated coefficient of discharge of the meter? *Ans.* 0.964

63. The loss of head from entrance to throat of a 10″ × 5″ Venturi meter is 1/16 times the throat velocity head. When the mercury in the differential gage attached to the meter deflects 4″, what is the indicated flow of water?   *Ans.* $Q = 2.23$ cfs

64. A 12″ × 6″ Venturi meter ($c = 0.985$) carries 2.00 cfs of water with a differential gage reading of 2.08 ft. What is the specific gravity of the gage liquid?   *Ans.* 1.75

65. Methane flows at a rate of 16.5 lb/sec through a 12″ × 6″ Venturi meter at a temperature of 60°F. The pressure at the meter inlet is 49.5 psi absolute. Using $k = 1.31$, $R = 96.3$, $\nu = 1.94 \times 10^{-4}$ ft²/sec at 1 atmosphere and $w = .0416$ lb/ft³ at 68°F and 1 atmosphere, calculate the expected deflection of the mercury in the differential gage.   *Ans.* 1.03 ft

66. Water flows through a 6″ pipe in which a long-radius flow nozzle 3″ in diameter is installed. For a deflection of 6″ of mercury in the differential gage, calculate the rate of flow in cfs. (Assume $c = 0.98$ from Diagram *D*.)   *Ans.* 1.00 cfs

67. Water at 80°F flows at the rate of 1.62 cfs through the nozzle in Problem 66. What is the deflection of the mercury in the differential gage? (Use Diagram *D*.)   *Ans.* 1.29 ft

68. Had a dust-proofing oil at 80°F been flowing at 1.62 cfs in Problem 67, what would have been the deflection of the mercury?   *Ans.* 1.22 ft

69. If air at 68°F flows through the same pipe and nozzle (Problem 66), how many pounds of air per second will flow if the absolute pressures in pipe and jet are 30.0 psi and 25.0 psi respectively? *Ans.* 3.81 lb/sec

70. What depth of water must exist behind a rectangular sharp-crested suppressed weir 5 ft long and 4 ft high when a flow of 10 cfs passes over it? (Use the Francis formula.)   *Ans.* 4.72 ft

71. A flow of 30 cfs occurs in a rectangular flume 4 ft deep and 6 ft wide. Find the height at which the crest of a sharp-edged suppressed weir should be placed in order that water will not overflow the sides of the flume. ($m = 3.33$)   *Ans.* 2.72 ft

72. A flow of 384 cfs passes over a suppressed weir which is 16 ft long. The total depth upstream from the weir must not exceed 8 ft. Determine the height to which the crest should be placed to carry this flow. ($m = 3.33$)   *Ans.* 4.39 ft

73. A suppressed weir ($m = 3.33$) under a constant head of 0.300 ft feeds a tank containing a 3″ diameter orifice. The weir, which is 2 ft long and 2.70 ft high, is located in a rectangular channel. The lost head through the orifice is 2.00 ft and $c_c = 0.65$. Determine the head to which the water will rise in the tank and the coefficient of velocity for the orifice.   *Ans.* $h = 20.3$ ft, $c_v = 0.95$

74. A contracted weir 4.00 ft long is placed in a rectangular channel 9 ft wide. The height of weir crest is 3.25 ft and the head is 15.0 in. Determine the flow, using $m = 3.40$.   *Ans.* 17.8 cfs

75. A triangular weir has a 90° notch. What head will produce 1200 gpm? ($m = 2.50$)   *Ans.* 1.027 ft

76. A 36″ pipeline, containing a 36″ × 12″ Venturi meter, supplies water to a rectangular canal. The pressure at the inlet of the Venturi is 30.0 psi and at the throat is 8.67 psi. A suppressed weir ($m = 3.33$), 3 ft high, placed in the canal, discharges under a 9″ head. What is the probable width of the canal?   *Ans.* 20.0 ft

77. Water flows over a suppressed weir ($m = 3.33$) which is 12 ft long and 2 ft high. For a head of 1.200 ft, find the flow in cfs.   *Ans.* 54.4 cfs

78. A tank, 12 ft long and 4 ft wide, contains 4 ft of water. How long will it take to lower the water to a 1 ft depth if a 3″ diameter orifice ($c = 0.60$) is opened in the bottom of the tank?   *Ans.* 406 sec

79. A rectangular tank, 16 ft by 4 ft, contains 4 ft of oil, specific gravity 0.75. If it takes 10 minutes and 10 seconds to empty the tank through a 4″ diameter orifice in the bottom, determine the average value of the coefficient of discharge.   Ans. 0.60

80. In Problem 79, for a coefficient of discharge of 0.60, what will be the depth after the orifice has been flowing for 5 minutes?   Ans. 1.03 ft

81. A tank with a trapezoidal cross section has a constant length of 5 ft. When the water is 8 ft deep above the 2″ diameter orifice ($c = 0.65$), the width of the water surface is 6 ft and, at a 3 ft head depth, the width of water surface is 4 ft. How long will it take to lower the water from the 8 ft depth to the 3 ft depth?   Ans. 482 sec

82. A suppressed weir is located in the end of a tank which is 10 ft square. If the initial head on the weir is 2.00 ft, how long will it take for 125 ft³ of water to run out of the tank? ($m = 3.33$)
Ans. 2.68 sec

83. A rectangular channel, 60 ft long by 10 ft wide, is discharging its flow over a 10 ft long suppressed weir under a head of 1.00 ft. If the supply is suddenly cut off, what will be the head on the weir in 36 seconds? ($m = 3.33$)   Ans. 0.25 ft

84. Two orifices in the side of a tank are one above the other vertically and are 6 ft apart. The total depth of water in the tank is 14 ft and the head on the upper orifice is 4 ft. For the same values of c_v, show that the jets will strike the horizontal plane on which the tank rests at the same point.

85. A 6″ diameter orifice discharges 12.00 cfs of water under a head of 144 ft. This water flows into a 12 ft wide rectangular channel at a depth of 3 ft, and then it flows over a contracted weir. The head on the weir is 1.000 ft. What is the length of the weir and the coefficient of the orifice?
Ans. 3.80 ft, $c = 0.635$

86. The head on a suppressed weir G, which is 12 ft long, is 1.105 ft and the velocity of approach may be neglected. For the system shown in Fig. 9-13, what is the pressure head at B? Sketch the hydraulic grade lines.   Ans. 193.2 ft

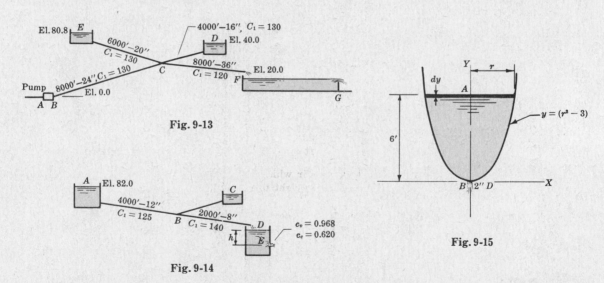

Fig. 9-13

Fig. 9-14

Fig. 9-15

87. In Fig. 9-14 above the elevation of the hydraulic grade line at B is 50.0 and the pipes BC and BD are arranged so that the flow from B divides equally. What is the elevation of the end of the pipe at D and what is the head that will be maintained on the 4″ diameter orifice E?
Ans. El. 23.8, h = 22.5 ft

88. For the tank shown in Fig. 9-15 above, using an average coefficient of discharge of 0.65 for the 2″ diameter orifice, how long will it take to lower the tank level 4 feet?   Ans. 390 sec

Chapter 10

Flow in Open Channels

OPEN CHANNEL

An open channel is a conduit in which the liquid flows with a free surface subjected to atmospheric pressure. The flow is caused by the slope of the channel and of the liquid surface. Accurate solution of flow problems is difficult and depends upon experimental data which must cover a wide range of conditions.

STEADY, UNIFORM FLOW

Steady, uniform flow encompasses two conditions of flow. Steady flow, as defined under pipe flow, refers to the condition in which the flow characteristics at any point do not change with time ($\partial V/\partial t = 0$, $\partial y/\partial t = 0$, etc.). Uniform flow refers to the condition in which the depth, slope, velocity and cross section remain constant over a given length of channel ($\partial y/\partial L = 0$, $\partial V/\partial L = 0$, etc.).

The energy grade line is parallel to the liquid surface (hydraulic grade line) and $V^2/2g$ above it. This is not true for steady, non-uniform flow.

NON-UNIFORM or VARIED FLOW

Non-uniform or varied flow occurs when the depth of flow changes along the length of the open channel, or $\partial y/\partial L \neq 0$. Varied flow may be steady or unsteady. It may also be classified as tranquil, rapid or critical.

LAMINAR FLOW

Laminar flow will occur in open channel flow for values of Reynolds number R_E of 2000 or less. The flow *may* be laminar up to $R_E = 10,000$. For open channel flow, $R_E = 4RV/v$, where R is the hydraulic radius.

The CHEZY FORMULA for *steady, uniform flow*, developed in Problem 1, is

$$V = C\sqrt{RS} \qquad\qquad (1)$$

where V = average velocity in ft/sec, C = coefficient,

R = hydraulic radius, S = slope of water surface or of the energy gradient or of the channel bottom; these lines are parallel for steady, uniform flow.

COEFFICIENT C can be obtained by using one of the following expressions.

$$C = \sqrt{\frac{8g}{f}} \qquad\qquad \text{(See Problem 1.)} \qquad\qquad (2)$$

$$C = \frac{41.65 + \dfrac{.00281}{S} + \dfrac{1.811}{n}}{1 + \dfrac{n}{\sqrt{R}}\left(41.65 + \dfrac{.00281}{S}\right)} \qquad\qquad \text{(Kutter)} \qquad\qquad (3)$$

160

$$C = \frac{1.486}{n} R^{1/6} \qquad\qquad \text{(Manning)} \qquad (4)$$

$$C = \frac{157.6}{1 + m/\sqrt{R}} \qquad\qquad \text{(Bazin)} \qquad (5)$$

$$C = -42 \log\left(\frac{C}{R_E} + \frac{\epsilon}{R}\right) \qquad\qquad \text{(Powell)} \qquad (6)$$

In expressions (2) through (5), n and m are roughness factors determined by experiments using water only. Some values are given in Table 9 of the Appendix. In general, the Manning formula is preferred. The Powell formula will be discussed in Problems 9 and 10.

DISCHARGE (Q) for steady uniform flow, in terms of the Manning formula, is

$$Q = AV = A\left(\frac{1.486}{n}\right) R^{2/3} S^{1/2} \qquad (7)$$

The conditions associated with steady uniform flow are called normal. Thus the terms normal depth and normal slope.

LOST HEAD (h_L), expressed in terms of the Manning formula, is

$$h_L = \left[\frac{Vn}{1.486\,R^{2/3}}\right]^2 L, \quad \text{using } S = h_L/L \qquad (8)$$

For non-uniform (varied) flow, mean values of V and R may be used with reasonable accuracy. For a long channel, short lengths should be employed in which the changes in depth are about the same magnitude.

VERTICAL DISTRIBUTION of VELOCITY

Vertical distribution of velocity in an open channel may be assumed as parabolic for laminar flow and logarithmic for turbulent flow.

For uniform *laminar* flow in wide open channels of mean depth y_m, the velocity distribution may be expressed as

$$v = \frac{gS}{\nu}(yy_m - \tfrac{1}{2}y^2) \qquad \text{or} \qquad v = \frac{wS}{\mu}(yy_m - \tfrac{1}{2}y^2) \qquad (9)$$

The mean velocity V, derived from this equation in Problem 3, becomes

$$V = \frac{gSy_m^2}{3\nu} \qquad \text{or} \qquad V = \frac{wSy_m^2}{3\mu} \qquad (10)$$

For uniform *turbulent* flow in wide open channels, the velocity distribution (developed in Problem 4) may be expressed as

$$v = 2.5\sqrt{\tau_o/\rho}\,\ln(y/y_o) \qquad \text{or} \qquad v = 5.75\sqrt{\tau_o/\rho}\,\log(y/y_o) \qquad (11)$$

SPECIFIC ENERGY

Specific energy (E) is defined as the energy per unit weight (ft lb/lb) relative to the bottom of the channel or

$$E = \text{depth} + \text{velocity head} = y + V^2/2g \qquad (12A)$$

A more exact expression of the kinetic energy term would be $\alpha V^2/2g$. See Chapter 6 for discussion of the kinetic-energy correction factor α.

In terms of the flow rate q per unit of channel width b (i.e., $q = Q/b$),

$$E = y + (1/2g)(q/y)^2 \qquad \text{or} \qquad q = \sqrt{2g(y^2E - y^3)} \qquad (12B)$$

For uniform flow, the specific energy remains constant from section to section. For non-uniform flow, the specific energy along the channel may increase or decrease.

CRITICAL DEPTH

Critical depth (y_c) for a constant unit flow q in a rectangular channel occurs when the specific energy is a minimum. As shown in Problems 27 and 28,

$$y_c = \sqrt[3]{q^2/g} = \tfrac{2}{3}E_c = V_c^2/g \qquad (13)$$

This expression may be rearranged to give

$$V_c = \sqrt{gy_c} \quad \text{or} \quad V_c/\sqrt{gy_c} = 1 \quad \text{for critical flow} \qquad (14)$$

Thus, if the Froude Number $N_F = V_c/\sqrt{gy_c} = 1$, critical flow occurs. If $N_F > 1$, super-critical flow (rapid flow) occurs; and if $N_F < 1$, subcritical flow (tranquil flow) occurs.

MAXIMUM UNIT FLOW

Maximum unit flow (q_{max}) in a rectangular channel for any given specific energy E, as developed in Problem 28, is

$$q_{max} = \sqrt{gy_c^3} = \sqrt{g(\tfrac{2}{3}E)^3} \qquad (15)$$

For CRITICAL FLOW in NON-RECTANGULAR CHANNELS, as developed in Prob. 27,

$$\frac{Q^2}{g} = \frac{A_c^3}{b'} \quad \text{or} \quad \frac{Q^2 b'}{gA_c^3} = 1 \qquad (16)$$

where b' is the width of the water surface. We may rearrange (16) by dividing by A_c^2, as follows:

$$V_c^2/g = A_c/b' \quad \text{or} \quad V_c = \sqrt{gA_c/b'} = \sqrt{gy_m} \qquad (17)$$

where the term A_c/b' is called the mean depth y_m.

NON-UNIFORM FLOW

For non-uniform flow, an open channel is usually divided into lengths L, called reaches, for study. To compute backwater curves, the energy equation (see Problem 39) yields

$$L \text{ in ft} = \frac{(V_2^2/2g + y_2) - (V_1^2/2g + y_1)}{S_o - S} = \frac{E_2 - E_1}{S_o - S} = \frac{E_1 - E_2}{S - S_o} \qquad (18)$$

where S_o = the slope of the channel bottom and S = the slope of the energy gradient.

For successive reaches where the changes in depth are about the same, energy gradient S may be written

$$S = \left(\frac{n V_{\text{mean}}}{1.486 R_{\text{mean}}^{2/3}}\right)^2 \quad \text{or} \quad \frac{V_{\text{mean}}^2}{C^2 R_{\text{mean}}} \qquad (19)$$

Surface profiles for gradually varied flow conditions in wide rectangular channels may be analyzed by using the expression

$$\frac{dy}{dL} = \frac{S_o - S}{(1 - V^2/gy)} \qquad (20)$$

The term dy/dL represents the slope of the water surface relative to the channel bottom. Thus if dy/dL is positive, the depth is increasing downstream. Problems 44 and 45 will develop the equation and a system of classifying surface profiles.

BROAD-CRESTED WEIRS can be used to measure the flow in a channel.

Unit flow $q = \sqrt{g}\,(\frac{2}{3}E)^{3/2}$, where E is the specific energy related to the weir crest or the upstream head plus the velocity head of approach. Due to friction, the true rate of discharge is from 90 to 92 percent of the value given by this formula. The approximate equation becomes $q = 3H^{3/2}$ (see Problem 52).

HYDRAULIC JUMP

Hydraulic jump occurs when a supercritical flow changes to a subcritical flow. In such cases, the elevation of the liquid surface increases suddenly in the direction of flow. For a constant flow in a rectangular channel, as derived in Problem 46,

$$\frac{q^2}{g} = y_1 y_2 \Big(\frac{y_1 + y_2}{2}\Big) \tag{21}$$

Solved Problems

1. Develop the general (Chezy) equation for steady, uniform flow in an open channel.

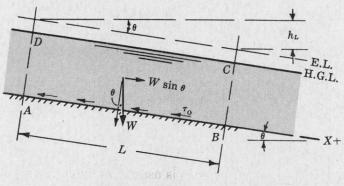

Fig. 10-1

Solution:

In the Fig. 10-1 above, consider the volume of liquid $ABCD$ of constant cross section A and of length L. The volume may be considered in equilibrium since the flow is steady (zero acceleration). Summing up the forces acting in the X-direction,

force on area AD − force on area BC + $W \sin\theta$ − resisting forces $= 0$

$$\overset{(h_L)}{w\bar{h}A - w\bar{h}A + wAL\sin\theta - \tau_o pL = 0}$$

where τ_o is the boundary shear stress (lb/ft^2) acting on an area L ft long by wetted perimeter p ft wide. Then

$$wAL\sin\theta = \tau_o pL \qquad \text{and} \qquad \tau_o = (wA\sin\theta)/p = wRS \tag{A}$$

since $R = A/p$ and $\sin\theta = \tan\theta = S$ for small values of θ.

It was seen in Chapter 7, Problem 5, that $\tau_o = (w/g)f(V^2/8)$. Then

$$wRS = (w/g)f(V^2/8) \qquad \text{or} \qquad V = \sqrt{(8g/f)RS} = C\sqrt{RS} \tag{*B}$$

For *laminar* flow, f may be taken as $64/R_E$. Hence

$$C = \sqrt{(8g/64)R_E} = 2.005\sqrt{R_E} \tag{C}$$

2. Show that the vertical distribution of velocity is parabolic in a wide, open channel for uniform laminar flow. (y_m = mean depth of the channel).

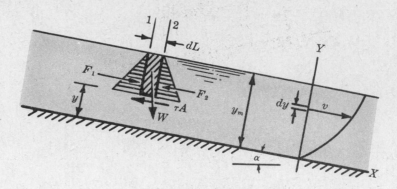

Fig. 10-2

Solution:

When velocity and depth of flow are relatively small, reflecting a Reynolds number < 2000, viscosity becomes the dominant flow factor. The resulting flow is laminar. (For open channels, R_E is defined as $4RV/\nu$). For the free body shown cross-hatched, using $\Sigma F_x = 0$, we obtain

$$F_1 - F_2 + w(y_m - y)dL\,dz \sin \alpha - \tau\,dL\,dz = 0$$

Since $F_1 = F_2$, we obtain

$$\tau = w(y_m - y) \sin \alpha$$

For laminar flow, $\tau = \mu\,dv/dy$, from which we obtain

$$dv = \frac{w}{\mu}(y_m - y) \sin \alpha\,dy = \frac{wS}{\mu}(y_m - y)\,dy \qquad (A)$$

For the small values of angle α associated with the slope of open channels, $\sin \alpha = \tan \alpha =$ slope S. Integrating (A) yields

$$v = \frac{wS}{\mu}(yy_m - \tfrac{1}{2}y^2) + C \qquad (B)$$

Since $v = 0$ when $y = 0$, the value of constant $C = 0$. Equation (B) is a quadratic equation representing a parabola.

3. What is the mean velocity V in Problem 2?

Solution:

$$\text{Mean velocity } V = \frac{Q}{A} = \frac{\int dQ}{\int dA} = \frac{\int v\,dA}{\int dA} = \frac{(wS/\mu)\int(yy_m - \tfrac{1}{2}y^2)dy\,dz}{\int dy\,dz = y_m\,dz}$$

where dz is a constant (dimension perpendicular to the paper).

$$V = \frac{wS\,dz}{\mu y_m\,dz}\int_0^{y_m}(yy_m - \tfrac{1}{2}y^2)dy = \frac{wSy_m^2}{3\mu}$$

4. For steady, uniform flow in wide open channels, establish a theoretical equation for the mean velocity for smooth surfaces.

Solution:

For turbulent flow in general, shear stress τ can be expressed as

$$\tau = \rho l^2(dv/dz)^2$$

where l is the mixing length and a function of z. (See Chapter 7.)

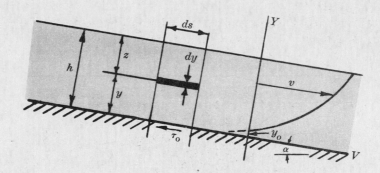

Fig. 10-3

Also, from Problem 1, expression (A), $\tau_o = wRS = whS$, since hydraulic radius R for wide channels equals the depth.

In the boundary layer, since y is very small, $z \cong h$ and $\tau \cong \tau_o$. Hence we may equate the values of τ_o, or

$$\rho l^2 (dv/dz)^2 = wzS \qquad \text{or} \qquad dv/dz = \pm\sqrt{gzS/l^2}$$

To integrate this expression, try a value of $l = k(h-z)(z/h)^{1/2}$. Then

$$-\frac{dv}{dz} = \sqrt{gS}\left[\frac{z^{1/2}}{k(h-z)(z/h)^{1/2}}\right] = \frac{\sqrt{gSh}}{k}\left(\frac{1}{h-z}\right)$$

Let $y = (h-z)$ and $dy = -dz$; then

$$+y\left(\frac{dv}{dy}\right) = \frac{\sqrt{gSh}}{k} \qquad \text{and} \qquad dv = \frac{\sqrt{gSh}}{k}\left(\frac{dy}{y}\right)$$

Since $\tau_o/\rho = whS/\rho = gSh$,

$$dv = \frac{1}{k}\sqrt{\tau_o/\rho}\left(\frac{dy}{y}\right) \qquad \text{or} \qquad v = \frac{1}{k}\sqrt{\tau_o/\rho}\,\ln y + C$$

For $y = y_o$, $v \cong 0$; then $C = (-1/k)\sqrt{\tau_o/\rho}\,\ln y_o$ and

$$v = \frac{1}{k}\sqrt{\tau_o/\rho}\,\ln(y/y_o) \tag{A}$$

Note: Neglecting the log curve to the left of y_o, while producing an approximation, gives satisfactory results well within the limits of expected accuracy, since y_o is very small. See Problem 5 for the value of y_o.

In this expression (A), $k \cong 0.40$ and is called the von Karman constant. Inasmuch as the term $\sqrt{\tau_o/\rho}$ has the dimensions of ft/sec, this term is called the shear velocity which is designated by v_*. Thus

$$v = 2.5v_* \ln(y/y_o) \tag{B}$$

From $Q = AV = (h \times 1)V = \int v(dy \times 1)$, we obtain the value of mean velocity V. Thus

$$V = \frac{\int v(dy \times 1)}{(h \times 1)} = \frac{2.5v_*}{h}\int_o^h (\ln y - \ln y_o)dy$$

Using L'Hospital's rule from the calculus, the mean velocity for smooth surfaces where a boundary layer exists can be evaluated as

$$V = 2.5v_*[\ln h - \ln y_o - 1] \tag{C}$$

It will be shown in Problem 5 that y_o equals $\nu/9v_*$. Therefore equations (B) and (C) may be written

$$v = 2.5v_* \ln(9v_*y/\nu) \tag{D}$$

and

$$V = 2.5v_*[\ln h - \ln(\nu/9v_*) - 1] \tag{E}$$

Frequently the average velocity in an open channel is taken as the velocity observed at a point 0.6 of the depth (measured from the surface). If we accept this value of $\bar{y}$, then we may write, from (B) above, average velocity V as

$$V = 2.5v_* \ln (0.4h/y_o)$$

From Problem 5, $y_o = \delta/103$. Then for wide channels this average velocity becomes, since hydraulic radius $R = h$,

$$V = 2.5v_* \ln 41.2R/\delta \tag{F}$$

5. **Determine the value of y_o in the preceding problem.**
   Solution:

   For smooth surfaces, in a boundary (laminar) layer,

   $$\tau_o = \mu(dv/dy) = \nu\rho(dv/dy) \qquad \text{or} \qquad dv/dy = (\tau_o/\rho)/\nu = v_*^2/\nu \ \text{(a constant)}$$

   Using δ as the thickness of the boundary layer,

   $$\int dv = (v_*^2/\nu) \int_o^\delta dy \qquad \text{or} \qquad v_\delta = v_*^2\delta/\nu = R_{E_*}v_* \tag{A}$$

   From experimentation, $R_{E_*} \cong 11.6$ (fairly constant). Hence,

   $$v_*^2\delta/\nu = 11.6v_* \qquad \text{or} \qquad \delta = 11.6\nu/v_* \tag{B}$$

   Putting $y = \delta$ into equation (B) of the preceding problem,

   $$v_\delta = 2.5v_* \ln \delta/y_o \tag{C}$$

   Combining (C) with (A), $\ln \delta/y_o = v_\delta/2.5v_* = R_{E_*}/2.5 \cong 4.64$,

   $$\delta/y_o = e^{4.64} = 103 \qquad \text{and} \qquad \delta = 103y_o \tag{D}$$

   Then from (B), $$y_o = \frac{\delta}{103} \cong \frac{11.6\nu}{103v_*} \cong \frac{\nu}{9v_*} \tag{E}$$

6. **Water at 60°F flows in a smooth and wide rectangular channel ($n = .009$) at a depth of 4 ft and on a slope of .0004. Compare the value of C obtained by using Manning's formula with that by using the expression $V = 2.5v_* \ln 41.2R/\delta$.**
   Solution:

   (a) Using Manning's formula, $C = (1.486/n)R^{1/6} = (1.486/.009)(4^{1/6}) = 176$.

   (b) Equating the Chezy formula for average velocity V with the given expression,

   $$C\sqrt{RS} = 2.5v_* \ln 41.2R/\delta$$

   Substituting $v_* = \sqrt{gSR}$ from Problem 4, we obtain

   $$C = 2.5\sqrt{g} \ln 41.2R/\delta \tag{A}$$

   For 60°F water, $\nu = 1.217 \times 10^{-5}$ and, using $\delta = 11.6\nu/v_*$ from (B) of Problem 5, we find $C = 177$.

7. (a) A wide rectangular channel flows 4 ft deep on a slope of 4 ft in 10,000 ft. Using the theoretical formula for velocity in Problem 4, calculate the values of theoretical velocities at depth increments of 1/10 depth, assuming the channel to be smooth. (b) Compare the average of the 0.2 and 0.8 depth values of velocity with the 0.6 depth velocity. (c) Compute the position of the average velocity below the surface of the water. Use kinematic viscosity as 1.50×10^{-5} ft²/sec.

Solution:

(a)   Since $v_* = \sqrt{\tau_0/\rho} = \sqrt{gRS} = \sqrt{ghS}$ and $y_0 = \nu/9v_*$,

$$v = 2.5v_* \ln y/y_0 = 2.5(2.303)\sqrt{ghS} \log 9v_*y/\nu$$

$$= 5.75\sqrt{32.2(4)(.0004)} \log \frac{9y\sqrt{32.2(4)(.0004)}}{1.5 \times 10^{-5}}$$

$$= 1.306 \log 136,100y \tag{A}$$

Using (A) we obtain the following values of velocity v:

Dist. Down (%)	y (ft)	$136,100y$	$\log 136,100y$	v (ft/sec)
0	4.0	544,400	5.736	7.48
10.	3.6	489,960	5.690	7.43
20.	3.2	435,520	5.639	7.36
30.	2.8	381,080	5.581	7.30
40.	2.4	326,640	5.514	7.20
50.	2.0	272,200	5.435	7.10
60.	1.6	217,760	5.338	6.97
70.	1.2	163,320	5.213	6.81
80.	0.8	108,880	5.037	6.58
90.	0.4	54,440	4.736	6.19
92.5	0.3	40,830	4.611	6.02
95.0	0.2	27,220	4.435	5.81
97.5	0.1	13,610	4.134	5.40
99.75	0.01	1,361	3.134	4.10

(b)   Average of the 0.2 and 0.8 depth values is $V = \frac{1}{2}(7.36 + 6.58) = 6.97$ ft/sec.

The 0.6 depth value is 6.97 ft/sec. Seldom does such exact agreement occur.

8. Assuming the Manning formula for C to be correct, what value of n will satisfy the "smooth" criterion in Problem 6?

Solution:

Equate the values of C, using expression (A) in Problem 6, as follows:

$$\frac{1.486\,R^{1/6}}{n} = 5.75\sqrt{g} \log\left(\frac{41.2R}{\delta}\right) = 5.75\sqrt{g} \log\left(\frac{41.2R\sqrt{gSR}}{11.6\nu}\right)$$

Substituting values and solving, $n = .0106$.

9. Using the Powell equation, what quantity of liquid will flow in a smooth rectangular channel 2 ft wide, on a slope of .010 if the depth is 1.00 ft? (Use $\nu = .00042$ ft²/sec.)

Solution:

Equation (6) is $C = -42 \log\left(\frac{C}{R_E} + \frac{\epsilon}{R}\right)$

For smooth channels, ϵ/R is small and can be neglected; then

$$C = 42 \log R_E/C \tag{A}$$

From the given data, R_E/C can be evaluated using $V = C\sqrt{RS}$:

$$R_E = 4RV/\nu = 4RC\sqrt{RS}/\nu$$

$$R_E/C = 4R^{3/2}S^{1/2}/\nu = 4(\tfrac{1}{2})^{3/2}(.010)^{1/2}/.00042 = 337$$

Then $C = 42 \log 337 = 106$ and

$$Q = CA\sqrt{RS} = 106(2)\sqrt{0.50(.010)} = 15.0 \text{ cfs}$$

10. Determine C by the Powell equation for a 2 ft by 1 ft rectangular channel, if $V = 5.50$ ft/sec, $\epsilon/R = .0020$ and $\nu = .00042$ ft²/sec.

Solution:

First calculate $R_E = 4RV/\nu = 4(0.50)(5.50)/.00042 = 26,200$. Then

$$C = -42 \log \left(\frac{C}{26,200} + .0020\right)$$

Solving by successive trials, we find that $C = 94$ is satisfactory.

Powell has plotted graphs of C vs R_E for various values of relative roughness ϵ/R. The graphs simplify the calculations. They also indicate a close analogy with the Colebrook formula for the flow in pipes.

11. (a) Show a correlation between roughness factor f and roughness factor n. (b) What is the average shear stress at the sides and bottom of a rectangular flume 12 ft wide, flowing 4 ft deep and laid on a slope of 1.60 ft/1000 ft?

Solution:

(a) Taking the Manning formula as a basis of correlation,

$$C = \sqrt{\frac{8g}{f}} = \frac{1.486\, R^{1/6}}{n}, \qquad \frac{1}{\sqrt{f}} = \frac{1.486 R^{1/6}}{n\sqrt{8g}}, \qquad f = \frac{8gn^2}{2.21\, R^{1/3}}$$

(b) From Problem 1,

$$\tau_o = wRS = w\left(\frac{\text{area}}{\text{wetted perimeter}}\right)(\text{slope}) = 62.4\left(\frac{12 \times 4}{4 + 12 + 4}\right)\left(\frac{1.60}{1000}\right) = 0.239 \text{ lb/ft}^2$$

12. What flow can be expected in a 4 ft wide rectangular, cement-lined channel laid on a slope of 4 ft in 10,000 ft, if the water flows 2 ft deep? Use both Kutter's C and Manning's C.

Solution:

(a) Using Kutter's C. From Table 9, $n = .015$. Hydraulic radius $R = 4(2)/8 = 1.00$ ft.

From Table 10, for $S = .0004$, $R = 1.00$ and $n = .015$, the value of $C = 98$.

$$Q = AV = AC\sqrt{RS} = (4 \times 2)(98)\sqrt{1.00 \times .0004} = 15.7 \text{ cfs}$$

(b) Using Manning's C,

$$Q = AV = A\frac{1.486}{n} R^{2/3} S^{1/2} = (4 \times 2)\frac{1.486}{.015}(1.00)^{2/3}(.0004)^{1/2} = 15.8 \text{ cfs}$$

13. In a hydraulics laboratory, a flow of 14.56 cfs was measured from a rectangular channel flowing 4 ft wide and 2 ft deep. If the slope of the channel was .00040, what is the roughness factor for the lining of the channel?

Solution:

(a) Using Kutter's formula,

$$Q = 14.56 = AC\sqrt{RS} = (4 \times 2)C\sqrt{[(4 \times 2)/8](.0004)} \qquad \text{and} \qquad C = 91$$

By interpolation in Table 10, $n = .016$.

(b) Using Manning's formula,

$$Q = 14.56 = A\frac{1.486}{n} R^{2/3} S^{1/2} = (4 \times 2)\frac{1.486}{n}(1)^{2/3}(.0004)^{1/2}, \quad n = .0163. \text{ Use } n = .016.$$

14. On what slope should a 24″ vitrified sewer pipe be laid in order that 6.00 cfs will flow when the sewer is half full? What slope if the sewer flows full? (Table 9 gives $n = .013$.)

Solution:

Hydraulic radius $R = \dfrac{\text{area}}{\text{wetted perimeter}} = \dfrac{\frac{1}{2}(\frac{1}{4}\pi d^2)}{\frac{1}{2}(\pi d)} = \frac{1}{4}d = \frac{1}{2}\text{ft}$.

(a) $Q = 6.00 = A\dfrac{1.486}{n}R^{2/3}S^{1/2} = \frac{1}{2}(\frac{1}{4}\pi)2^2 \times (1.486/.013)(\frac{1}{2})^{2/3}S^{1/2}$, $\sqrt{S} = .0532$ and $S = .00283$.

(b) $R = \frac{1}{4}d = \frac{1}{2}\text{ft}$, as before, and $A = \frac{1}{4}\pi(2^2)$. Then $\sqrt{S} = .0266$ and $S = .00071$.

15. A trapezoidal channel, bottom width 20 ft and side slopes 1 to 1, flows 4 ft deep on a slope of .0009. For a value of $n = 0.025$, what is the uniform discharge?

Solution:

Area $A = 20(4) + 2(\frac{1}{2})(4)(4) = 96.0$ ft². $\quad R = 96.0/[20 + 2(4\sqrt{2})] = 3.07$ ft.

$$Q = (1.486/n)AR^{2/3}S^{1/2} = 96.0(1.486/.025)(3.07)^{2/3}(.03) = 361 \text{ ft}^3/\text{sec}$$

16. Two concrete pipes ($C = 100$) must carry the flow from an open channel of half-square section 6 ft wide and 3 ft deep ($C = 120$). The slope of both structures is .00090. (a) Determine the diameter of the pipes. (b) Find the depth of water in the rectangular channel after it has become stabilized, if the slope is changed to .00160, using $C = 120$.

Solution:

(a)
$$Q_{\text{channel}} = Q_{\text{pipes}}$$
$$AC\sqrt{RS} = 2AC\sqrt{RS}$$
$$(6 \times 3)(120)\sqrt{\frac{6 \times 3}{12}(.00090)} = 2(\tfrac{1}{4}\pi d^2)(100)\sqrt{\frac{d}{4}(.00090)}$$
$$79.3 = 2.36d^{5/2} \quad\text{and}\quad d = 4.08 \text{ ft}$$

(b) For depth y, area $A = 6y$ and hydraulic radius $R = \dfrac{6y}{6 + 2y}$. For the same Q,

$$79.3 = (6y)(120)\sqrt{\frac{6y}{6 + 2y}(.00160)}, \qquad 6y\sqrt{\frac{6y}{6 + 2y}} = 16.55, \qquad y^3 - 2.54y = 7.62$$

Solving by successive trials: Try $y = 2.5$ ft, $(15.6 - 6.3) \neq 7.6$ (decrease y).
Try $y = 2.4$ ft, $(13.8 - 6.1) = 7.6$ (satisfactory).

Thus, depth to nearest 1/10 ft is 2.4 ft.

17. An average vitrified sewer pipe is laid on a slope of .00020 and is to carry 83.5 cfs when the pipe flows 0.90 full. What size pipe should be used?

Solution:

From Table 9, $n = .015$.

Calculate the hydraulic radius R. (Refer to Fig. 10-4.)

$$R = \frac{A}{p} = \frac{\text{circle} - (\text{sector } AOCE - \text{triangle } AOCD)}{\text{arc } ABC}$$

Angle $\theta = \cos^{-1}(0.40d/0.50d) = \cos^{-1}0.800$, $\quad \theta = 36°52'$.

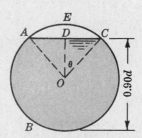

Fig. 10-4

Area of sector $AOCE = [2(36°52')/360°](\frac{1}{4}\pi d^2) = 0.1612d^2$.

Length of arc $ABC = \pi d - [2(36°52')/360°](\pi d) = 2.498d$.

Area of triangle $AOCD = 2(\frac{1}{2})(0.40d)(0.40d \tan 36°52') = 0.1200d^2$.

$$R = \frac{\frac{1}{4}\pi d^2 - (0.1612d^2 - 0.1200d^2)}{2.498d} = \frac{0.7442d^2}{2.498d} = 0.298d$$

(a) Using Kutter's C (assumed as 100 for first calculation),

$$Q = CA\sqrt{RS}, \qquad 83.5 = 100(0.7442d^2)\sqrt{0.298d(.00020)}, \qquad d^{5/2} = 145.7, \qquad d = 7.34 \text{ ft}$$

Checking on C, $R = 0.298 \times 7.34 = 2.19$ ft. Table 10 gives $C = 112$. Recalculating,

$$d^{5/2} = 145.7(100/112) = 130.0 \qquad \text{or} \qquad d = 7.01 \text{ ft} \qquad (C \text{ checks satisfactorily.})$$

(b) Using Manning's C (and above information),

$$Q = \frac{1.486}{n} AR^{2/3} S^{1/2}$$

$$83.5 = \frac{1.486}{0.015}(0.7442d^2)(0.298d)^{2/3}(.00020)^{1/2}, \qquad d^{8/3} = 180, \qquad d = 7.00 \text{ ft}$$

18. How deep will water flow at the rate of 240 cfs in a rectangular channel 20 ft wide, laid on a slope of .00010? Use $n = .0149$.

Solution:

Employing the Manning formula,

$$Q = \frac{1.486}{n} AR^{2/3} S^{1/2}, \qquad 240 = \frac{1.486}{.0149}(20y)\Big(\frac{20y}{20+2y}\Big)^{2/3}(.01), \qquad 12 = y\Big(\frac{20y}{20+2y}\Big)^{2/3}$$

Solving by successive trials, we find that $y = 5.25$ ft is satisfactory. The water will flow at a depth of 5.25 ft, called the normal depth.

19. How wide must a rectangular channel be constructed in order to carry 500 cfs at a depth of 6 ft on a slope of .00040? Use $n = .010$.

Solution:

Using the Manning formula, with $A = 6b$ and $R = 6b/(b+12)$, and solving by successive trials, we find the required width $b = 13.1$ ft.

20. Develop the Manning equation discharge factors K and K' enumerated in Tables 11 and 12 of the Appendix.

Solution:

The discharge factors to be used in the Manning formula may be evaluated as follows. Any cross sectional area can be expressed as $A = F_1 y^2$, where F_1 is a dimensionless factor and y^2 is the square of the depth. In similar fashion, the hydraulic radius R can be expressed as $R = F_2 y$. Then the Manning formula becomes

$$Q = \frac{1.486}{n}(F_1 y^2)(F_2 y)^{2/3} S^{1/2} \qquad \text{or} \qquad \frac{Qn}{y^{8/3} S^{1/2}} = 1.486 F_1 F_2^{2/3} = K \tag{1}$$

Similarly, in terms of a base width b, $A = F_3 b^2$, $R = F_4 b$ and

$$\frac{Qn}{b^{8/3} S^{1/2}} = 1.486 F_3 F_4^{2/3} = K' \tag{2}$$

Tables 11 and 12 give values of K and K' for representative trapezoids. Values of K and K' may be calculated for any shape of cross section.

21. What are the discharge factors K and K' for a rectangular channel 20 ft wide by 4 ft deep. Compare with the values in Tables 11 and 12.

Solution:

(a) $A = F_1 y^2$, $80 = F_1(16)$, $F_1 = 5.0$. $R = F_2 y$, $80/28 = F_2(4)$, $F_2 = 0.713$. $K = 1.486 F_1 F_2^{2/3} = 5.94$.

   Table 11 indicates that for $y/b = 4/20 = 0.20$, $K = 5.94$. (Check)

(b) $A = F_3 b^2$, $80 = F_3(400)$, $F_3 = 0.20$. $R = F_4 b$, $80/28 = F_4(20)$, $F_4 = 0.143$. $K' = 1.486 F_3 F_4^{2/3} = 0.812$.

   Table 12 indicates that for $y/b = 4/20 = 0.20$, $K' = 0.812$. (Check)

22. Solve Problem 18, using the discharge factors in Table 12.

Solution:

From Problem 20, equation (2),

$$\frac{Qn}{b^{8/3} S^{1/2}} = K', \qquad \frac{240(.0149)}{(20)^{8/3}(.00010)^{1/2}} = 0.121 = K'$$

   Table 12 indicates that for trapezoids with vertical sides, a K' of 0.121 represents a depth-to-width ratio (y/b) between 0.26 and 0.28". Interpolating, $y/b = 0.262$. Then $y = 0.262(20) = 5.24$ ft, as calculated in Problem 18.

23. Solve Problem 19 using the discharge factors in Table 11.

Solution:

From Problem 20, equation (1),

$$\frac{Qn}{y^{8/3} S^{1/2}} = K, \qquad \frac{500(.010)}{(6)^{8/3}(.00040)^{1/2}} = 2.11 = K$$

   $K = 2.11$ lies between 2.22 and 2.09. By interpolation, the y/b value for $K = 2.11$ is 0.458. Then $b = 6/0.458 = 13.1$ ft, as found in Problem 19.

24. A channel with a trapezoidal cross section is to carry 900 cfs. If slope $S = .000144$, $n = 0.015$, base width $b = 20$ ft and the side slopes are 1 vertical to $1\frac{1}{2}$ horizontal, determine the normal depth of flow y_N by formula and by use of Tables.

Solution:

(a) By formula,

$$900 = \frac{1.486}{0.015}(20y_N + 1.5 y_N^2)\left(\frac{20 y_N + 1.5 y_N^2}{20 + 2 y_N \sqrt{3.25}}\right)^{2/3}(.000144)^{1/2}$$

or

$$758 = \frac{(20 y_N + 1.5 y_N^2)^{5/3}}{(20 + 2 y_N \sqrt{3.25})^{2/3}}$$

Try $y_N = 8.0$ ft: $758 \overset{?}{=} \dfrac{(160 + 96)^{5/3}}{(20 + 16 \sqrt{3.25})^{2/3}}$ or $758 \neq 768$ (close enough).

   The depth of flow can be evaluated by successive trials to the accuracy desired. The normal depth is slightly less than 8.0 ft.

(b) Preparatory to using Table 12 in the Appendix,

$$\frac{Qn}{b^{8/3} S^{1/2}} = \frac{900(.015)}{(20)^{8/3}(.000144)^{1/2}} = 0.382 = K'$$

In Table 12, for $1\frac{1}{2}$ horizontal to 1 vertical,

$$y/b = 0.38, \; K' = 0.353 \qquad \text{and} \qquad y/b = 0.40, \; K' = 0.389$$

Interpolating for $K' = 0.382$ gives $y/b = 0.396$.   Then $y_N = 0.396(20) = 7.92$ ft.

25. For a given cross sectional area, determine the best dimensions for a trapezoidal channel.

Solution:

Examination of the Chezy equation indicates that, for a given cross sectional area and slope, the rate of flow through a channel of given roughness will be a maximum when the hydraulic radius is a maximum. It follows that the hydraulic radius will be a maximum when the wetted perimeter is a minimum. Referring to Fig. 10-5,

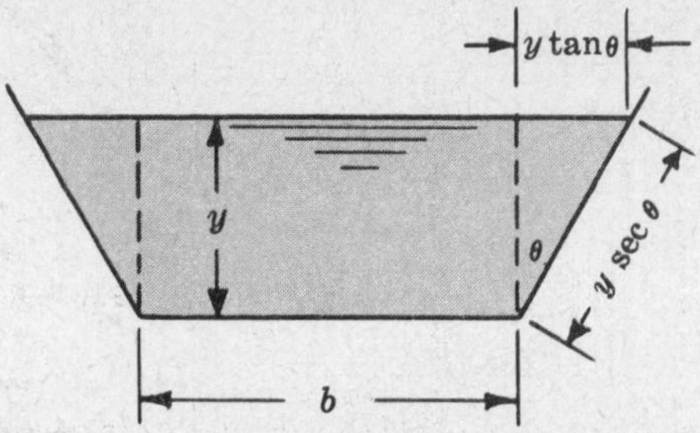

$$A \; = \; by \; + \; 2(\tfrac{1}{2}y)(y \tan \theta)$$

or $\qquad\qquad b = A/y - y \tan \theta$

$p = b + 2y \sec \theta \quad$ or $\quad p = A/y - y \tan \theta + 2y \sec \theta$

Fig. 10-5

Differentiating p with respect to y and equating to zero,

$$dp/dy \; = \; -A/y^2 \; - \; \tan \theta \; + \; 2 \sec \theta \; = \; 0 \qquad \text{or} \qquad A \; = \; (2 \sec \theta - \tan \theta)y^2$$

$$\text{Maximum } R \; = \; \frac{A}{p} \; = \; \frac{(2 \sec \theta - \tan \theta)y^2}{(2 \sec \theta - \tan \theta)y^2/y \; - \; y \tan \theta \; + \; 2y \sec \theta} \; = \; \frac{y}{2}$$

Notes:

(1) For all trapezoidal channels, the best hydraulic section is obtained when $R = y/2$. The symmetrical section will be a half-hexagon.

(2) For a rectangular channel (when $\theta = 0°$), $A = 2y^2$ and also $A = by$, yielding $y = b/2$, in addition to $R = y/2$. Thus the best depth is half the width with the hydraulic radius half the depth.

(3) The circle has the least perimeter for a given area. A semicircular open channel will discharge more water than any other shape (for the same area, slope and factor n).

26. (a) Determine the most efficient section of a trapezoidal channel, $n = .025$, to carry 450 cfs. To prevent scouring, the maximum velocity is to be 3.00 ft/sec and the side slopes of the trapezoidal channel are 1 vertical to 2 horizontal. (b) What slope S of the channel is required? Refer to the figure of Problem 25.

Solution:

$(a) \qquad\qquad R \; = \; \dfrac{y}{2} \; = \; \dfrac{A}{p} \; = \; \dfrac{by + 2(\tfrac{1}{2}y)(2y)}{b + 2y\sqrt{5}} \qquad \text{or} \qquad b = 2y\sqrt{5} - 4y \qquad\qquad (1)$

$\qquad\qquad A \; = \; Q/V \; = \; 450/3 \; = \; by + 2y^2 \qquad \text{or} \qquad b = (150 - 2y^2)/y \qquad\qquad (2)$

Equating (1) and (2), we obtain $y = 7.79$ ft. Substituting in (2), $b = 3.67$ ft.

For this trapezoid, $b = 3.67$ ft and $y = 7.79$ ft.

$(b) \qquad\qquad V = (1.486/n)R^{2/3}S^{1/2}, \qquad 3.00 = (1.486/.025)(7.79/2)^{2/3}S^{1/2}, \qquad S = .000418$

27. Develop the expression for critical depth, critical specific energy and critical velocity (a) for rectangular channels and (b) for any channel.

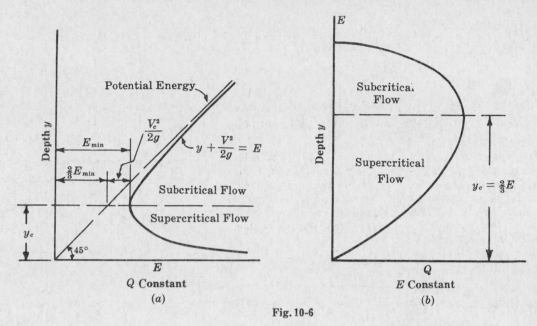

Fig. 10-6

Solution:

(a) **Rectangular Channels.**

As defined,
$$E = y + \frac{V^2}{2g} = y + \frac{1}{2g}\left(\frac{Q/b}{y}\right)^2 = y + \frac{1}{2g}\left(\frac{q}{y}\right)^2 \tag{1}$$

The critical depth for a given flow Q occurs when E is a minimum. Following the usual calculus procedure,

$$\frac{dE}{dy} = \frac{d}{dy}\left[y + \frac{1}{2g}\left(\frac{q}{y}\right)^2\right] = 1 - \frac{q^2}{gy^3} = 0, \qquad q^2 = gy_c^3, \qquad y_c = \sqrt[3]{q^2/g} \tag{2}$$

Eliminating q in (1), using values in (2),

$$E_c = y_c + \frac{gy_c^3}{2gy_c^2} = \frac{3}{2}y_c \tag{3}$$

Since $q = yV$ ($b =$ unity), expression (2) gives

$$y_c^3 = \frac{q^2}{g} = \frac{y_c^2 V_c^2}{g}, \qquad V_c = \sqrt{gy_c}, \qquad \frac{V_c^2}{2g} = \frac{y_c}{2} \tag{4}$$

(b) **Any Channel.**
$$E = y + \frac{V^2}{2g} = y + \frac{1}{2g}\left(\frac{Q}{A}\right)^2$$

For a constant Q, and since area A varies with depth y,

$$\frac{dE}{dy} = 1 + \frac{Q^2}{2g}\left(-\frac{2}{A^3}\cdot\frac{dA}{dy}\right) = 1 - \frac{Q^2}{A^3 g}\frac{dA}{dy} = 0$$

Area dA is defined as water surface width $b' \times dy$. Substitute in the above equation and obtain

$$\frac{Q^2 b'}{gA_c^3} = 1 \qquad \text{or} \qquad \frac{Q^2}{g} = \frac{A_c^3}{b'} \tag{5}$$

This equation must be satisfied for critical flow conditions. The right-hand side is a function of depth y, and generally a trial-and-error solution is necessary to determine the value of y_c which satisfies equation (5).

By dividing Q^2 by A_c^2, or in terms of average velocity, (5) may be written

$$V_c^2/g \;=\; A_c/b' \qquad \text{or} \qquad V_c \;=\; \sqrt{gA_c/b'} \tag{6}$$

Using the mean depth y_m equal to the area A divided by the surface dimension b', equation (5) may be written

$$Q \;=\; A\sqrt{gA/b'} \;=\; A\sqrt{gy_m} \tag{7}$$

Also, $\qquad\qquad V_c \;=\; \sqrt{gA_c/b'} \;=\; \sqrt{gy_m} \qquad \text{or} \qquad V_c^2/gy_m = 1 \tag{8}$

The minimum specific energy is, using (8),

$$E_{\min} \;=\; y_c + V_c^2/2g \;=\; y_c + \tfrac{1}{2}y_m \tag{9}$$

For a rectangular channel $A_c = b'y_c$, and (6) reduces to equation (4) above.

See the above figure for equation (1) plotted twice, with Q held constant, and with E held constant. When the flow is near the critical, a rippling, unstable surface results. It is not desirable to design channels with slopes near the critical.

---

28. Derive the expression for the maximum unit flow q in a rectangular channel for a given specific energy E.

Solution:

Solving (1) of Problem 27 for q gives $q = y\sqrt{2g}(E - y)^{1/2}$. Differentiating with respect to y and equating to zero, we obtain $y_c = \tfrac{2}{3}E$. Equation (2) of Problem 27 now becomes

$$q_{\max}^2 \;=\; g(\tfrac{2}{3}E_c)^3 \;=\; gy_c^3 \qquad \text{or} \qquad q_{\max} \;=\; \sqrt{gy_c^3}$$

Summarizing, for rectangular channels, the characteristics of critical flow are:

(a) $E_{\min} = \tfrac{3}{2}\sqrt[3]{q^2/g}$

(b) $q_{\max} = \sqrt{gy_c^3} = \sqrt{g(\tfrac{2}{3}E_c)^3}$

(c) $y_c = \tfrac{2}{3}E_c = V_c^2/g = \sqrt[3]{q^2/g}$

(d) $V_c/\sqrt{gy_c} = N_F = 1$

(e) Tranquil or subcritical flow occurs when $N_F < 1$ and $y/y_c > 1$.

(f) Rapid or supercritical flow occurs when $N_F > 1$ and $y/y_c < 1$.

---

29. A rectangular channel carries 200 cfs. Find the critical depth y_c and critical velocity V_c for (a) a width of 12 ft and (b) a width of 9 ft. (c) What slope will produce the critical velocity in (a) if $n = .020$?

Solution:

(a) $y_c = \sqrt[3]{q^2/g} = \sqrt[3]{(200/12)^2/32.2} = 2.05$ ft, $\qquad V_c = \sqrt{gy_c} = \sqrt{32.2 \times 2.05} = 8.12$ ft/sec.

(b) $y_c = \sqrt[3]{q^2/g} = \sqrt[3]{(200/9)^2/32.2} = 2.48$ ft, $\qquad V_c = \sqrt{gy_c} = \sqrt{32.2 \times 2.48} = 8.94$ ft/sec.

(c) $V_c = \dfrac{1.486}{n}R^{2/3}S^{1/2}, \qquad 8.12 = \dfrac{1.486}{0.020}\Big(\dfrac{12 \times 2.05}{16.1}\Big)^{2/3}S^{1/2}, \qquad S = .00680.$

---

30. A trapezoidal channel with side slopes of 2 horizontal to 1 vertical is to carry a flow of 590 cfs. For a bottom width of 12 ft, calculate (a) the critical depth and (b) the critical velocity.

Solution:

(a) The area $A_c = 12y_c + 2(\frac{1}{2}y_c \times 2y_c) = 12y_c + 2y_c^2$, and surface width $b' = 12 + 4y_c$.

Expression (5) of Problem 27 yields $\dfrac{(590)^2}{32.2} = \dfrac{(12y_c + 2y_c^2)^3}{12 + 4y_c}$.

Solving this equation by trial, $y_c = 3.45$ ft.

(b) The critical velocity V_c is determined by using equation (6) of Problem 27.

$$V_c = \sqrt{\frac{gA_c}{b'}} = \sqrt{\frac{32.2(41.40 + 23.80)}{12 + 13.80}} = 9.02 \text{ ft/sec}$$

As a check, using $y = y_c = 3.45$, $V_c = Q/A_c = 590/[12(3.45) + 2(3.45)^2] = 9.02$ ft/sec.

31. A trapezoidal channel has a bottom width of 20 ft, side slopes of 1 to 1, and flows at a depth of 3.00 ft. For $n = .015$, and a discharge of 360 cfs, calculate (a) the normal slope, (b) the critical slope and critical depth for 360 cfs, and (c) the critical slope at the normal depth of 3.00 ft.

Solution:

(a) $Q = A\dfrac{1.486}{n}R^{2/3}S_N^{1/2}$, $360 = (60 + 9)\Big(\dfrac{1.486}{0.015}\Big)\Big(\dfrac{69.0}{20 + 6\sqrt{2}}\Big)^{2/3}S_N^{1/2}$, $S_N = .000852$

(b) $V = \dfrac{Q}{A} = \dfrac{360}{20y + y^2}$ and $V_c = \sqrt{\dfrac{gA_c}{b'}} = \sqrt{\dfrac{32.2(20y_c + y_c^2)}{20 + 2y_c}}$

Equating the velocity terms, squaring, and simplifying, we obtain

$$\frac{[y_c(20 + y_c)]^3}{10 + y_c} = 8050$$

which when solved by successive trials gives critical depth $y_c = 2.1$ ft.

The critical slope S_c is calculated using the Manning equation:

$$360 = [20(2.1) + (2.1)^2]\Big(\dfrac{1.486}{0.015}\Big)\Big(\dfrac{20(2.1) + (2.1)^2}{20 + 2(2.1\sqrt{2})}\Big)^{2/3}S_c^{1/2},   S_c = .0028$$

This slope will maintain uniform, critical flow at a critical depth of 2.1 ft and with $Q = 360$ cfs.

(c) From (a), for $y_N = 3.00$ ft, $R = 2.42$ ft and $A = 69.0$ ft². Also, using (6) of Problem 27,

$$V_c = \sqrt{gA/b'} = \sqrt{32.2(69.0)/[20 + 2(3)]} = 9.25 \text{ ft/sec}$$

Substituting these values in the Manning equation for velocity gives

$$9.25 = (1.486/.015)(2.42)^{2/3}S_c^{1/2},   S_c = .00225$$

This slope will produce uniform, critical flow in the trapezoidal channel at a depth of 3.00 ft. Note that in this case the flow $Q = AV = 69.0(9.25) = 637$ cfs.

32. A rectangular channel, 30 ft wide, carries 270 cfs when flowing 3.00 ft deep. (a) What is the specific energy? (b) Is the flow subcritical or supercritical?

Solution:

(a) $E = y + \dfrac{V^2}{2g} = y + \dfrac{1}{2g}\Big(\dfrac{Q}{A}\Big)^2 = 3.00 + \dfrac{1}{64.4}\Big(\dfrac{270}{30 \times 3}\Big)^2 = 3.14$ ft (ft lb/lb).

(b) $y_c = \sqrt[3]{q^2/g} = \sqrt[3]{(270/30)^2/32.2} = 1.36$ ft

The flow is subcritical since the depth of flow exceeds the critical depth. (See Problem 28.)

33. A trapezoidal channel has a bottom width of 20 ft and side slopes of 2 horizontal to 1 vertical. When the depth of water is 3.50 ft, the flow is 370 cfs. (a) What is the specific energy? (b) Is the flow subcritical or supercritical?

Solution:

(a) Area $A = 20(3.50) + 2(\tfrac{1}{2})(3.50)(7.00) = 94.5 \text{ ft}^2$.

$$E = y + \frac{1}{2g}\left(\frac{Q}{A}\right)^2 = 3.50 + \frac{1}{64.4}\left(\frac{370}{94.5}\right)^2 = 3.74 \text{ ft}$$

(b) Using $\dfrac{Q^2}{g} = \dfrac{A_c^3}{b'}$, $\quad \dfrac{(370)^2}{32.2} = \dfrac{(20y_c + 2y_c^2)^3}{20 + 4y_c}$. $\quad$ Solving by trial, $y_c = 2.05$ ft.

$\qquad$ Actual depth exceeds critical depth and flow is subcritical.

34. The discharge through a rectangular channel ($n = .012$) 15 ft wide is 400 cfs when the slope is 1 ft in 100 ft. Is the flow subcritical or supercritical?

Solution:

(1) Investigate critical conditions for the channel.

$$q_{max} = 400/15 = \sqrt{gy_c^3} \quad \text{and} \quad y_c = 2.81 \text{ ft}$$

(2) Critical slope for above critical depth can be found by the Chezy-Manning formula.

$$Q = A\frac{1.486}{n}R^{2/3}S_c^{1/2}$$

$$400 = (15 \times 2.81)\left(\frac{1.486}{0.012}\right)\left(\frac{15 \times 2.81}{15 + 2(2.81)}\right)^{2/3}S_c^{1/2}, \qquad S_c = .0023+$$

Since the stated slope *exceeds* the critical slope, the flow is supercritical.

35. A rectangular channel, 10 ft wide, carries 400 cfs. (a) Tabulate (as preliminary to preparing a diagram) depth of flow against specific energy for depths from 1 ft to 8 ft. (b) Determine the minimum specific energy. (c) What type of flow exists when the depth is 2 ft and when it is 8 ft? (d) For $C = 100$, what slopes are necessary to maintain the depths in (c)?

Solution:

(a) From $E = y + \dfrac{V^2}{2g} = y + \dfrac{(Q/A)^2}{2g}$ we obtain:

$$\text{For } y = 1 \text{ ft}, \qquad E = 1.00 + \frac{(400/10)^2}{2g} = 25.9 \text{ ft lb/lb}$$

$$
\begin{array}{llll}
= 2 & = 2.00 + 6.21 & = & 8.21 \\
= 3 & = 3.00 + 2.76 & = & 5.76 \\
= 4 & = 4.00 + 1.55 & = & 5.55 \\
= 5 & = 5.00 + 0.99 & = & 5.99 \\
= 6 & = 6.00 + 0.69 & = & 6.69 \\
= 7 & = 7.00 + 0.51 & = & 7.51 \\
= 8 & = 8.00 + 0.39 & = & 8.39 \text{ ft lb/lb}
\end{array}
$$

(b) The minimum value of E lies between 5.76 and 5.55 ft lb/lb.

$\qquad$ Using equation (2) of Problem 27, $y_c = \sqrt[3]{q^2/g} = \sqrt[3]{(400/10)^2/32.2} = 3.68$ ft.

$\qquad$ Then $E_{min} = E_c = \tfrac{3}{2}y_c = \tfrac{3}{2}(3.68) = 5.52$ ft lb/lb.

$\qquad$ Note that $E = 8.21$ at 2.00 ft depth and 8.39 at 8.00 ft depth. Figure (a) of Problem 27 indicates this fact, i.e., two depths for a given specific energy when flow Q is constant.

(c) For 2 ft depth (below critical depth) the flow is supercritical and for the 8 ft depth the flow is subcritical.

(d)
$$Q = CA\sqrt{RS}$$

For $y = 2$ ft, $A = 20$ ft^2 and $R = 20/14 = 1.43$ ft, $400 = 100(20)\sqrt{1.43S}$ and $S = .0280$.

For $y = 8$ ft, $A = 80$ ft^2 and $R = 80/26 = 3.08$ ft, $400 = 100(80)\sqrt{3.08S}$ and $S = .000812$.

36. A rectangular flume $(n = .012)$ is laid on a slope of .0036 and carries 580 cfs. For critical flow conditions, what width is required?
Solution:

From Problem 28, $q_{max} = \sqrt{gy_c^3}$. Hence $580/b = \sqrt{32.2y_c^3}$.

Using successive trials, check calculated flow against given flow.

Trial 1. Let $b = 8.0$ ft, $y_c = \sqrt[3]{(580/8.0)^2/32.2} = 5.47$ ft.

Then $R = A/p = (8.0 \times 5.47)/18.94 = 2.31$ ft

and $Q = AV = (8.0 \times 5.47)\left[\dfrac{1.486}{0.012}(2.31)^{2/3}(.0036)^{1/2}\right] = 568$ cfs.

Trial 2. Since the flow must be increased, let $b = 8.50$ ft.

Then $y_c = \sqrt[3]{(580/8.50)^2/32.2} = 5.25$ ft, $R = (8.50 \times 5.25)/19.00 = 2.35$ ft,

and $Q = AV = (8.50 \times 5.25)\left[\dfrac{1.486}{0.012}(2.35)^{2/3}(.0036)^{1/2}\right] = 585$ cfs.

The result is probably close enough.

37. For a constant specific energy of 6.60 ft lb/lb, what maximum flow may occur in a rectangular channel 10.0 ft wide?
Solution:

Critical depth $y_c = \frac{2}{3}E = \frac{2}{3}(6.60) = 4.40$ ft.   (See equation (1) of Problem 28.)

Critical velocity $V_c = \sqrt{gy_c} = \sqrt{32.2 \times 4.40} = 11.90$ ft/sec   and

maximum $Q = AV = (10.0 \times 4.40)(11.90) = 524$ cfs

Using $q_{max} = \sqrt{gy_c^3}$ (equation (2) of Problem 28), we have

maximum $Q = bq_{max} = 10.0\sqrt{32.2(4.40)^3} = 523$ cfs

38. A rectangular channel, 20 ft wide, $n = .025$, flows 5 ft deep on a slope of 14.7 ft in 10,000 ft. A suppressed weir C, 2.45 ft high is built across the channel $(m = 3.45)$. Taking the elevation of the bottom of the channel just upstream from the weir to be 100.00, estimate (using one reach) the elevation of the water surface at a point A, 1000 ft upstream from the weir.
Solution:

Calculate the new elevation of water surface at B in Fig. 10-7 below (before drop-down). Note that the flow is non-uniform since the depths, velocities and areas are not constant after the weir is installed.

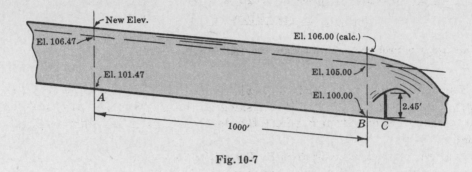

Fig. 10-7

$$Q = (20 \times 5)(1.486/.025)(100/30)^{2/3}(.00147)^{1/2} = 507 \text{ cfs}$$

For an estimated depth of 6 ft just upstream from the weir,

$$\text{Velocity of approach } V = Q/A = 507/(20 \times 6) = 4.23 \text{ ft/sec}$$

The weir formula gives $\quad 507 = 3.45 \times 20\left[\left(H + \frac{(4.23)^2}{2g}\right)^{3/2} - \left(\frac{(4.23)^2}{2g}\right)^{3/2}\right].$ Then

$$(H + 0.278)^{3/2} = 7.34 + 0.15 = 7.49 \quad \text{and} \quad H = 3.55 \text{ ft.}$$
$$\text{Height } Z = 2.45 \text{ ft}$$
$$\text{Depth } y = \overline{6.00} \text{ ft} \quad \text{(assumption checks)}$$

New elevation at A must lie between 106.47 and 107.47. Try an elevation of 106.90 (and check in the Bernoulli equation).

New area at $A = 20(106.90 - 101.47) = 108.6 \text{ ft}^2$, and $V = 507/108.6 = 4.67 \text{ ft/sec.}$

Mean velocity $= \frac{1}{2}(4.23 + 4.67) = 4.45 \text{ ft/sec.}$

Mean hydraulic radius $R = \frac{1}{2}(120.0 + 108.6)/[\frac{1}{2}(32.0 + 30.9)] = 3.64.$

$$\text{Lost Head } h_L = \left(\frac{V}{1.486R^{2/3}}\right)^2 L = \left(\frac{4.45 \times .025}{1.486(3.64)^{2/3}}\right)^2(1000) = 1.00 \text{ ft.}$$

Now apply the Bernoulli equation, A to B, datum B,

$$[106.90 + (4.67)^2/2g] - 1.00 = [106.00 + (4.23)^2/2g]$$

which reduces to $\qquad\qquad 106.24 = 106.28 \quad \text{(approximately)}$

The difference of .04 ft is within the error in roughness factor n alone. Further refinement does not seem justified. Use elevation of 106.90.

39. Develop a formula for the length-energy-slope relationship for non-uniform flow problems similar to the preceding case.

Solution:

By applying the energy equation, section 1 to section 2 in the direction of flow, using the datum below the channel bottom, we obtain

$$\text{energy at } 1 - \text{lost head} = \text{energy at } 2$$
$$(z_1 + y_1 + V_1^2/2g) - h_L = (z_2 + y_2 + V_2^2/2g)$$

The slope of the energy line S is h_L/L; then $h_L = SL$. The slope of the channel bottom S_0 is $(z_1 - z_2)/L$; then $z_1 - z_2 = S_0L$. Rearranging and substituting,

$$S_0L + (y_1 - y_2) + (V_1^2/2g - V_2^2/2g) = SL$$

This expression is usually solved for length L in open channel studies. Thus

$$L \text{ in ft} = \frac{(y_1 + V_1^2/2g) - (y_2 + V_2^2/2g)}{S - S_0} = \frac{E_1 - E_2}{S - S_0} \qquad (A)$$

The next few problems will illustrate the use of expression (A).

40. A rectangular flume ($n = .013$) is 6 ft wide and carries 66 cfs of water. At a certain section F the depth is 3.20 ft. If the slope of the channel bed is constant at .000400, determine the distance from F where the depth is 2.70 ft. (Use one reach.)

Solution:

   Assume the depth is *upstream* from F. Use subscripts 1 and 2, as usual.

   $A_1 = 6(2.70) = 16.20 \text{ ft}^2$, $V_1 = 66/16.20 = 4.07 \text{ ft/sec}$, $R_1 = 16.20/11.40 = 1.42 \text{ ft}$

   $A_2 = 6(3.20) = 19.20 \text{ ft}^2$, $V_2 = 66/19.20 = 3.44 \text{ ft/sec}$, $R_2 = 19.20/12.40 = 1.55 \text{ ft}$

Hence, $V_{mean} = 3.755$ and $R_{mean} = 1.485$.   Then for non-uniform flow,

$$L = \frac{(V_2^2/2g + y_2) - (V_1^2/2g + y_1)}{S_0 - S} = \frac{(0.184 + 3.20) - (0.257 + 2.70)}{.000400 - \left(\frac{.013 \times 3.755}{1.486(1.485)^{2/3}}\right)^2} = -1855 \text{ ft}$$

   The minus sign signifies that the section with the 2.70 ft depth is downstream from F, not upstream as assumed.

   This problem illustrates the method to be employed. A more accurate answer could be obtained by assuming intermediate depths of 3.00 ft and 2.85 ft (or exact depths by interpolating values), calculating ΔL values and adding these. In this manner a *backwater* curve may be calculated. The backwater curve is not a straight line.

41. A rectangular channel, 40 ft wide, carries 900 cfs of water. The slope of the channel is .00283. At Section 1 the depth is 4.50 ft and at Section 2, 300 ft downstream, the depth is 5.00 ft. What is the average value of roughness factor n?

Solution:

   $A_2 = 40(5.00) = 200 \text{ ft}^2$, $V_2 = 900/200 = 4.50 \text{ ft/sec}$, $R_2 = 200/50 = 4.00 \text{ ft}$

   $A_1 = 40(4.50) = 180 \text{ ft}^2$, $V_1 = 900/180 = 5.00 \text{ ft/sec}$, $R_1 = 180/49 = 3.67 \text{ ft}$

Hence, $V_{mean} = 4.75 \text{ ft/sec}$ and $R_{mean} = 3.835 \text{ ft}$.   For non-uniform flow,

$$L = \frac{(V_2^2/2g + y_2) - (V_1^2/2g + y_1)}{S_0 - \left(\frac{nV}{1.486R^{2/3}}\right)^2}, \qquad 300 = \frac{(0.315 + 5.00) - (0.388 + 4.50)}{.00283 - \left(\frac{n \times 4.75}{1.486(3.835)^{2/3}}\right)^2}$$

and $n = .0288$.

42. A rectangular channel, 20 ft wide, has a slope of 1 ft per 1000 ft. The depth at Section 1 is 8.50 ft and at Section 2, 2000 ft downstream, the depth is 10.25 ft. If $n = .011$, determine the probable flow in cfs.

Solution:

   Using the stream bed at Section 2 as datum,

$$\text{energy at 1} = y_1 + V_1^2/2g + z_1 = 8.50 + V_1^2/2g + 2.00$$
$$\text{energy at 2} = y_2 + V_2^2/2g + z_2 = 10.25 + V_2^2/2g + 0$$

   The drop in the energy line = energy at 1 − energy at 2. Since the value is unknown, a slope assumption will be made.

$$\text{Slope } S = \frac{\text{lost head}}{L} = \frac{(10.50 - 10.25) + (V_1^2/2g - V_2^2/2g)}{2000} \qquad (1)$$

   Assume $S = .000144$. Also, the values of A_{mean} and R_{mean} are necessary.

$$A_1 = 20(8.50) = 170 \text{ ft}^2, \quad R_1 = 170/37 = 4.60 \text{ ft}$$
$$A_2 = 20(10.25) = 205 \text{ ft}^2, \quad R_2 = 205/40.5 = 5.07 \text{ ft}$$

Hence, $A_{mean} = 187.5 \text{ ft}^2$ and $R_{mean} = 4.835 \text{ ft}$.

(1) First Approximation.

$$Q = A_m(1.486/n)R_m^{2/3}S^{1/2} = 187.5(1.486/.011)(4.835)^{2/3}(.000144)^{1/2} = 867 \text{ cfs}$$

Check on slope S in equation (1) above, as follows:

$$V_1 = 867/170 = 5.10, \quad V_1^2/2g = 0.404$$
$$V_2 = 867/205 = 4.23, \quad V_2^2/2g = 0.278$$

$$S = \frac{(10.50 - 10.25) + 0.126}{2000} = .000188$$

The energy gradient drops 0.376 ft in 2000 ft, which is more than was assumed.

(2) Second Approximation.

Try $S = .000210$. Then $Q = 867 \times \left(\frac{.000210}{.000144}\right)^{1/2} = 1050 \text{ cfs}$.

Again checking, $\quad V_1 = 1050/170 = 6.18 \text{ ft/sec}, \quad V_1^2/2g = 0.593 \text{ ft}$
$$V_2 = 1050/205 = 5.12 \text{ ft/sec}, \quad V_2^2/2g = 0.408 \text{ ft}$$

$$S = \frac{(10.50 - 10.25) + 0.185}{2000} = .000218$$

This slope checks (reasonably) the assumption made. Hence, approximate $Q = 1050$ cfs.

43. A reservoir feeds a rectangular channel, 15 ft wide, $n = .015$. At entrance, the depth of water in the reservoir is 6.22 ft above the channel bottom. (Refer to Fig. 10-8 below.) The flume is 800 ft long and drops 0.72 ft in this length. The depth behind a weir at the discharge end of the channel is 4.12 ft. Determine, using one reach, the capacity of the channel assuming the loss at entrance to be $0.25V_1^2/2g$.

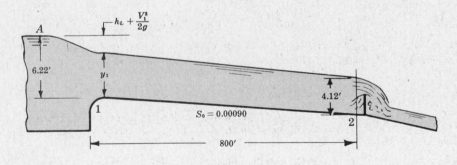

Fig. 10-8

Solution:

The Bernoulli equation, A to 1, datum 1, gives

$$(0 + \text{negl.} + 6.22) - 0.25V_1^2/2g = (0 + V_1^2/2g + y_1) \tag{1}$$

and

$$L = \frac{(V_2^2/2g + y_2) - (V_1^2/2g + y_1)}{S_o - \left(\frac{nV_m}{1.486R_m^{2/3}}\right)^2} \tag{2}$$

Solve these equations by successive trials until L approximates or equals 800 ft.

Try $y_1 = 5.0$ ft; then from (1), $V_1^2/2g = (6.22 - 5.00)/1.25 = 0.976$ ft, $V_1 = 7.93$ ft/sec and $q = y_1V_1 = 5.0(7.93) = 39.6$ cfs, $V_2 = 39.6/4.12 = 9.62$ ft/sec.

$$V_{\text{mean}} = \tfrac{1}{2}(7.93 + 9.62) = 8.78 \text{ ft/sec}$$

and $\qquad R_{\text{mean}} = \tfrac{1}{2}(R_1 + R_2) = \tfrac{1}{2}[(15 \times 5)/25 + (15 \times 4.12)/23.24] = 2.83$ ft

Substituting in equation (2) above, we find $L = 392$ ft.

Increase the value of y_1 to 5.50 ft and repeat the calculation. Results in tabular form are:

y_1	V_1	q_1	V_2	V_m	R_m	L	Notes
5.50	6.09	33.5	8.13	7.11	2.92	2800 ft	$\therefore$ decrease y_1
5.20	7.25	37.8	9.17	8.21	2.87	760 ft	close enough

The capacity of the channel $= 37.8 \times 15 = 567$ cfs.

Should greater accuracy be required, start at the lower end and, for unit flow $q = 37.8$ cfs, find the length of the reach to a point where the depth is about 10% more than 4.12 or about 4.50 ft, thence to a depth of 4.90 ft, and so on. Should the sum of the lengths exceed 800 ft, decrease the value of y_1, obtaining a resulting increase in q.

44. Derive the expression which gives the slope of the liquid surface in wide rectangular channels for gradually varied flow.

Solution:

The total energy per pound of fluid with respect to an arbitrary datum plane is

$$H = y + V^2/2g + z$$

where the kinetic-energy correction factor α is taken as unity. Differentiating this expression with respect to L, the distance along the channel, yields

$$\frac{dH}{dL} = \frac{dy}{dL} + \frac{d(V^2/2g)}{dL} + \frac{dz}{dL} \qquad (A)$$

For rectangular channels (or for wide channels of mean depth y_m), $V^2 = (q/y)^2$ and

$$\frac{d(q^2/2gy^2)}{dL} = -\frac{2q^2}{2gy^3}\left(\frac{dy}{dL}\right) = -\frac{V^2}{gy}\left(\frac{dy}{dL}\right)$$

Substituting in (A), using $dH/dL = -S$ or the slope of the energy line, and $dz/dL = -S_o$ or the slope of the channel bottom, we obtain

$$-S = \frac{dy}{dL} - \frac{V^2}{gy}\left(\frac{dy}{dL}\right) - S_o \qquad \text{or} \qquad \frac{dy}{dL} = \frac{S_o - S}{(1 - V^2/gy)} = \frac{S_o - S}{1 - N_F^2} \qquad (B)$$

The term dy/dL represents the slope of the water surface relative to the channel bottom. When the channel slopes down in the direction of flow, S_o is positive. Similarly, S is positive (always). For uniform flow, $S = S_o$ and $dy/dL = 0$.

Another form of equation (B) can be obtained as follows. Manning's formula is

$$Q = (1.486/n)AR^{2/3}S^{1/2}$$

Solving this equation for the slope of the energy line, using $q = Q/b$, $A = by$ and $R = y$ for wide rectangular channels, yields

$$\frac{dH}{dL} = S = \frac{n^2(q^2 b^2/b^2 y^2)}{(1.486)^2 y^{4/3}}$$

Similarly, the slope of the channel bottom, in terms of normal depth y_N and coefficient n_N can be written

$$\frac{dz}{dL} = S_o = \frac{n_N^2(q^2 b^2/b^2 y_N^2)}{(1.486)^2 y_N^{4/3}}$$

Then the first part of equation (B) becomes

$$-\frac{n^2(q^2 b^2/b^2 y^2)}{(1.486)^2 y^{4/3}} = (1 - V^2/gy)\frac{dy}{dL} - \frac{n_N^2(q^2 b^2/b^2 y_N^2)}{(1.486)^2 y_N^{4/3}}$$

But $V^2 = q^2/y^2$, $n \cong n_N$ and $q^2/g = y_c^3$. Then

$$\frac{-n^2 q^2}{(1.486)^2 \, y^{10/3}} \;=\; \frac{dy}{dL}(1 - y_c^3/y^3) \;-\; \frac{n^2 q^2}{(1.486)^2 \, y_N^{10/3}} \tag{C}$$

$$\frac{dy}{dL} \;=\; \frac{(nq/1.486)^2 \, [1/y_N^{10/3} - 1/y^{10/3}]}{1 - (y_c/y)^3} \tag{D}$$

Using $Q/b = q = y_N[(1.486/n)y_N^{2/3}S_o^{1/2}]$ or $(nq/1.486)^2 = y_N^{10/3}S_o$,   equation (D) becomes

$$\frac{dy}{dL} \;=\; S_o\left[\frac{1 - (y_N/y)^{10/3}}{1 - (y_c/y)^3}\right] \tag{E}$$

There are limiting conditions to the surface profiles. For example, as y approaches y_c, the denominator of (E) approaches zero. Thus dy/dL becomes infinite and the curves cross the critical depth line perpendicular to it. Hence surface profiles in the vicinity of $y = y_c$ are only approximate.

Similarly, when y approaches y_N, the numerator approaches zero. Thus the curves approach the normal depth y_N asymptotically.

Finally, as y approaches zero, the surface profile approaches the channel bed perpendicularly which is impossible under the assumption concerning gradually varied flow.

45. Summarize the system for classifying surface profiles for gradually-varied flow in wide channels.

Solution:

There are a number of different conditions in a channel that give rise to some twelve different types of non-uniform (varied) flow. In expression (E) in Problem 44, for positive values of dy/dL, the depth y increases downstream along the channel, and for negative values of dy/dL the depth y will decrease downstream along the channel.

In the table below a summary of the twelve different types of varied flow is presented. Several of these will be discussed and the reader may analyze the remaining types of flow in a similar fashion.

The "mild" classification results from the channel slope S_o being such that normal depth $y_N > y_c$. If the depth y is greater than y_N and y_c, the curve is called "type 1"; if the depth y is between y_N and y_c, type 2; and if depth y is smaller than y_N and y_c, type 3.

It will be noticed that, for the type 1 curves, since the velocity is decreasing due to the increased depth, the water surface must approach a horizontal asymptote (see M_1, C_1 and S_1). Similarly, curves which approach the normal depth line do so asymptotically. As pointed out previously, curves which approach the critical depth line cross it vertically, inasmuch as the denominator of expression (E) in Problem 44 becomes zero for such cases. Therefore, curves for critical slopes are exceptions to the preceding statements, since it is impossible to have a water surface curve both tangent and perpendicular to the critical depth line.

On each profile in the following table the vertical scale is greatly magnified with respect to the horizontal scale. As indicated in the numerical problems for M_1 curves, such profiles may be thousands of feet in extent.

The tabulation below gives the relations between slopes and depths, the sign of dy/dL, the type of profile, the symbol for the profile, the type of flow, and a sketch representing the form of the profile. The values of y within each profile can be observed to be greater than or less than y_N and/or y_c by studying each sketch.

Channel Slope	Depth Relations	$\left(\dfrac{dy}{dL}\right)$	Type of Profile	Symbol	Type of Flow	Form of Profile
Mild $0 < S < S_c$	$y > y_N > y_c$	$+$	Backwater	M_1	Subcritical	
	$y_N > y > y_c$	$-$	Dropdown	M_2	Subcritical	
	$y_N > y_c > y$	$+$	Backwater	M_3	Supercritical	
Horizontal $S = 0$ $y_N = \infty$	$y > y_c$	$-$	Dropdown	H_2	Subcritical	
	$y_c > y$	$+$	Backwater	H_3	Supercritical	
Critical $S_N = S_c$ $y_N = y_c$	$y > y_c = y_N$	$+$	Backwater	C_1	Subcritical	
	$y_c = y = y_N$		Parallel to bed	C_2	Uniform, Critical	
	$y_c = y_N > y$	$+$	Backwater	C_3	Supercritical	
Steep $S > S_c > 0$	$y > y_c > y_N$	$+$	Backwater	S_1	Subcritical	
	$y_c > y > y_N$	$-$	Dropdown	S_2	Supercritical	
	$y_c > y_N > y$	$+$	Backwater	S_3	Supercritical	
Adverse $S < 0$ $y_N = \infty$	$y > y_c$	$-$	Dropdown	A_2	Subcritical	
	$y_c > y$	$+$	Backwater	A_3	Supercritical	

46. For a rectangular channel, develop an expression for the relation between the depths before and after a hydraulic jump. Refer to the figure below.

   Solution:

For the free body between Sections 1 and 2, considering a unit width of channel and unit flow q,

$$P_1 = w\bar{h}A = w(\tfrac{1}{2}y_1)y_1 = \tfrac{1}{2}wy_1^2 \qquad \text{and similarly} \qquad P_2 = \tfrac{1}{2}wy_2^2$$

From the principle of impulse and momentum,

$$\Delta P_x\, dt = \Delta \text{ linear momentum } = \frac{W}{g}(\Delta V_x)$$

$$\tfrac{1}{2}w(y_2^2 - y_1^2)dt = \frac{wq\, dt}{g}(V_1 - V_2)$$

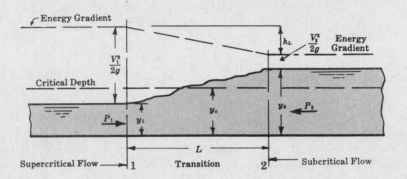

Fig. 10-9

Since $V_2 y_2 = V_1 y_1$ and $V_1 = q/y_1$, the above equation becomes

$$q^2/g = \tfrac{1}{2}y_1 y_2(y_1 + y_2) \tag{1}$$

Since $q^2/g = y_c^3$,

$$y_c^3 = \tfrac{1}{2}y_1 y_2(y_1 + y_2) \tag{2}$$

The length of the jump has been found to vary between $4.3y_2$ and $5.2y_2$.

For the relation between L/y_2 and Froude number $V_1/\sqrt{gy_1}$, see Page 73 of "Engineering Hydraulics", Hunter Rouse, John Wiley & Sons, 1950.

The hydraulic jump is an energy dissipator. In designing hydraulic jump stilling basins, knowledge of jump length and depth y_2 is important. Good energy dissipation occurs when $V_1^2/gy_1 = 20$ to 80.

47. A rectangular channel, 20 ft wide, carries 400 cfs and discharges onto a 20 ft wide apron with no slope with a mean velocity of 20 ft/sec. What is the height of the hydraulic jump? What energy is absorbed (lost) in the jump?

   Solution:

(a) $V_1 = 20$ ft/sec, $q = 400/20 = 20$ cfs/ft width, and $y_1 = q/V_1 = 20/20 = 1.0$ ft. Then

$$q^2/g = \tfrac{1}{2}y_1 y_2(y_1 + y_2), \qquad (20)^2/32.2 = \tfrac{1}{2}y_2(1 + y_2), \qquad 12.4 = 0.50y_2 + 0.50y_2^2$$

from which $y_2 = -5.50$ ft, $+4.50$ ft. The negative root being extraneous, $y_2 = 4.5$ ft and the height of the hydraulic jump is $(4.5 - 1.0) = 3.5$ ft.

Note that $y_c = \sqrt[3]{(20)^2/32.2}$ or $\sqrt[3]{\tfrac{1}{2}y_1 y_2(y_1 + y_2)} = 2.32$ ft.

Hence the flow at 1 ft depth is supercritical and the flow at 4.5 ft depth is subcritical.

(b) Before jump, $E_1 = V_1^2/2g + y_1 = (20)^2/2g + 1.00 = 7.21$ ft lb/lb.

$\quad\quad\quad$ After jump, $E_2 = V_2^2/2g + y_2 = [400/(20 \times 4.50)]^2/2g + 4.50 = 4.81$ ft lb/lb.

$\quad\quad\quad$ Lost energy per second $= wQH = 62.4(400)(7.21 - 4.81) = 59,900$ ft lb/sec.

48. A rectangular channel, 16 ft wide, carries a flow of 192 cfs. The depth of water on the downstream side of the hydraulic jump is 4.20 ft. (a) What is the depth upstream? (b) What is the loss of head?

Solution:

(a) $\quad\quad q^2/g = \frac{1}{2}y_1y_2(y_1 + y_2), \quad\quad\quad (192/16)^2/32.2 = 2.10y_1(y_1 + 4.20), \quad\quad y_1 = 0.455$ ft

(b) $\quad\quad\quad\quad\quad\quad A_1 = 16(0.455) = 7.29$ ft^2, $\quad V_1 = 192/7.29 = 26.3$ ft/sec

$\quad\quad\quad\quad\quad\quad\quad A_2 = 16(4.20) \quad = 67.2$ ft^2, $\quad V_2 = 192/67.2 = 2.86$ ft/sec

$\quad\quad\quad\quad\quad\quad\quad E_1 = V_1^2/2g + y_1 = (26.3)^2/2g + 0.455 = 11.18$ ft lb/lb

$\quad\quad\quad\quad\quad\quad\quad E_2 = V_2^2/2g + y_2 = (2.86)^2/2g + 4.20 \quad = \quad 4.33$ ft lb/lb

$\quad\quad$ Energy lost $= 11.18 - 4.33 = 6.85$ ft lb/lb or ft.

49. After flowing over the concrete spillway of a dam, 9000 cfs then passes over a level concrete apron ($n = .013$). The velocity of the water at the bottom of the spillway is 42.0 ft/sec and the width of the apron is 180 ft. Conditions will produce a hydraulic jump, the depth in the channel below the apron being 10.0 ft. In order that the jump be contained on the apron, (a) how long should the apron be built? (b) How much energy is lost from the foot of the spillway to the downstream side of the jump?

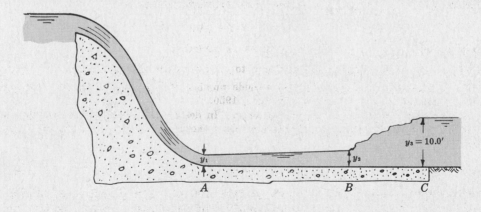

Fig. 10-10

Solution:

(a) Referring to Fig. 10-10, first calculate the depth y_2 at the upstream end of the jump.

$\quad\quad\quad\quad q^2/g = \frac{1}{2}y_2y_3(y_2 + y_3), \quad\quad (9000/180)^2/32.2 = \frac{1}{2}(10y_2)(y_2 + 10) \quad\quad y_2 = 1.37$ ft

Also, $\quad\quad\quad\quad\quad\quad\quad\quad\quad y_1 = q/V_1 = (9000/180)/42.0 = 1.19$ ft

$\quad\quad\quad$ Now calculate length L_{AB} of the retarded flow.

$V_1 = 42.0$ ft/sec, $\quad\quad V_1^2/2g = 27.4$ ft, $\quad\quad R_1 = (180 \times 1.19)/182.38 = 1.174$ ft

$V_2 = q/y_2 = 50.0/1.37 = 36.5$ ft/sec, $\quad\quad V_2^2/2g = 20.7$ ft, $\quad\quad R_2 = (180 \times 1.37)/182.74 = 1.350$ ft

Then $V_{\text{mean}} = 39.25$ ft/sec, $R_{\text{mean}} = 1.262$ ft, and

$$L_{AB} = \frac{(V_2^2/2g + y_2) - (V_1^2/2g + y_1)}{S_o - S} = \frac{(20.7 + 1.37) - (27.4 + 1.19)}{0 - \left(\frac{.013 \times 39.25}{1.486(1.262)^{2/3}}\right)^2} = 75.5 \text{ ft}$$

The length of the jump L_J from B to C is from $4.3y_3$ to $5.2y_3$ ft. Assuming the conservative value of $5.0y_3$,

$$L_J = 5.0 \times 10.0 = 50.0 \text{ ft}$$

Hence, total length $ABC = 76 + 50 = 126$ ft (approximately).

(b) Energy at $A = y_1 + V_1^2/2g = 1.19 + 27.4 = 28.59$ ft lb/lb.
Energy at $C = y_3 + V_3^2/2g = 10.0 + (5.0)^2/2g = 10.39$ ft lb/lb.
Total energy lost $= wQH = 62.4(9000)(28.59 - 10.39) = 10.2 \times 10^6$ ft lb/sec.

50. In order that the hydraulic jump below a spillway be not swept downstream, establish the relationship between the variables indicated in Fig. 10-11. (Prof. E. A. Elevatorski suggested the dimensionless parameters which follow. See "Civil Engineering", August 1958.)

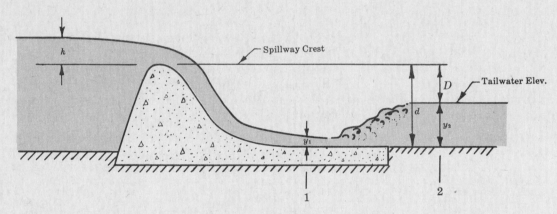

Fig. 10-11

Solution:

The energy equation is applied between a section upstream of the dam where h is measured and section 1, neglecting the velocity head of approach, i.e.,

$$(h + d) + 0 + \text{negl.} - \text{losses (neglected)} = 0 + 0 + V_1^2/2g$$

or $V_1 = \sqrt{2g(h + d)}$.

Since $q = y_1 V_1$ (cfs/ft of width), $y_1 = \dfrac{q}{V_1} = \dfrac{q}{\sqrt{2g(d + h)}}$

or
$$y_1 = \frac{q}{\sqrt{2g}\,(d/h + 1)^{1/2}\,h^{1/2}} \qquad (A)$$

From Problem 46, the hydraulic jump relationship is

$$\frac{y_2^2 - y_1^2}{2} = \frac{qV_1}{g}\left(\frac{y_2 - y_1}{y_2}\right) \qquad \text{or} \qquad gy_2^2 + gy_1y_2 = 2qV_1$$

Solving,
$$y_2 = \frac{-y_1 \pm \sqrt{y_1^2 + 8qV_1/g}}{2}$$

Dividing by y_1 produces a dimensionless form

$$\frac{y_2}{y_1} = -\tfrac{1}{2} \pm \tfrac{1}{2}\sqrt{1 + 8qV_1/y_1^2 g} = \tfrac{1}{2}[\sqrt{1 + 8q^2/gy_1^3} - 1] \qquad (B)$$

Since $y_2 = (d - D)$, $y_2/y_1 = (d - D)/y_1$ is substituted in (B) together with the value of y_1 from (A) to yield

$$\frac{d - D}{y_1} = \tfrac{1}{2}[\sqrt{1 + 8q^2/gy_1^3} - 1]$$

$$\frac{2(d - D)\sqrt{2g}\,(d/h + 1)^{1/2} h^{1/2}}{q} + 1 = \sqrt{1 + \frac{8(2^{3/2})(g^{3/2})(d/h + 1)^{3/2} h^{3/2}}{qg}}$$

The equation is put in dimensionless form by multiplying the left-hand side by h/h, dividing through by $\sqrt{8}$, and collecting terms:

$$\left(\frac{h^{3/2} g^{1/2}}{q}\right)\left(\frac{d - D}{h}\right)\left(\frac{d}{h} + 1\right)^{1/2} + 0.353 = \sqrt{\tfrac{1}{8} + 2.828\left(\frac{g^{1/2} h^{3/2}}{q}\right)\left(\frac{d}{h} + 1\right)^{3/2}} \qquad (C)$$

The dimensionless terms in (C) may be written

$$\pi_1 = \frac{h^{3/2} g^{1/2}}{q}, \qquad \pi_2 = \frac{D}{h}, \qquad \pi_3 = \frac{d}{h}$$

Equation (C) then becomes

$$\pi_1(\pi_3 - \pi_2)(\pi_3 + 1)^{1/2} + 0.353 = \sqrt{\tfrac{1}{8} + 2.828\pi_1(\pi_3 + 1)^{3/2}} \qquad (D)$$

Prof. Elevatorski has prepared a graph of Equation (D) to allow for ready solution. For calculated values of π_1 and π_2, the graph gives the value of π_3. (See "Civil Engineering", August 1958.)

Prof. Elevatorski, in commenting on the omission of the energy loss on the spillway face, states "by neglecting the loss due to friction, a slight excess of tailwater will be produced in the stilling basin. A slightly-drowned jump produces a better all-around energy dissipator as compared to one designed for the y_2 depth".

51. Determine the elevation of the spillway apron if $q = 50$ cfs/ft, $h = 9$ ft, $D = 63$ ft, and the spillway crest is at Elevation 200.0.

Solution:

Employing the dimensionless ratios derived in the preceding problem,

$$\pi_1 = g^{1/2} h^{3/2}/q = 5.67(9^{3/2})/50 = 3.06, \qquad \pi_2 = D/h = 63/9 = 7.00, \qquad \pi_3 = d/h = d/9$$

Equation (D) of Problem 50 may then be written

$$3.06(d/9 - 7.00)(d/9 + 1)^{1/2} + 0.353 = \sqrt{0.125 + 2.828(3.06)(d/9 + 1)^{3/2}}$$

Solving by successive trials for $\pi_3 = d/9$, we find $\pi_3 = 8.65$, or $d = 77.85$ ft. The elevation of the spillway apron is $(200 - 77.9) = 122.1$ ft above datum.

52. Establish the equation for flow over a broad-crested weir assuming no lost head.

Solution:

At the section where critical flow occurs, $q = V_c y_c$. But $y_c = V_c^2/g = \tfrac{2}{3}E$, and $V_c = \sqrt{g(\tfrac{2}{3}E_c)}$. Hence the theoretical value of flow q becomes

$$q = \sqrt{g(\tfrac{2}{3}E_c)} \times \tfrac{2}{3}E_c = 3.09 E_c^{3/2}$$

However, the value of E_c is difficult to measure accurately, because the critical depth is difficult to locate. The practical equation becomes

$$q = CH^{3/2} \cong 3H^{3/2}$$

The weir should be calibrated in place to obtain accurate results.

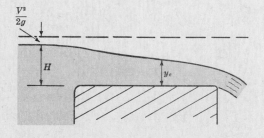

Fig. 10-12

53. Develop an expression for a critical flow meter and illustrate the use of the formula.

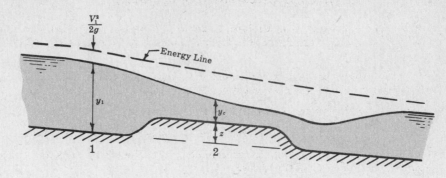

Fig. 10-13

Solution:

An excellent method of measuring flow in open channels is by means of a constriction. The measurement of the critical depth is not required. The depth y_1 is measured a short distance upstream from the constriction. The raised floor should be about $3y_c$ long and of such height as to have the critical velocity occur on it.

For a rectangular channel of constant width, the Bernoulli equation is applied between sections 1 and 2, in which the lost head in accelerated flow is taken as one-tenth of the difference in velocity heads, i.e.,

$$y_1 + \frac{V_1^2}{2g} - \frac{1}{10}\left(\frac{V_c^2}{2g} - \frac{V_1^2}{2g}\right) = \left(y_c + \frac{V_c^2}{2g} + z\right)$$

which equation neglects the slight drop in the channel bed between 1 and 2. Recognizing that $E_c = y_c + V_c^2/2g$, we rearrange as follows:

$$(y_1 + 1.10V_1^2/2g) = [z + 1.0E_c + \tfrac{1}{10}(\tfrac{1}{3}E_c)]$$
$$(y_1 - z + 1.10V_1^2/2g) = 1.033E_c = 1.033(\tfrac{3}{2}\sqrt[3]{q^2/g})$$

or

$$q = 2.94(y_1 - z + 1.10V_1^2/2g)^{3/2} \qquad (A)$$

Since $q = V_1 y_1$,

$$q = 2.94(y_1 - z + .0171q^2/y_1^2)^{3/2} \qquad (B)$$

To illustrate the use of expression (B), consider a rectangular channel 10 ft wide with a critical depth meter having dimension $z = 1.10$ ft. If the measured depth y_1 is 2.42 ft, what is the discharge Q?

As a first approximation, neglect the last term in (B). Then

$$q = 2.94(2.42 - 1.10)^{3/2} = 4.41 \text{ cfs/ft width}$$

Now, using the entire equation (B), by successive trials we find $q = 4.73$. Hence

$$Q = q(10) = 4.73(10) = 47.3 \text{ cfs}$$

Supplementary Problems

54. Using y_N as the depth in the figure in Problem 1, derive an expression for laminar flow along a flat plate of infinite width, considering the free body in Problem 1 to be one unit wide.
    Ans. $y_N^2 = 3\nu V/gS$

55. The Darcy friction factor f is usually associated with pipes. However, for the preceding problem, evaluate Darcy's factor f using the answer given for that problem.   *Ans.* $96/R_E$

56. Show that mean velocity V can be expressed as $0.263 v_* R^{1/6}/n$.

57. Show that Manning's n and Darcy's f are related to each other by $n = .093 f^{1/2} R^{1/6}$.

58. Calculate the mean velocity in the rectangular channel of Problem 7 by summing up the area under the depth-velocity curve.   *Ans.* 6.95 ft/sec

59. Upon what slope should the flume shown in Fig. 10-14 below be laid in order to carry 522.5 cfs? ($C = 100$)   *Ans.* .00373

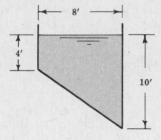

Fig. 10-14

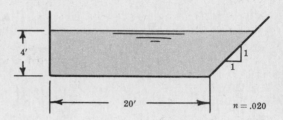

Fig. 10-15

60. The canal shown in Fig. 10-15 above is laid on a slope of .00016. When it reaches a railroad embankment, the flow is to be carried by 2 concrete pipes ($n = .012$) laid on a slope of 2.5 ft in 1000 ft. What size pipes should be used?   *Ans.* 4.16 ft

61. A flow of 78.5 cfs is carried in a flume which is a half-square. The flume is 4000 ft long and it drops 2 ft in that length. Using Manning's formula and $n = .012$, determine the dimensions.
    Ans. 6.4 ft × 3.2 ft

62. Water flows at a depth of 6.25 ft in a rectangular canal 8 ft wide. The average velocity is 1.90 ft/sec. What is the probable slope on which the canal is laid if $C = 100$?   *Ans.* .000148

63. A canal cut in rock ($n = .030$) is trapezoidal in section with a bottom width of 20 ft and side slopes of 1 on 1. The allowable average velocity is 2.50 ft/sec. What slope will produce 200 cfs?
    Ans. .000675

64. What is the flow of water in a new 24″ vitrified sewer pipe flowing half-full on a slope of .0025?
    Ans. 5.66 cfs

65. A canal ($n = .017$) has a slope of .00040 and is 10,000 ft long. Assuming the hydraulic radius is 4.80 ft, what correction in grade must be made to produce the same flow if the roughness factor changes to .020?   *Ans.* new $S = .000552$

66. How deep will water flow in a 90° V-shaped flume, $n = .013$, laid on a grade of .00040 if it is to carry 90.0 cfs?   *Ans.* $y = 5.15$ ft

67. A given amount of lumber is to be used to build a V-notch triangular flume. What vertex angle should be used for maximum flow on a given slope?   *Ans.* 90°

68. Water flows 3 ft deep in a rectangular canal 20 ft wide, $n = .013$, $S = .0144$. How deep would the same quantity flow on a slope of .00144?   *Ans.* 6.6 ft

69. A flume discharges 42.0 cfs on a slope of 0.50 ft in 1000 ft. The section is rectangular and roughness factor $n = .012$. Determine the best dimensions, i.e., the dimensions for minimum wetted perimeter.
    Ans. 2.54 ft × 5.08 ft

70. A lined, rectangular canal, 16 ft wide, carries a flow of 408 cfs at a depth of 2.83 ft. Find n if the slope of the canal is 3.22 ft in 1600 ft. (Use Manning's formula.)   *Ans.* $n = .0121$

71. Find the average shear stress over the wetted perimeter in Problem 70.   *Ans.* 0.263 lb/ft²

72. Using the Manning formula, show that the theoretical depth for maximum velocity in a circular conduit is 0.81 of the diameter.

73. Design the most efficient trapezoidal channel to carry 600 cfs at a maximum velocity of 3.00 ft/sec. Use $n = .025$ and side slopes of 1 vertical on 2 horizontal.   *Ans.* $b = 4.22$ ft, $y = 9$ ft

74. Calculate the slope of the channel in the preceding problem.   *Ans.* .000345

75. Which canal structure, shown in Fig. 10-16 below, will carry the greater flow, if both are laid on the same slope?   *Ans.* (b) Trapezoidal section

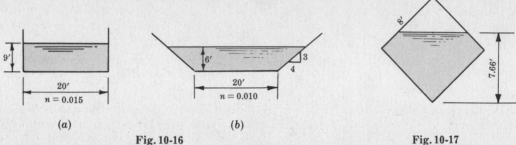

(a)   (b)
Fig. 10-16   **Fig. 10-17**

76. A square box culvert is 8 ft on a side and is installed as shown in Fig. 10-17 above. What is the hydraulic radius for a depth of 7.66 ft?   *Ans.* 2.33 ft

77. What is the radius of the semi-circular flume B, shown in Fig. 10-18 below, if its slope $S = .0200$ and $C = 90$?   *Ans.* $r = 1.80$ ft

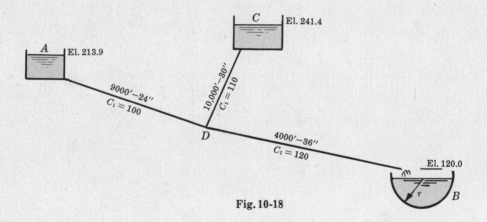

Fig. 10-18

78. Calculate the specific energy when a flow of 220 cfs is produced in a rectangular channel 10 ft wide at a depth of 3.0 ft.   *Ans.* 3.84 ft

79. Calculate the specific energy when 310 cfs flows in a trapezoidal channel, base width 8 ft with side slopes 1 on 1, at a depth of 3.90 ft.   *Ans.* 4.60 ft

80. A sewer pipe 6 ft in diameter carries 80.6 cfs when flowing 4 ft deep. What is the specific energy?   *Ans.* 4.25 ft

81. At what depths may the water flow at 220 cfs in Problem 78 for a specific energy of 5.00 ft lb/lb? What is the critical depth?   *Ans.* 1.46 ft and 4.65 ft, 2.47 ft

82. In a rectangular channel 10 ft wide the flow is 265 cfs. Is the flow subcritical or supercritical at depths of 2.00 ft, 3.00 ft and 4.00 ft?   *Ans.* supercritical, subcritical, subcritical

83. In a rectangular channel 10 ft wide the flow is 265 cfs when the velocity is 8.00 ft/sec. State the nature of the flow.   *Ans.* subcritical

84. For a critical depth of 3.22 ft in a rectangular channel 10 ft wide, compute the discharge.
Ans. 328 cfs

85. Determine the critical slope of a rectangular channel 20 ft wide, $n = .012$, when the flow is 990 cfs.
Ans. .00207

86. A trapezoidal channel with side slopes of 1 on 1 carries a flow of 720 cfs. For a bottom width of 16 ft, calculate the critical velocity. *Ans.* 9.98 ft/sec

87. A rectangular canal, 6000 ft long, 60 ft wide and 10 ft deep, carries 1800 cfs ($C = 75$). The cleaning of the canal raises C to 100. If the depth at the upper end remains 10 ft deep, find the depth at the lower end for the same flow (using one reach). *Ans.* $y_2 = 10.64$ ft

88. A rectangular channel, $n = .016$, is laid on a slope of .0064 and carries 600 cfs. For critical flow conditions, what width is required? *Ans.* 8.5 ft

89. A rectangular channel, $n = .012$, 10 ft wide laid on a slope of .0049, carries 160 cfs. The channel is to be contracted to produce critical flow. What width of contracted section will accomplish this, neglecting any loss in the gradual reduction in width? *Ans.* 4.5 ft

90. In a rectangular channel 12 ft wide, $C = 100$, $S = .0225$, the flow is 500 cfs. The slope of the channel changes to .00250. How far below the point of change in slope is the depth 2.75 ft, using one reach? *Ans.* 104 ft

91. Using the data in Problem 90, (*a*) calculate the critical depth in the flatter channel, (*b*) calculate the depth required for uniform flow in the flatter channel, (*c*) calculate the depth just before the hydraulic jump occurs, using equation from Problem 46. (Note that this depth occurs 104 ft from change in grade, from Problem 90.) *Ans.* 3.78 ft, 5.05 ft, 2.75 ft

92. A broad-crested weir is 1.25 ft high above the bottom of a rectangular channel, 10 ft wide. The measured head above the weir crest is 1.95 ft. Determine the approximate flow in the channel. (Use $c = 0.92$.) *Ans.* 83.5 cfs

93. Show that the critical depth in a triangular channel is $2V_c^2/g$.

94. Show that the critical depth in a triangular channel may be expressed as 4/5 of the minimum specific energy.

95. Show that the critical depth in a parabolic channel is 3/4 of the minimum specific energy if the dimensions of the channel are y_c deep and b' wide at the water surface.

96. For a rectangular channel, show that the discharge q per foot of width equals $3.087E_{min}^{3/2}$.

97. For a triangular channel, show that discharge $Q = 1.148(b'/y_c)E_{min}^{5/2}$.

98. For a parabolic channel, show that discharge $Q = 2.005b'E_{min}^{3/2}$.

Chapter 11

Forces Developed by Moving Fluids

INTRODUCTION

Knowledge of the forces exerted by moving fluids is of significant importance in the analysis and design of such objects as pumps, turbines, airplanes, rockets, propellers, ships, automobile bodies, buildings and many hydraulic devices. The energy relationship is not sufficient to solve most of these problems. One additional tool of mechanics, the momentum principle, is most important. The boundary layer theory provides a further basis for analysis. Extensive and continuing experiments add data regarding laws of variation of fundamental coefficients.

IMPULSE-MOMENTUM PRINCIPLE, from kinetic mechanics, states that

$$\text{Linear impulse} = \text{change in linear momentum}$$

or

$$(\Sigma F)t = M(\Delta V)$$

The quantities in the equation are vector quantities and must be added and subtracted accordingly. Components are usually most convenient, and, to avoid possible mistakes in sign, the following forms are suggested:

(a) In the X-direction,

$$\text{initial linear momentum} \pm \text{linear impulse} = \text{final linear momentum}$$

$$MV_{x_1} \pm \Sigma F_x \cdot t = MV_{x_2} \tag{1}$$

(b) In the Y-direction,

$$MV_{y_1} \pm \Sigma F_y \cdot t = MV_{y_2} \tag{2}$$

where M = mass having its momentum changed in time t.

These expressions may be written, with appropriate subscripts x, y or z, in the following form:

$$\Sigma F_x = \rho Q(V_2 - V_1)_x, \quad \text{etc.} \tag{3}$$

MOMENTUM CORRECTION FACTOR β, evaluated in Problem 1, is

$$\beta = \frac{1}{A} \int_A (v/V)^2 \, dA \tag{4}$$

For laminar flow in pipes, $\beta = 1.33$. For turbulent flow in pipes, β varies from 1.01 to 1.07. In most cases β can be considered as unity.

DRAG

Drag is the component of the resultant force exerted by a fluid on a body *parallel* to the relative motion of the fluid. The usual equation is

$$\text{Drag in lb} = C_D \rho A \frac{V^2}{2} \tag{5}$$

LIFT

Lift is the component of the resultant force exerted by a fluid on a body *perpendicular* to the relative motion of the fluid. The usual equation is

$$\text{Lift in lb} = C_L \rho A \frac{V^2}{2} \tag{6}$$

where C_D = the drag coefficient, which is dimensionless

C_L = the lift coefficient, which is dimensionless

ρ = the density of the fluid, in slugs/ft^3

A = some characteristic area in ft^2, usually the area projected on a plane perpendicular to the relative motion of the fluid

V = relative velocity of the fluid with respect to the body, in ft/sec.

TOTAL DRAG FORCE

Total drag force consists of the friction drag and the pressure drag. However, seldom are both these effects of appreciable magnitude simultaneously. For objects which exhibit no lift, profile drag is synonymous with total drag. The following tabulation will illustrate.

Object	Friction drag		Pressure drag		Total drag
1. Spheres.	negligible	+	Pressure drag	=	total drag
2. Cylinders (axis perpendicular to velocity).	negligible	+	Pressure drag	=	total drag
3. Disks and thin plates (perpendicular to velocity).	zero	+	Pressure drag	=	total drag
4. Thin plates (parallel to velocity).	Friction drag	+	negligible to zero	=	total drag
5. Well-streamlined objects.	Friction drag	+	small to negligible	=	total drag

DRAG COEFFICIENTS

Drag coefficients are dependent upon Reynolds Number at low and intermediate velocities, but are independent at high velocity. However, at high velocities the drag coefficient is related to the Mach Number which has negligible effect at low velocities. Diagrams *F*, *G* and *H* illustrate the variations for certain geometric shapes. Problems 24 and 40 discuss these relations.

For flat plates and airfoils, the drag coefficients are usually tabulated for the plate area and for the chord-length product respectively.

LIFT COEFFICIENTS

Kutta gives theoretical maximum values of lift coefficients for thin flat plates not normal to the relative velocity of the fluid as

$$C_L = 2\pi \sin \alpha \tag{7}$$

where α = the angle of attack or the angle the plate makes with the relative velocity of the fluid. In normal range of operation, present-day airfoil sections have values about 90% of this theoretical maximum. Angle α should not exceed about 25°.

MACH NUMBER

Mach number is the dimensionless ratio of the velocity of the fluid to the acoustic velocity (sometimes called celerity).

$$\text{Mach Number} = N_M = \frac{V}{c} = \frac{V}{\sqrt{E/\rho}} \tag{8}$$

For gases, $c = \sqrt{kgRT}$ (see Chapter 1).

Values of V/c up to the critical value of 1.0 indicate subsonic flow; at 1.0, sonic flow; and values above 1.0 indicate supersonic flow (see Diagram H).

BOUNDARY LAYER THEORY

Boundary layer theory was first developed by Prandtl. He showed that, for a moving fluid all friction losses occur within a thin layer adjacent to a solid boundary (called the boundary layer), and that flow outside this layer may be considered frictionless. The velocity near the boundary is affected by boundary shear. In general, the boundary layer is very thin at the upstream boundaries of an immersed object but it increases in thickness due to the continual action of shear stress.

At low Reynolds numbers, the entire boundary layer is governed by viscous forces and laminar flow occurs therein. For intermediate values of Reynolds number the boundary layer is laminar near the boundary surface and turbulent beyond. For high Reynolds numbers the entire boundary layer is turbulent.

FLAT PLATES

For flat plates of length L feet, held parallel to the relative motion of a fluid, the following equations are applicable.

1. Boundary Layer Laminar (up to Reynolds number about 500,000).

(a) Mean drag coefficient $(C_D) = \dfrac{1.328}{\sqrt{R_E}} = \dfrac{1.328}{\sqrt{VL/\nu}}$ (9)

(b) Boundary layer thickness δ (in ft) at any distance x is given by

$$\frac{\delta}{x} = \frac{5.20}{\sqrt{R_{E_x}}} = \frac{5.20}{\sqrt{Vx/\nu}} \tag{10}$$

(c) Shear stress τ_0 in lb/ft² is estimated by

$$\tau_0 = 0.33\,\rho V^{3/2}\sqrt{\nu/x} = 0.33\,(\mu V/x)\sqrt{R_{E_x}} = \frac{0.33\,\rho V^2}{\sqrt{R_{E_x}}} \tag{11}$$

where V = velocity of the fluid approaching the boundary (ambient velocity)
  x = distance from the leading edge in ft
  L = total length of plate in ft
  R_{E_x} = local Reynolds number, for distance x.

It can be seen that the thickness of the boundary layer will increase as the square root of dimension x increases and also as the square root of the kinematic viscosity increases, while δ will decrease as the square root of the velocity increases. Similarly, the boundary shear τ_0 will increase as the square root of ρ and μ increases, will decrease as the square root of x increases, and will increase as the three-halves power of V increases.

2. **Boundary Layer Turbulent** (smooth boundary)

(a) Mean drag coefficient $(C_D) = \dfrac{0.074}{R_E^{0.20}}$ for $2 \times 10^5 < R_E < 10^7$ (12)

$$= \frac{0.455}{(\log_{10} R_E)^{2.58}} \text{ for } 10^6 < R_E < 10^9 \qquad (13)$$

For a rough boundary, the drag coefficient varies with the relative roughness ϵ/L and not with Reynolds number.

K. E. Schoenherr suggests the formula $1/\sqrt{C_D} = 4.13 \log (C_D R_{E_x})$ which equation is considered more accurate than expressions (12) and (13), particularly for Reynolds numbers greater than 2×10^7.

(b) Boundary layer thickness δ is estimated by

$$\frac{\delta}{x} = \frac{0.38}{R_{E_x}^{0.20}} \qquad \text{for } 5 \times 10^4 < R_E < 10^6 \qquad (14)$$

$$= \frac{0.22}{R_{E_x}^{0.167}} \qquad \text{for } 10^6 < R_E < 5 \times 10^8 \qquad (15)$$

(c) Shear stress is estimated by

$$\tau_0 = \frac{0.023\rho V^2}{(\delta V/\nu)^{1/4}} = 0.0587 \frac{V^2}{2} \rho \left(\frac{\nu}{xV}\right)^{1/5} \qquad (16)$$

3. **Boundary Layer in Transition** from laminar to turbulent on plate surface (R_E from about 500,000 to about 20,000,000).

(a) Mean drag coefficient $(C_D) = \dfrac{0.455}{(\log_{10} R_E)^{2.58}} - \dfrac{1700}{R_E}$ (17)

Diagram G illustrates the variation of C_D with Reynolds number for these three conditions of flow.

WATER HAMMER

Water hammer is the term used to express the resulting shock caused by the sudden decrease in the motion (velocity) of a fluid. In a pipeline the time of travel of the pressure wave up and back (round trip) is given by

$$\text{Time (round trip) in sec} = 2 \times \frac{\text{length of pipe in ft}}{\text{celerity of pressure wave in ft/sec}}$$

or $$T = \frac{2L}{c} \qquad (18)$$

The increase in pressure caused by the sudden closing of a valve is calculated by

$$\text{Change in pressure in psf} = \text{density} \times \text{celerity} \times \text{change in velocity}$$

or $$dp = \rho c\, dV \quad \text{or} \quad dh = c\, dV/g \qquad (19)$$

where dh is the change in pressure head.

For rigid pipes, the velocity (celerity) of the pressure wave is

$$c = \sqrt{\frac{\text{bulk modulus of fluid in psf}}{\text{density of fluid}}} = \sqrt{\frac{E_B}{\rho}} \qquad (20)$$

For non-rigid pipes the expression is

$$c = \sqrt{\frac{E_B}{\rho[1 + (E_B/E)(d/t)]}} \qquad (21)$$

where E = modulus of elasticity of pipe walls, psf
  d = inner diameter of pipe in inches
  t = thickness of wall of pipe in inches.

SUPERSONIC SPEEDS

Supersonic speeds completely change the nature of the flow. The drag coefficient is related to Mach number N_M (see Diagram H) since viscosity has a small effect upon the drag. The pressure disturbance created forms a cone with the apex at the nose of the body or projectile. The cone represents the wave front or *shock wave*, which may be photographed. The cone angle or *Mach angle* α is given by

$$\sin \alpha = \frac{\text{celerity}}{\text{velocity}} = \frac{1}{V/c} = \frac{1}{N_M} \qquad (22)$$

Solved Problems

1. Determine the momentum correction factor β which should be applied when using average velocity V in the momentum calculation (two-dimensional flow).

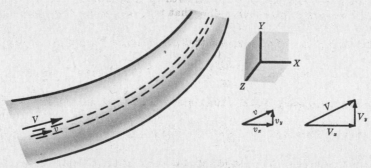

Fig. 11-1

Solution:

The mass discharge dM through the streamtube shown in Fig. 11-1 is $\rho\, dQ$. The correct momentum in the X-direction is

$$\text{Mom}_x = \int dM\, v_x = \int \rho\, dQ\, v_x = \int \rho v_x(v\, dA)$$

Using the average velocity for the cross section, the correct momentum would be

$$\text{Mom}_x = \beta(MV_x) = \beta(\rho Q V_x) = \beta\rho(AV)V_x$$

Equating the two values of correct momentum produces

$$\beta = \frac{\int \rho v_x (v\, dA)}{\rho A V(V_x)} = \frac{1}{A} \int_A (v/V)^2\, dA$$

since, from the velocity vector diagrams above $v_x/V_x = v/V$.

2. Evaluate the momentum correction factor if the velocity profile satisfies the equation $v = v_{max}[(r_o^2 - r^2)/r_o^2]$. (See Chap. 6, Prob. 17 for sketch.)

Solution:

From Prob. 17, Chap. 6, the average velocity was found to be $\frac{1}{2}v_{max}$. Using this value for average velocity V, we obtain

$$\beta = \frac{1}{A} \int_A \left(\frac{v}{V}\right)^2 dA = \frac{1}{\pi r_o^2} \int_o^{r_o} \left[\frac{v_{max}(r_o^2 - r^2)/r_o^2}{\frac{1}{2}v_{max}}\right]^2 (2\pi r\, dr)$$

$$= \frac{8}{r_o^6}(\tfrac{1}{2}r_o^6 - \tfrac{1}{2}r_o^6 + \tfrac{1}{6}r_o^6) = \tfrac{4}{3} = 1.33$$

3. A jet of water, $3''$ in diameter, moving to the right impinges on a flat plate held normal to its axis. (a) For a velocity of 80.0 ft/sec, what force will keep the plate in equilibrium? (b) Compare the average dynamic pressure on the plate with the maximum (stagnation) pressure if the plate is 20 times the area of the jet.

Solution:

Let the X-axis lie along the jet's path. Thus the plate destroys the initial momentum of the water in the X-direction. Taking M as the mass of water which has its momentum reduced to zero in dt seconds and F_x to the left as the force exerted by the plate on the water, we have:

(a)   Initial linear momentum $-$ linear impulse $=$ final linear momentum

$$M(80.0) - F_x\, dt = M(0)$$

$$\frac{wQ}{g} dt\, (80.0) - F_x\, dt = 0$$

and $$F_x = \frac{\overset{A}{62.4[(\pi/4)(3/12)^2]}\overset{V}{(80.0)} \times \overset{V}{80.0}}{32.2} = 608 \text{ lb (to the left for equilibrium).}$$

There is no Y-component of force involved in this problem, the Y-components on the plate balancing (and canceling) each other. Note that time dt cancels and might well have been chosen as 1 second.

For the plate it is well to recognize that this impulse-momentum expression can be arranged as follows:

$$F = MV = \frac{wQ}{g}V = \frac{w}{g}(AV)V = \rho A V^2 \quad \text{(lb)} \qquad (1)$$

(b) For the average pressure, divide the total dynamic force by the area over which it is acting.

$$\text{Average pressure} = \frac{\text{force}}{\text{area}} = \frac{\rho A V^2}{20A} = \frac{\rho V^2}{20} = \frac{w}{10}\left(\frac{V^2}{2g}\right) \quad \text{(psf)}$$

From Problems 1 and 5 of Chapter 9, the stagnation pressure $= p_s = w(V^2/2g)$ (psf). Thus the average pressure is 1/10 the stagnation pressure for this case.

4. A curved plate deflects a $3''$ diameter stream of water through an angle of $45°$. For a velocity in the jet of 130 ft/sec to the right, compute the value of the components of the force developed against the curved plate (assuming no friction).

Solution:

The components will be chosen along the jet's initial path and normal thereto. The water has its momentum changed by the force exerted by the plate.

(a) For the X-direction, taking $+$ to the right, assuming F_x positive,

$$\text{Initial linear momentum} + \text{linear impulse} = \text{final linear momentum.}$$

$$MV_{x_1} + F_x\,dt = MV_{x_2}$$

$$\frac{wQ\,dt}{g}V_{x_1} + F_x\,dt = \frac{wQ\,dt}{g}V_{x_2}$$

Rearranging and noting that $V_{x_2} = +V_{x_1}\cos 45°$, we obtain

$$F_x = \frac{62.4[(\pi/4)(3/12)^2](130)}{32.2}(130 \times 0.707 - 130) = -471 \text{ lb}$$

where the negative sign indicates that F_x is to the left (assumed to the right). Had F_x been assumed to the left, the solution would yield $+471$ lb, the sign signifying that the assumption was correct.

The effect of the water on the plate is opposite and equal to the effect of the plate on the water. Hence, X-component on plate $= 471$ lb to the right.

(b) For the Y-direction, taking *up* positive,

$$MV_{y_1} + F_y\,dt = MV_{y_2}$$

$$0 + F_y\,dt = \frac{62.4(0.0491)(130)dt}{32.2}(0.707 \times 130)$$

and $F_y = +1136$ lb upward on water. Hence, Y-component on plate $= 1136$ lb downward.

5. The force exerted by a $1''$ diameter stream of water against a flat plate held normal to the stream's axis is 145 lb. What is the flow in cfs?
    Solution:

    From equation (1) of Problem 3,

$$F_x = \frac{62.4QV}{32.2} = \rho AV^2$$

$$145 = \frac{62.4[(\pi/4)(1/12)^2]V^2}{32.2} \text{ and } V = 117.2 \text{ ft/sec.}$$

Then $Q = AV = [(\pi/4)(1/12)^2](117.2) = 0.639$ cfs.

6. Had the plate in Problem 3 been moving to the right with a velocity of 30.0 ft/sec, what force would the jet exert on the plate?
    Solution:

    Using $t = 1$ second,   initial $MV_{x_1} + F_x(1) =$ final MV_{x_2}.

    In this case the mass of water having its momentum changed is not identical with the mass for the stationary plate. For the fixed plate, in one second a mass of

$$(w/g)(\text{volume}) = (w/g)(A \times 80.0)$$

has its momentum changed. For the moving plate, in one second the mass striking the plate is

$$M = (w/g)[A(80.0 - 30.0)]$$

where $(80.0 - 30.0)$ is the relative velocity of the water with respect to the plate.

Then $$F_x = \frac{62.4[(\pi/4)(3/12)^2](80.0 - 30.0)}{32.2}(30.0 - 80.0)$$

and $F_x =$ force of plate on water $= -238$ lb, to the left. Hence the force of the water on the plate is 238 lb to the right.

Had the plate been moving to the left at 30.0 ft/sec, more water would have had its momentum changed in any time t. The value for V_{x_2} is -30.0 ft/sec. The magnitude of the force would be

$$F_x = \frac{62.4(0.0491)[80.0-(-30.0)]}{32.2}(-30.0-80.0) = -1150 \text{ lb to the left on the water}$$

7. The fixed surface shown in Fig. 11-2 divides the jet so that 1.00 cfs goes in each direction. For an initial velocity of 48.0 ft/sec, find the values of the X and Y components to keep the surface in equilibrium (assuming no friction).

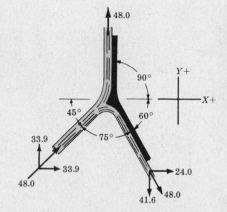

Fig. 11-2

Solution:

(a) In the X-direction, using $t = 1$ second,

$$MV_{x_1} - F_x(1) = \tfrac{1}{2}MV_{x_2} + \tfrac{1}{2}MV_{x_2}'$$

$$\frac{62.4(2.00)}{32.2}(33.9) - F_x = \frac{62.4}{32.2}\left(\frac{2.00}{2}\right)(0+24.0)$$

and $F_x = +131.4 - 46.6 = +84.8$ lb to the left.

(b) In the Y-direction,

$$MV_{y_1} - F_y(1) = \tfrac{1}{2}MV_{y_2} - \tfrac{1}{2}MV_{y_2}'$$

$$\frac{62.4(2.00)}{32.2}(33.9) - F_y = \frac{62.4}{32.2}\left(\frac{2.00}{2}\right)(+48.0-41.6)$$

and $F_y = +131.4 - 12.4 = 119.0$ lb downward.

8. A 3″ diameter jet has a velocity of 110.0 ft/sec. It strikes a blade moving in the same direction at 70.0 ft/sec. The deflection angle of the blade is 150°. Assuming no friction, calculate the X and Y components of the force exerted by the water on the blade. (Refer to Fig. 11-3(a) below.)

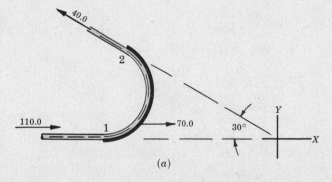

(a)

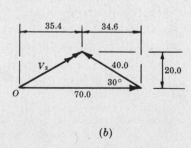

(b)

Fig. 11-3

Solution:

The relative velocity $V_{x_1} = 110.0 - 70.0 = 40.0$ ft/sec to the right.

The velocity of the water at $2 = V_{\text{water/blade}} \leftrightarrow V_{\text{blade}}$ (see Fig. 11-3(b) above)

from which $V_{2_x} = 35.4$ ft/sec to the right   and $V_{2_y} = 20.0$ ft/sec upward.

Employ the principle of impulse-momentum in the X-direction.

(a)   Initial $MV_x - F_x(1) = $ final MV_x
                                     $M(110) - F_x = M(+35.4)$

and $F_x = \dfrac{62.4}{32.2}\left[\dfrac{\pi}{4}\left(\dfrac{3}{12}\right)^2 \times 40.0\right](110.0 - 35.4) = 284$ lb, to the left on the water.

(b)   Initial $MV_y - F_y(1) = $ final MV_y
                                     $M(0) - F_y = M(+20.0)$

and $F_y = \dfrac{62.4}{32.2}\left[\dfrac{\pi}{4}\left(\dfrac{3}{12}\right)^2 \times 40.0\right](0 - 20.0) = -76.0$ lb, upward on the water.

The components of the force exerted by the water on the blade are 284 lb to the right and 76.0 lb downward.

9. Had friction in Problem 8 reduced the velocity of the water with respect to the blade from 40.0 ft/sec to 35.0 ft/sec, (a) what would be the components of the force exerted by the blade on the water and (b) what would be the final absolute velocity of the water?

Solution:

The components of the absolute velocity at (2) are found by solving a vector triangle similar to Figure (b) of Problem 8, using 70.0 horizontally and 35.0 upward to the left at 30°. Thus

$$V_{2_x} = 39.7 \text{ ft/sec to the right} \quad \text{and} \quad V_{2_y} = 17.5 \text{ ft/sec upward}$$

(a) Then $F_x = \dfrac{62.4}{32.2}\left[\dfrac{\pi}{4}\left(\dfrac{3}{12}\right)^2 \times 40.0\right](110.0 - 39.7) = 267$ lb, to the left on the water

$F_y = \dfrac{62.4}{32.2}\left[\dfrac{\pi}{4}\left(\dfrac{3}{12}\right)^2 \times 40.0\right](0 - 17.5) = -66.5$ lb, upward on the water.

(b) From the components shown above, the absolute velocity of the water leaving the blade is

$$V_2 = \sqrt{(39.7)^2 + (17.5)^2} = 43.4 \text{ ft/sec upward and to the right at an angle}$$
$$\theta_x = \tan^{-1} 17.5/39.7 = 23.8°.$$

10. For a given velocity of jet, determine the conditions which will produce the maximum amount of work (or power) on a series of moving blades (neglecting friction along blades).

Solution:

Consider first the velocity of the blades v which will produce maximum power. Referring to the adjacent figure, the expression for power in the X-direction will be developed, taking motion of the blade along the X-axis. Since the entire jet strikes either one blade or several blades, the mass flowing per second is having its momentum changed or $M = (w/g)AV$.

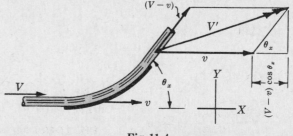

Fig. 11-4

Power = work done per sec = force × distance traveled per sec in direction of force.

(1) Determine the force by the momentum principle. The final absolute velocity in the X-direction is
$$V'_x = v + (V - v)\cos\theta_x$$

and   Initial momentum − linear impulse = final momentum
$$MV - F_x(1) = M[v + (V - v)\cos\theta_x]$$
or $$F_x = (wAV/g)[(V - v)(1 - \cos\theta_x)]$$

Then $\qquad\qquad$ Power $P = (wAV/g)[(V-v)(1-\cos\theta_x)]v$ $\qquad\qquad$ (1)

Since $(V-v)v$ is the variable which must attain its maximum value for maximum power, equating the first derivative to zero we obtain

$$dP/dv = (wAV/g)(1-\cos\theta_x)(V-2v) = 0$$

Thus $v = V/2$, or the blades should have a velocity equal to half of the jet velocity.

(2) Examination of expression (1) above for given values of V and v indicates that maximum power results when $\theta_x = 180°$ (by inspection). Since this angle is impractical, an angle of about $170°$ has proven to be satisfactory. The reduction in power is small percentagewise.

(3) In the Y-direction, the unbalanced force is made equal to zero by using cusped blades, diverting one-half the mass of water toward each end of the Y-axis.

11. (a) Referring to Fig. 11-5 below, at what angle should a jet of water moving at 50.0 ft/sec impinge upon a series of blades moving at 20.0 ft/sec in order that the water will strike tangent to the blades (no shock)? (b) What power is developed if the flow is 2.86 cfs? (c) What is the efficiency of the blades?

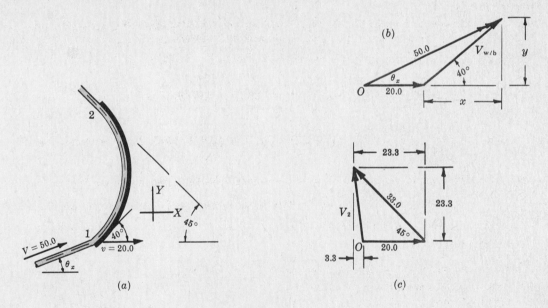

Fig. 11-5

Solution:

(a) $\qquad\qquad$ Velocity of water = velocity of water/blade $\rightarrow$ velocity of blade

or $\qquad\qquad$ 50.0 at $\angle\theta_x$ = ? at $40°$ $\rightarrow$ $20.0 \rightarrow$

From vector diagram, Fig. 11-5(b) above, $50\cos\theta_x = 20.0 + x$, $50\sin\theta_x = y$, and $\tan 40° = y/x = 0.8391$. Solving these equations, $\theta_x = 25°5'$.

(b) Solving Fig. (b) above for the velocity of the water with respect to the blades,

$\qquad y = 50\sin\theta_x = 50\sin 25°5' = 21.20$ ft/sec $\quad$ and $\quad V_{w/b} = y/(\sin 40°) = 33.0$ ft/sec.

Also, absolute $V_{x_2} = 3.3$ ft/sec to the left, from Fig. (c) above. Then

$$\begin{array}{cc} M & V_1\cos\theta_x \end{array}$$

force $F_x = \dfrac{62.4 \times 2.86}{32.2}[50 \times 0.906 - (-3.3)] = 270$ lb $\quad$ and $\quad$ power $E_x = 270 \times 20 = 5400$ ft lb/sec.

(c) Efficiency $= \dfrac{5400}{\frac{1}{2}M(50)^2} = \dfrac{5400}{6920} = 78.1\%$.

12. A 24″ pipe carrying 31.4 cfs of oil (sp gr 0.85) has a 90° bend in a horizontal plane. The loss of head in the bend is 3.50 ft of oil and the pressure at entrance is 42.5 psi. Determine the resultant force exerted by the oil on the bend.

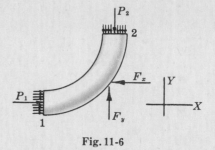

Fig. 11-6

Solution:

Referring to Fig. 11-6, the free body diagram indicates the static and dynamic forces acting on the mass of oil in the bend. These forces are calculated as follows.

(a) $P_1 = p_1 A = 42.5 \times \frac{1}{4}\pi(24)^2 = 19,200$ lb.

(b) $P_2 = p_2 A$, where $p_2 = p_1 -$ loss in psi, from the Bernoulli equation since $z_1 = z_2$ and $V_1 = V_2$.

Then $P_2 = (42.5 - 0.85 \times 62.4 \times 3.50/144) \times \frac{1}{4}\pi(24)^2 = 18,600$ lb

(c) Using the impulse-momentum principle, and knowing that $V_1 = V_2 = Q/A = 10.0$ ft/sec,

$$MV_{x_1} + \Sigma \text{ (forces in X-direction) } \times 1 = MV_{x_2}$$

$$19,200 - F_x = (0.85 \times 62.4 \times 31.4/32.2)(0 - 10.0) = -517 \text{ lb}$$

and $F_x = +19,720$ lb, to the left on the oil.

(d) Similarly, for $t = 1$ second,

$$MV_{y_1} + \Sigma \text{ (forces in Y-direction) } \times 1 = MV_{y_2}$$

$$F_y - 18,600 = (0.85 \times 62.4 \times 31.4/32.2)(10.0 - 0) = +517 \text{ lb}$$

and $F_y = +19,120$ lb, upward on the oil.

On the pipe bend, the resultant force R acts to the right and downward and is

$$R = \sqrt{(19,720)^2 + (19,120)^2} = 27,450 \text{ lb} \quad \text{at} \quad \theta_x = \tan^{-1} 19120/19720 = 44.2°$$

13. The 24″ pipe of Problem 12 is connected to a 12″ pipe by a standard *reducer* fitting. For the same flow of 31.4 cfs of oil, and a pressure of 40.0 psi, what force is exerted by the oil on the reducer, neglecting any lost head?

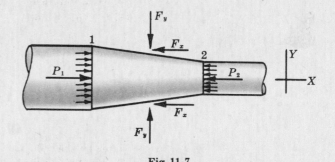

Fig. 11-7

Solution:

Since $V_1 = 10.0$ ft/sec, $V_2 = (2/1)^2 \times 10.0 = 40.0$ ft/sec. Also, the Bernoulli equation between Section 1 at entrance and Section 2 at exit yields

$$\left(\frac{p_1}{w} + \frac{(10)^2}{2g} + 0\right) - \text{negligible lost head} = \left(\frac{p_2}{w} + \frac{(40)^2}{2g} + 0\right)$$

Solving, $\dfrac{p_2}{w} = \dfrac{40.0 \times 144}{0.85 \times 62.4} + \dfrac{100}{2g} - \dfrac{1600}{2g} = 85.4$ ft of oil   and $p_2' = 31.5$ psi.

The figure above represents the forces acting on the mass of oil in the reducer.

$$P_1 = p_1 A_1 = 40.0 \times \frac{1}{4}\pi(24)^2 = 18,100 \text{ lb (to the right)}$$
$$P_2 = p_2 A_2 = 31.5 \times \frac{1}{4}\pi(12)^2 = 3560 \text{ lb (to the left)}$$

In the X-direction the momentum of the oil is changed. Then

$$MV_{x_1} + \Sigma \text{ (forces in } X\text{-direction)} \times 1 = MV_{x_2}$$
$$(18,100 - 3560 - F_x)1 = (0.85 \times 62.4 \times 31.4/32.2)(40.0 - 10.0)$$

and $F_x = 13,000$ lb acting to the left on the oil.

The forces in the Y-direction will balance each other and $F_y = 0$.

Hence the force exerted by the oil on the reducer is 13,000 lb to the right.

14. A 45° reducing bend, 24″ diameter upstream, 12″ diameter downstream, has water flowing through it at the rate of 15.7 cfs under a pressure of 21.0 psi. Neglecting any loss in the bend, calculate the force exerted by the water on the reducing bend.

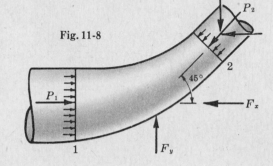

Fig. 11-8

Solution:

$$V_1 = 15.7/A_1 = 5.00 \text{ ft/sec}$$
and $$V_2 = 20.0 \text{ ft/sec}$$

The Bernoulli equation, Section 1 to Section 2, produces

$$\left(\frac{21.0 \times 144}{62.4} + \frac{25}{2g} + 0\right) - \text{negligible lost head} = \left(\frac{p_2}{w} + \frac{400}{2g} + 0\right)$$

from which $p_2/w = 42.7$ ft and $p_2' = 18.5$ psi.

In Fig. 11-8 is shown the mass of water acted upon by static and dynamic forces.

$$P_1 = p_1 A_1 = 21.0 \times \tfrac{1}{4}\pi(24)^2 = 9500 \text{ lb}$$
$$P_2 = p_2 A_2 = 18.5 \times \tfrac{1}{4}\pi(12)^2 = 2090 \text{ lb}$$
$$P_{2_x} = P_{2_y} = 2090 \times 0.707 = 1478 \text{ lb}$$

In the X-direction,

$$MV_{x_1} + \Sigma \text{ (forces in } X\text{-direction)} \times 1 = MV_{x_2}$$
$$(9500 - 1478 - F_x)1 = (62.4 \times 15.7/32.2)(20.0 \times 0.707 - 5.00)$$

and $F_x = 7740$ lb to the left.

In the Y-direction,

$$(+F_y - 1478)1 = (62.4 \times 15.7/32.2)(20.0 \times 0.707 - 0)$$

and $F_y = 1910$ lb upward.

The force exerted by the water on the reducing bend is $F = \sqrt{(7740)^2 + (1910)^2} = 7980$ lb to the right and downward, at an angle $\theta_x = \tan^{-1} 1910/7740 = 13°52'$.

15. Referring to the Fig. 11-9 below, a 2″ diameter stream of water strikes a 4 ft square door which is at an angle of 30° with the stream's direction. The velocity of the water in the stream is 60.0 ft/sec and the jet strikes the door at its center of gravity. Neglecting friction, what normal force applied at the edge of the door will maintain equilibrium?

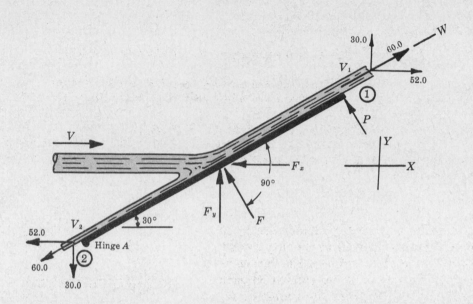

Fig. 11-9

Solution:

The force exerted by the door on the water will be normal to the door (no friction). Hence, since no forces act in the W-direction in the figure above, there will be no change in momentum in that direction. Thus, using W components,

$$\text{Initial momentum} \pm 0 = \text{final momentum}$$
$$+M(V\cos 30°) = +M_1 V_1 - M_2 V_2$$
$$(w/g)(A_{jet}V)(V\cos 30°) = (w/g)(A_1 V_1)V_1 - (w/g)(A_2 V_2)V_2$$

But $V = V_1 = V_2$ (friction neglected). Then

$$A_{jet}\cos 30° = A_1 - A_2 \quad \text{and, from the equation of continuity,} \quad A_{jet} = A_1 + A_2$$

Solving, $A_1 = A_{jet}(1 + \cos 30°)/2 = A_{jet} \times 0.933$ and $A_2 = A_{jet}(1 - \cos 30°)/2 = A_{jet} \times 0.067$.

The stream divides as indicated and the momentum equation produces, for the X-direction,

$$\left[\frac{62.4}{32.2}(\tfrac{1}{4}\pi)(\tfrac{1}{6})^2 60\right]60 - F_x(1) = \left[\frac{62.4}{32.2}(\tfrac{1}{4}\pi)(\tfrac{1}{6})^2 0.933(60)\right]52.0 + \left[\frac{62.4}{32.2}(\tfrac{1}{4}\pi)(\tfrac{1}{6})^2 0.067(60)\right](-52.0)$$

and $F_x = 38.0$ lb.

Similarly, in the Y-direction,

$$M(0) + F_y(1) = \left[\frac{62.4}{32.2}(.0218)(0.933)60\right]30.0 + \left[\frac{62.4}{32.2}(.0218)(.067)60\right](-30.0)$$

and $F_y = 65.9$ lb.

For the door as the free body, $\Sigma M_{hinge} = 0$ and

$$+ 38.0(1) + 65.9(2 \times 0.866) - P(4) = 0 \quad \text{or} \quad P = 38.1 \text{ lb}$$

16. Determine the reaction of a jet flowing through an orifice on the containing tank.

Solution:

In the adjoining figure a mass of liquid $ABCD$ is taken as a free body. The only horizontal forces acting are F_1 and F_2, which change the momentum of the water.

$$(F_1 - F_2) \times 1 = M(V_2 - V_1), \text{ where } V_1 \text{ can be considered negligible.}$$

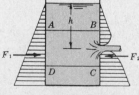

Fig. 11-10

Reaction $F = F_1 - F_2 = \dfrac{wQ}{g} V_2 = \dfrac{wA_2V_2}{g} V_2$.

But $A_2 = c_c A_o$ and $V_2 = c_v \sqrt{2gh}$.

Hence $F = \dfrac{w(c_c A_o)}{g} c_v^2 (2gh) = (c\, c_v) w A_o (2h)$ (to the right on the liquid).

(1) For average values of $c = 0.60$ and $c_v = 0.98$, the reaction $F = 1.176 wh A_o$. Hence the force acting to the left on the tank is about 18% more than the static force on a plug which would just fill the orifice.

(2) For ideal flow (no friction, no contraction), $F = 2(whA_o)$.
    This force is equal to twice the force on a plug which would just fill the orifice.

(3) For a nozzle ($c_c = 1.00$), the reaction $F = c_v^2 wA(2h)$ where h would be the effective head causing the flow.

17. The jets from a garden sprinkler are 1″ in diameter and are normal to the 2 ft radius. If the pressure at the base of the nozzles is 50 psi, what force must be applied on each sprinkler pipe, 1 ft from the center of rotation, to maintain equilibrium? (Use $c_v = 0.80$ and $c_c = 1.00$.)

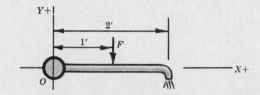

Fig. 11-11

Solution:

The reaction of the sprinkler jet may be calculated from the momentum principle. Inasmuch as the force which causes a change in momentum in the X-direction acts along the X-axis, no torque is exerted. We are interested, therefore, in the change in momentum in the Y-direction. But the initial momentum in the Y-direction is zero. The jet velocity

$$V_Y = c_v \sqrt{2gh} = 0.80 \sqrt{2g(50 \times 2.31 + \text{negligible velocity head})} = 68.8 \text{ ft/sec}$$

Thus $$F_Y\, dt = M(V_Y) = \left[\frac{62.4}{32.2} \times \tfrac{1}{4}\pi (\tfrac{1}{12})^2 \times 68.8\, dt \right](-68.8)$$

or $F_Y = -50.2$ lb downward on the water. Hence the force of the jet on the sprinkler is $+50.2$ lb upward. Then

$$\Sigma M_o = 0, \qquad F(1) - 2(50.2) = 0, \qquad F = 100.4 \text{ lb for equilibrium}$$

18. Develop basic thrust equations for propulsive devices.

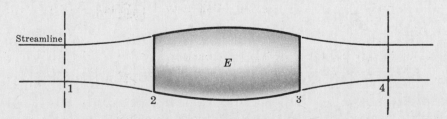

Fig. 11-12

Solution:

In Fig. 11-12 above, consider the air-breathing engine E that utilizes W lb of air per second. At section 1, the velocity V_1 of the air entering the engine is taken as the flight velocity. Also the air enters at atmospheric pressure (where no shock waves occur). In the engine E the air is compressed and heated by combustion. The air leaves the nozzle at section 3 at a high exit velocity and hence with greatly increased momentum.

In most air-breathing engines, the weight of air per second at exit is greater than the weight of air per second entering the engine due to the addition of fuel. This increase is about 2%. The weight of air at exit is usually measured at Section 3.

The thrust can be evaluated in terms of the change in momentum, as follows:

$$\text{Thrust } F \;=\; \frac{W_{\text{exit}} V_4}{g} \;-\; \frac{W_1 V_1}{g} \tag{A}$$

In those cases where the pressure at Section 3 may be greater than atmospheric pressure, an additional acceleration of the gas is provided. The additional force is the difference in pressure times the area at Section 3. Thus, for change in momentum from Sections 1 to 3, we obtain

$$F \;=\; \frac{W_{\text{exit}} V_3}{g} \;+\; A_3(p_3 - p_4) \;-\; \frac{W_1 V_1}{g} \tag{B}$$

Should the effective exhaust velocity be required, equations (A) and (B) can be solved simultaneously to produce

$$V_4 \;=\; V_3 \;+\; \frac{gA_3}{W_3}(p_3 - p_4) \tag{C}$$

It will be noted that if $p_3 = p_4$, $V_4 = V_3$.

The term $W_1 V_1/g$ is called the negative thrust or ram drag. The gross thrust (produced by the nozzle) is $W_3 V_4/g$ in (A) and $W_3 V_3/g + A_3(p_3 - p_4)$ in (B).

For a rocket, equation (A) is used to evaluate the thrust, recognizing that $V_1 = 0$ for such a device.

19. A jet engine is being tested in the laboratory. The engine consumes 50 lb of air per sec and 0.5 lb of fuel per sec. If the exit velocity of the gases is 1500 ft/sec, what is the thrust?

Solution:

Using formula (A) in Problem 18, thrust $F \;=\; (50.5 \times 1500 - 50 \times 0)/32.2 \;=\; 2350$ lb.

20. A jet engine operates at 600 ft/sec and consumes air at the rate of 50.0 lb/sec. At what velocity should the air be discharged in order to develop a thrust of 1500 lb?

Solution:

Thrust $F \;=\; 1500 \;=\; (50.0/32.2)(V_{\text{exit}} - 600)$ from which $V_{\text{exit}} = 1565$ ft/sec.

21. A turbojet engine is tested in the laboratory under conditions simulating an altitude where atmospheric pressure $p = 785.3$ psfa, temperature $T = 429.5°$ Rankine and specific weight $w = 0.0343$ lb/ft^3. If the exit area of the engine is 1.50 ft^2 and the exit pressure is atmospheric, what is the Mach number N_M if the gross thrust is 1470 lb? (Use $k = 1.33$.)

Solution:

Since in equation (B) of Problem 18, $p_3 = p_4$ and $V_1 = 0$,

$$F = W_e V_e/g = (w A_e V_e) V_e/g, \qquad 1470 = .0343(1.50) V_e^2/g, \qquad V_e = 959 \text{ ft/sec}$$

Mach number $N_M = V_e/c = V_e/\sqrt{kgRT} = 959/\sqrt{1.33(32.2)(53.3)(429.5)} = 0.97.$

22. In Problem 21, what would be the gross thrust if the exit pressure becomes 10 psia and the Mach number is 1.00? (Use $k = 1.33$.)

Solution:

In order to evaluate the exit velocity for the new exit conditions, the temperature at exit must be calculated from

$T_e/429.5 = (10 \times 144/785.3)^{(k-1)/k}$ from which $T_e = 500°$ Rankine

Then $V_e = N_M c = N_M\sqrt{kgRT} = 1.00\sqrt{1.33(32.2)(53.3)(500)} = 1068$ ft/sec.

Furthermore, the specific weight at exit must be calculated from

$(w_1/w_2)^k = p_1/p_2,$ $(w_e/.0343)^{1.33} = 10 \times 144/785.3,$ $w_e = .0540$ lb/ft³

Using (B) of Prob. 18, $F = .0540(1.50)(1068)^2/32.2 + 1.50(1440 - 785.3) - 0 = 4042$ lb.

23. A rocket device burns its propellant at a rate of 15.2 lb/sec. The exhaust gases leave the rocket at a relative velocity of 3220 ft/sec and at atmospheric pressure. The exhaust nozzle has an area of 50.0 in² and the gross weight of the rocket is 500 lb. At the given instant, 2500 horsepower is developed by the rocket engine. What is the rocket velocity?

Solution:

For a rocket, no air enters the device and the section 1 terms in formula (B) of Prob. 18 are zero. Also, since the exit pressure is atmospheric, $p_3 = p_4$. Thus, the thrust

$$F_T = (W_e/g)V_e = (15.2/32.2)(3220) = 1520 \text{ lb}$$

and since $2500 \text{ hp} = F_T V_{rocket}/550,$ $V_{rocket} = 905$ ft/sec.

24. Assuming that the drag force is a function of density, viscosity, elasticity and velocity of the fluid, and a characteristic area, show that the drag force is a function of the Mach number and the Reynolds number (see also Chap. 5, Problems 9 and 16).

Solution:

As illustrated in Chapter 5, a dimensional analysis study will provide the desired relation, as follows.
$$F_D = f_1(\rho, \mu, E, V, A)$$
or
$$F_D = C\rho^a \mu^b E^c V^d L^{2e}$$

Then, dimensionally, $F^1 L^0 T^0 = (F^a T^{2a} L^{-4a})(F^b T^b L^{-2b})(F^c L^{-2c})(L^d T^{-d}) L^{2e}$

and $1 = a+b+c,$ $0 = -4a-2b-2c+d+2e,$ $0 = 2a+b-d$

Solving in terms of b and c yields

$$a = 1-b-c, \qquad d = 2-b-2c, \qquad e = 1-b/2$$

Substituting, $F_D = C\rho^{1-b-c}\mu^b E^c V^{2-b-2c} L^{2-b}$

Expressing this equation in the commonly-used form produces

$$F = CA\rho V^2 \left(\frac{\mu}{L\rho V}\right)^b \left(\frac{E}{\rho V^2}\right)^c$$
or
$$F = A\rho V^2 f_2(R_E, N_M)$$

This equation indicates that the drag coefficient of objects of given configuration and alignment will depend upon their Reynolds and Mach numbers only.

For incompressible fluids, the Reynolds Number is dominant, and the effect of the Mach Number N_M is small to negligible; thus the drag coefficient becomes a function of Reynolds Number R_E only. (See Diagrams F and G in the Appendix.) Indeed, for small values of N_M, a fluid may be considered as incompressible insofar as the coefficient of drag is concerned.

When the Mach Number N_M is equal to or greater than 1.0 (with velocities of the fluid equal to or greater than the velocity of sound) the drag coefficient is a function of N_M only. (See Diagram H in the Appendix.) However, there are often instances where the coefficient of drag depends on both R_E and N_M.

A similar derivation may be presented regarding the coefficient of lift, and the conclusions stated above are equally applicable to the coefficient of lift. Use of the Buckingham Pi Theorem is suggested.

25. A 50 mph wind strikes a 6 ft by 8 ft sign normal to its surface. For a standard barometer, what force acts against the sign? ($w = .0752$ lb/ft^3)

Solution:

For a small jet of fluid striking a large plate at rest, we have seen that the force exerted by the fluid is

$$\text{Force}_x = \Delta(MV_x) = (w/g)(AV_x)V_x = \rho A V_x^2$$

The stationary plate under consideration affects a large amount of air. The momentum is not reduced to zero in the X-direction as was the case for the jet of water. Tests on plates moving through fluids at different velocities indicate that the drag coefficient varies with the length to width ratio and that its value is essentially constant above Reynolds number of 1000. (See Diagram F, Appendix.) It is immaterial whether an object moves in the fluid at rest or the fluid moves past a stationary object; drag coefficients and drag forces are the same for each case. It is the relative velocity that is important.

The coefficient (C_D) is employed in the following equation: Force $F = C_D \rho A \dfrac{V^2}{2}$.

This is sometimes written to include velocity head, as follows: Force $F = C_D w A \dfrac{V^2}{2g}$.

Using $C_D = 1.20$ from Diagram F,   Force $F = 1.20\left(\dfrac{.0752}{32.2}\right)(48)\dfrac{(50 \times 5280/3600)^2}{2} = 362$ lb.

26. A flat plate, 4 ft by 4 ft, moves at 22 ft/sec normal to its plane. At standard pressure and 68°F air temperature, determine the resistance of the plate (a) moving through air and (b) moving through water at 60°F.

Solution:

(a)   Diagram F indicates $C_D = 1.16$ for length/width $= 1$.

$$\text{Drag force} = C_D \rho A \frac{V^2}{2} = 1.16\left(\frac{.0752}{32.2}\right)(4 \times 4)\frac{(22)^2}{2} = 10.5 \text{ lb.}$$

(b)   Drag force $= C_D \rho A \dfrac{V^2}{2} = 1.16(1.94)(4 \times 4)\dfrac{(22)^2}{2} = 8700$ lb.

27. A long copper wire, $\frac{1}{2}''$ in diameter, is stretched taut and is exposed to a wind at a velocity of 90.0 ft/sec, normal to the wire. Compute the drag force per foot of length.

Solution:

For air at 68°F, Table 1 gives $\rho = .00233$ slug/ft^3 and $\nu = 16.0 \times 10^{-5}$ ft^2/sec. Then

$$R_E = \frac{Vd}{\nu} = \frac{90 \times 1/24}{16.0} \times 10^5 = 23{,}400$$

From Diagram F, $C_D = 1.30$. Then

$$\text{Drag force} = C_D \rho A \frac{V^2}{2} = 1.30(.00233)(1 \times \tfrac{1}{24})\frac{(90.0)^2}{2} = 0.510 \text{ lb per ft of length}$$

28. A 3 ft by 4 ft plate moves at 44 ft/sec in still air at an angle of 12° with the horizontal. Using a coefficient of drag $C_D = 0.17$ and a coefficient of lift $C_L = 0.72$, determine (a) the resultant force exerted by the air on the plate, (b) the frictional force and (c) the horsepower required to keep the plate moving. (Use $w = .0752$ lb/ft^3.)

Solution:

(a)   Drag force $= C_D\left(\dfrac{w}{g}\right)A\,\dfrac{V^2}{2}$

$= 0.17\left(\dfrac{.0752}{32.2}\right)(12)\dfrac{(44)^2}{2} = 4.61$ lb.

Lift force $= C_L\left(\dfrac{w}{g}\right)A\,\dfrac{V^2}{2}$

$= 0.72\left(\dfrac{.0752}{32.2}\right)(12)\dfrac{(44)^2}{2} = 19.5$ lb.

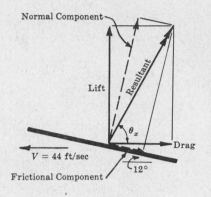

Referring to Fig. 11-13, the resultant of the drag and lift components is

Fig. 11-13

$R = \sqrt{(4.61)^2 + (19.5)^2} = 20.1$ lb   acting on the plate at $\theta_x = \tan^{-1} 19.5/4.61 = 76°42'$

(b)   The resultant force might also have been resolved into a normal component and a frictional component (shown dotted in the figure). From the vector triangle,

$$\text{frictional component} = R\cos(\theta_x + 12°) = 20.1(.0227) = 0.46 \text{ lb.}$$

(c)   Horsepower $=$ (force in direction of motion $\times$ velocity)/550 $= (4.61 \times 44)/550 = 0.369$.

29. If an airplane weighs 4000 lb and has a wing area of 300 ft², what *angle of attack* must the wings make with the horizontal at a speed of 100 mph? Assume the coefficient of lift varies linearly from 0.35 at 0° to 0.80 at 6° and use $w = .0752$ lb/ft³ for air.
   Solution:

For equilibrium in the vertical direction, $\Sigma Y = 0$. Hence, lift $-$ weight $= 0$, or

$$\text{Weight} = C_L wA\,\dfrac{V^2}{2g}, \qquad 4000 = C_L(.0752)(300)\dfrac{(100\times5280/3600)^2}{2g}, \qquad C_L = 0.53$$

By interpolation between 0° and 6°, angle of attack $= 2.4°$.

30. What wing area is required to support a 5000 lb plane when flying at an *angle of attack* of 5° at 88 ft/sec? Use coefficients given in Problem 29.
   Solution:

From given data (or from a curve), $C_L = 0.725$ for 5° angle by interpolation. As in Problem 29,

$$\text{Weight} = \text{lift}, \qquad 5000 = 0.725(.0752/32.2)A(88)^2/2, \qquad A = 762 \text{ ft}^2$$

31. An airfoil of 400 ft² area has an angle of attack of 6° and is traveling at 80 ft/sec. If the coefficient of drag varies linearly from .040 at 4° to 0.120 at 14°, what horsepower is required to maintain the velocity in air at 40°F and 13.0 psi absolute?
   Solution:

$$w = \frac{p}{RT} = \frac{13.0 \times 144}{53.3(460 + 40)} = .0702 \text{ lb/ft}^3 \text{ for air}$$

For 6° angle of attack, $C_D = .056$, by interpolation.

$$\text{Drag force} = C_D \rho A V^2/2 = .056(.0702/32.2)(400)(80)^2/2 = 156.5 \text{ lb}$$
$$\text{Power} = (156.5 \text{ lb})(80 \text{ ft/sec})/550 = 22.8 \text{ horsepower}$$

32. In the preceding problem, for a coefficient of lift of 0.70 and a chord length of 5 ft, determine (a) the lift force and (b) the Reynolds and Mach numbers.

Solution:

(a)   Lift force $F_L = C_L\rho AV^2/2 = 0.70(.0702/g)(400)(80)^2/2 = 1950$ lb.

(b)   The characteristic length in the Reynolds number is the chord length. Then

$$R_E = \frac{VL\rho}{\mu} = \frac{80 \times 5 \times .0702}{(3.62 \times 10^{-7})(32.2)} = 2{,}412{,}000$$

It will be remembered that the absolute coefficient of viscosity does not change with pressure variations.

$$N_M = V/\sqrt{E/\rho} = V/\sqrt{kgRT} = 80/\sqrt{(1.4)(32.2)(53.3)(500)} = .073$$

33. An air wing of 270 ft² area moves at 84.0 ft/sec. If 14.0 horsepower is required to keep the wing in motion, what angle of attack is indicated using the same variation of the coefficient of drag as in Problem 31? Use $w = .0702$ as in Problem 31.

Solution:
$$14.0 \text{ hp} = (\text{force})(84.0 \text{ ft/sec})/550, \qquad \text{force} = 91.7 \text{ lb}$$
$$\text{Force} = C_D\rho AV^2/2, \qquad 91.7 = C_D(.0702/32.2)(270)(84.0)^2/2, \qquad C_D = .0441$$

Using the data regarding angle of attack and C_D, interpolation yields 4.5° angle of attack.

34. Consider the area on one side of a moving van to be 600 ft². Determine the resultant force acting on the side of the van when the wind is blowing at 10 mph normal to the area (a) when the van is at rest and (b) when the van is moving at 30 mph normal to the direction of the wind. In (a) use $C_D = 1.30$, and in (b) use $C_D = 0.25$ and $C_L = 0.60$. ($\rho = .00237$ slug/ft³)

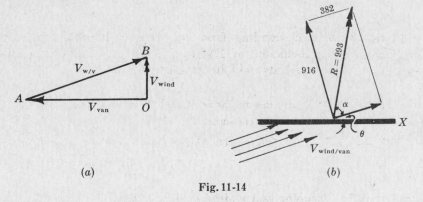

Fig. 11-14

Solution:

(a)   The force acting normal to the area $= C_D(\rho/2)AV^2$. Then

$$\text{Resultant force} = 1.30(.00237/2)(600)(10 \times 5280/3600)^2 = 198 \text{ lb normal to area}$$

(b)   It will be necessary to calculate the relative velocity of the wind with respect to the van. From kinetic mechanics,
$$V_{\text{wind}} = V_{\text{wind/van}} + V_{\text{van}}$$

Figure (a) above indicates this vector relationship, i.e.,

$$OB = OA + AB = 30.0 + V_{w/v}$$

Thus the relative velocity $= \sqrt{(30)^2 + (10)^2} = 31.6$ mph to the right and upward at an angle $\theta = \tan^{-1} 10/30 = 18.4°$.

The component of the resultant force normal to the relative velocity of wind with respect to van is

$$\text{Lift force} = C_L(\rho/2)AV^2 = 0.60(.00237/2)(600)(31.6 \times 5280/3600)^2$$
$$= 916 \text{ lb normal to the relative velocity}$$

The component of the resultant force parallel to the relative motion of wind to van is

$$\text{Drag force} = C_D(\rho/2)AV^2 = 0.25(.00237/2)(600)(31.6 \times 5280/3600)^2$$
$$= 382 \text{ lb parallel to the relative velocity}$$

Referring to Fig. 11-14(b), the resultant force $= \sqrt{(916)^2 + (382)^2} = 993$ lb at an angle $\alpha = \tan^{-1} 916/382 = 67.4°$. Hence the angle with the longitudinal axis (X-axis) is $18.4° + 67.4° = 85.8°$.

35. A kite weighs 2.50 lb and has an area of 8.00 ft². The tension in the kite string is 6.60 lb when the string makes an angle of 45° with the horizontal. For a wind of 20 mph, what are the coefficients of lift and drag if the kite assumes an angle of 8° with the horizontal? Consider the kite essentially a flat plate and $w_{\text{air}} = .0752$ lb/ft³.

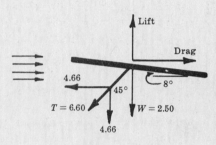

Fig. 11-15

Solution:

Fig. 11-15 indicates the forces acting on the kite taken as a free body. The components of the tension are 4.66 lb each.

From $\Sigma X = 0$,   drag = 4.66 lb.

From $\Sigma Y = 0$,   lift = 4.66 + 2.50 = 7.16 lb.

Drag force $= C_D\rho AV^2/2$, $4.66 = C_D(.0752/32.2)(8.00)(20 \times 5280/3600)^2/2$, $C_D = 0.58$.

Lift force $= C_L\rho AV^2/2$, $7.16 = C_L(.0752/32.2)(8.00)(20 \times 5280/3600)^2/2$, $C_L = 0.90$.

36. A man weighing 170 lb is descending from an airplane using an 18 ft diameter parachute. Assuming a drag coefficient of 1.00 and neglecting the weight of the parachute, what maximum terminal velocity will be attained?

Solution:

The forces on the parachute are the weight down and the drag force up.

For equilibrium, $\Sigma Y = 0$ (velocity constant),

$$W = C_D\rho AV^2/2, \qquad 170 = 1.00(.0752/32.2)(\pi 9^2)V^2/2, \qquad V = 23.9 \text{ ft/sec}$$

37. A steel ball, $\frac{1}{8}$ inch in diameter and weighing 0.284 lb/in³, is falling in an oil of specific gravity 0.908 and kinematic viscosity .00157 ft²/sec. What is the terminal velocity of the ball?

Solution:

The forces acting on the steel ball are the weight down, the buoyant force up, and the drag force up. For constant velocity, $\Sigma Y = 0$, and, transposing,

$$\text{weight of sphere} - \text{buoyant force} = \text{drag force}$$
or $$w_S(\text{volume}) - w_o(\text{volume}) = C_D\rho AV^2/2$$

Using lb/in³ $\times$ in³ = weight,

$$\frac{4}{3}\pi\left(\frac{1}{16}\right)^3\left(0.284 - \frac{0.908 \times 62.4}{1728}\right) = C_D\left(\frac{0.908 \times 62.4}{32.2}\right)\pi\left(\frac{1}{12 \times 16}\right)^2\frac{V^2}{2}$$

Assuming a value of C_D of 3.00 (see Diagram F, spheres) and solving,

$$V^2 = 3.43/C_D = 1.143 \quad \text{and} \quad V = 1.070 \text{ ft/sec}$$

Check C_D assumed by calculating Reynolds number and using Diagram F.

$$R_E = \frac{Vd}{\nu} = \frac{1.070 \times 1/(8 \times 12)}{.00157} = 7.11 \quad \text{and} \quad C_D = 5.6 \text{ (increase } C_D)$$

Recalculating and checking, using $C_D = 6.5$ to anticipate the effect of increased value of C_D,

$$V^2 = 3.43/6.5 = 0.528, \quad V = 0.726, \quad R_E = 4.82, \quad C_D = 7.2 \text{ (increase } C_D)$$

Trying $C_D = 7.8$,

$$V^2 = 3.43/7.8 = 0.440, \quad V = 0.664, \quad R_E = 4.40, \quad C_D = 7.8 \text{ (checks)}$$

Therefore, the terminal velocity $= 0.66$ ft/sec.

Had Reynolds number been less than 0.60, the equation for the drag force could be written

$$C_D \rho A V^2/2 = (24/R_E)\rho A V^2/2 = (24\nu/Vd)\rho(\pi d^2/4)V^2/2.$$

Since $\mu = \rho\nu$, drag force $= 3\pi\mu dV$.

38. A 1 in. diameter sphere of lead, weighing 710 lb/ft^3, is moving downward in an oil at a constant velocity of 1.17 ft/sec. Calculate the absolute viscosity of the oil if the specific gravity is 0.93.

Solution:

As in the preceding problem, but using lb/ft^3 × ft^3 = weight,

$$(w_S - w_o)(\text{volume}) = C_D \rho A V^2/2$$

Then $(710 - 0.93 \times 62.4)(4\pi/3)(1/24)^3 = C_D(0.93 \times 62.4/32.2)\pi(1/24)^2(1.17)^2/2$ and $C_D = 29.4$.

From Diagram F, for $C_D = 29.4$, $R_E = 0.85$ and

$$0.85 = Vd/\nu = (1.17)(1/12)/\nu, \quad \nu = 0.115 \text{ ft}^2/\text{sec}$$

Thus $\qquad \mu = \nu\rho = 0.115(0.93 \times 62.4)/32.2 = 0.207 \text{ lb sec/ft}^2$

39. A sphere, $\frac{1}{2}''$ in diameter, rises in an oil at the maximum velocity of 0.12 ft/sec. What is the specific weight of the sphere if the density of the oil is 1.78 slug/ft^3 and the absolute viscosity is .000710 lb sec/ft^2?

Solution:

For constant velocity upward, $\Sigma Y = 0$ and

$$\text{buoyant force} - \text{weight} - \text{drag} = 0$$

$$(4\pi/3)(1/48)^3(1.78 \times 32.2 - w_s) = C_D(1.78)\pi(1/48)^2(0.12)^2/2$$

$$(57.32 - w_s) = 0.461 C_D \qquad\qquad (1)$$

The coefficient of drag can be evaluated using Diagram F and Reynolds number.

$$\text{Reynolds number} = \frac{Vd\rho}{\mu} = \frac{0.12 \times 1/24 \times 1.78}{.000710} = 12.5$$

Then from Diagram F, $C_D = 3.9$ (for sphere) and, from (1),

$$w_s = 57.32 - 0.461 \times 3.9 = 55.5 \text{ lb/ft}^3$$

40. For laminar flow at low Reynolds numbers, show that the coefficient of drag for a sphere is equal to 24 divided by Reynolds number (shown graphically on Diagram F in the Appendix).

Solution:

The drag force $F_D = C_D \rho A V^2/2$, as previously noted.

For laminar flow the drag force depends upon the viscosity and the velocity of the fluid and the diameter d of the sphere. Thus

$$F_D = f(\mu, V, d) = C \mu^a V^b d^c$$

Then

$$F^1 L^0 T^0 = (F^a T^a L^{-2a})(L^b T^{-b})(L^c)$$

and

$$1 = a, \qquad 0 = -2a + b + c, \qquad 0 = a - b$$

from which $a = 1$, $b = 1$ and $c = 1$. Thus, drag force $F_D = C(\mu V d)$. G. G. Stokes has shown mathematically that $C = 3\pi$, which fact has been confirmed by many experiments.

We may now equate the two expressions for drag force, substituting $\frac{1}{4}\pi d^2$ for projected area A, and then solving for C_D.

$$3\pi\mu V d = C_D \rho(\tfrac{1}{4}\pi d^2)V^2/2 \qquad \text{and} \qquad C_D = \frac{24\mu}{V d\rho} = \frac{24}{R_E}$$

41. For laminar flow of a fluid passed a thin plate, develop an expression for the thickness δ of the boundary layer, assuming that the velocity distribution equation is

$$v = V\left(\frac{2y}{\delta} - \frac{y^2}{\delta^2}\right).$$

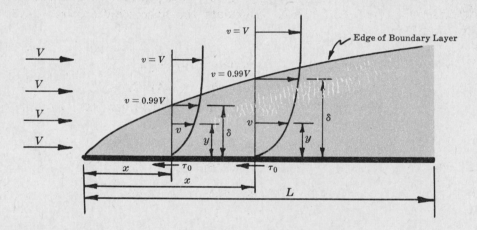

Fig. 11-16

Solution:

It is assumed that steady flow occurs ($\partial v/\partial t = 0$), that the velocity outside the boundary layer is uniformly V, that δ is very small with respect to distance x, and that $dp/dy = 0 = dp/dx$, both outside and within the boundary layer. Furthermore, by definition the edge of the boundary layer is considered as the locus of points where the velocity is 0.99 of the undisturbed velocity V.

The mass passing through any section of the boundary layer per unit width is $\int_o^\delta \rho v(dy \times 1)$ and the change in velocity at any point is $(V - v)$. Inasmuch as the pressure forces on the section cancel each other, thereby not contributing to the change in momentum, the change in momentum is caused by the shearing force $\tau_o\, dA$ or $\tau_o(dx \times 1)$. From the above, the change in momentum in unit time is

$$\int_o^\delta \rho(V - v)v(dy \times 1)$$

This expression is equal to the shearing force acting for a unit time, or

$$\text{drag force/unit width,} \quad F_D' = \int_o^x \tau_o(dx \times 1) = \int_o^\delta \rho(V - v)v(dy \times 1)$$

Substituting the parabolic velocity relation in the equation gives

$$F'_D = \int_0^\delta \rho(V - 2yV/\delta + y^2V/\delta^2)(V)(2y/\delta - y^2/\delta^2)\,dy$$

$$= \rho V^2 \int_0^\delta (1 - 2y/\delta + y^2/\delta^2)(2y/\delta - y^2/\delta^2)\,dy = \tfrac{2}{15}\rho V^2 \delta \qquad (A)$$

In order to obtain a useful expression for δ, consider that $\tau_o\,dx$ = the differential unit drag force dF'_D and that the flow is laminar. Then, in $\tau_o = \mu(dv/dy)_o$, the term

$$\left(\frac{dv}{dy}\right)_o = \frac{d}{dy}[V(2y/\delta - y^2/\delta^2)] = \frac{2V}{\delta}(1 - y/\delta) \qquad (B)$$

Substituting the values above in $\mu(dv/dy)_o = dF'_D/dx$ and realizing that the shear stress is τ_o when $y = 0$, we obtain $\mu(2V/\delta) = \tfrac{2}{15}\rho V^2(d\delta/dx)$ or

$$\int_0^\delta \delta\,d\delta = \frac{15\mu}{\rho V}\int_0^x dx$$

which yields

$$\delta^2 = \frac{30\mu x}{\rho V} \qquad \text{or} \qquad \frac{\delta}{x} = \sqrt{\frac{30\nu}{xV}} = \frac{5.48}{\sqrt{R_{E_x}}} \qquad (C)$$

Blasius' more exact solution gives 5.20 as the numerator of (C).

42. For laminar flow, derive the expression (a) for the shear stress at the boundary (plate) in the preceding problem and (b) for the local drag coefficient C_D.

Solution:

(a) From (B), Prob. 41, when $y = 0$, $\tau_o = 2\mu V/\delta$. Then, using the value of δ in (C) above,

$$\tau_o = \frac{2\mu V}{\sqrt{30\mu x/\rho V}} = 0.365\sqrt{\frac{\rho V^3 \mu}{x}} = 0.365\frac{\rho V^2}{\sqrt{R_{E_x}}} \qquad (A)$$

Results of experimentation give the more exact formula as

$$\tau_o = 0.33\sqrt{\frac{\rho V^3 \mu}{x}} = 0.33\frac{\rho V^2}{\sqrt{R_{E_x}}} \qquad (B)$$

(b) The local drag coefficient C_{D_x} is obtained by equating $\tau_o A$ and the local drag force, i.e.,

$$F_D = \tau_o A = C_{D_x}\rho A V^2/2$$

or

$$C_{D_x} = \frac{2\tau_o}{\rho V^2} = \frac{0.66\rho V^2}{\rho V^2\sqrt{R_{E_x}}} = \frac{0.66}{\sqrt{R_{E_x}}} \qquad (C)$$

It can be seen that the total drag force on one side of a plate is the sum of all the $(\tau_o\,dA)$ terms or

$$F_D = \int_0^L \tau_o(dx \cdot 1) = \int_0^L 0.33\sqrt{\rho V^3 \mu}\,(x^{-1/2}\,dx) = 0.33(2L^{1/2})\sqrt{\rho V^3 \mu}$$

In the usual form, $F_D = C_D\rho A V^2/2$. For this case $A = L \times 1$; hence

$$C_D\rho L V^2/2 = 0.33(2)\sqrt{\rho V^3 \mu L} \qquad \text{and} \qquad C_D = 1.32\sqrt{\frac{\mu}{\rho VL}} = \frac{1.32}{\sqrt{R_E}} \qquad (D)$$

43. A 4 ft by 4 ft thin plate is held parallel to a stream of air moving at 10 ft/sec (standard conditions). Calculate (a) the surface drag of the plate, (b) the thickness of the boundary layer at the trailing edge and (c) the shear stress at the trailing edge.

Solution:

(a) Since the "skin friction" drag coefficient varies with Reynolds number, R_E should be found.

$$R_E = VL/\nu = 10(4)/(16.0 \times 10^{-5}) = 250{,}000 \quad \text{(laminar range)}$$

Assuming laminar boundary layer conditions over the entire plate,

$$\text{coefficient } C_D = 1.328/\sqrt{R_E} = 1.328/\sqrt{250{,}000} = .00266$$

$$\text{Drag force } D \text{ (two sides)} = 2C_D \rho A V^2/2 = (.00266)(.0752/32.2)(4 \times 4)(10)^2$$
$$= .0099 \text{ lb.}$$

(b) $\dfrac{\delta}{x} = \dfrac{5.20}{\sqrt{R_{E_x}}}$ and $\delta = \dfrac{5.20(4)}{\sqrt{250{,}000}} = .0416 \text{ ft.}$

(c) $\tau = 0.33 \dfrac{\mu V}{x} \sqrt{R_{E_x}} = 0.33 \dfrac{(3.75 \times 10^{-7})10}{4} \sqrt{250{,}000} = .000155 \text{ lb/ft}^2.$

44. A smooth plate, 10 ft by 4 ft, moves through air (60°F) at a relative velocity of 4 ft/sec parallel to the plate surface and to its length. Calculate the drag force on one side of the plate (a) assuming laminar conditions, and (b) assuming turbulent conditions over the entire plate. (c) For laminar conditions, estimate the thickness of the boundary layer at the middle of the plate and at the trailing edge.

Solution:

(a) Calculate Reynolds number: $R_E = VL/\nu = 4(10)/(15.8 \times 10^{-5}) = 253{,}000.$

For laminar conditions, $C_D = \dfrac{1.328}{\sqrt{R_E}} = \dfrac{1.328}{\sqrt{253{,}000}} = .00264$ (also see Diagram G).

Drag force $= C_D \rho A V^2/2 = .00264(.00237)(10 \times 4)(4^2)/2 = .00200 \text{ lb}$

(b) For turbulent conditions, with $R_E < 10^7$, $C_D = \dfrac{.074}{R_E^{0.20}}.$ (see equation (12).)

Then $C_D = \dfrac{.074}{(253{,}000)^{0.20}} = \dfrac{.074}{12.04} = .00614$ (also see Diagram G).

Drag force $= C_D \rho A V^2/2 = .00614(.00237)(10 \times 4)(4^2)/2 = .00465 \text{ lb}$

(c) For $x = 5$ ft, $R_{E_x} = 4(5)/(15.8 \times 10^{-5}) = 126{,}500.$

Note that Reynolds number is calculated for $L = x$ ft. This value of Reynolds number is referred to as the local Reynolds number. Then

$$\delta = \frac{5.20x}{\sqrt{R_{E_x}}} = \frac{5.20(5)}{\sqrt{126{,}500}} = .073 \text{ ft}$$

For $x = 10$ ft, $R_{E_x} = 253{,}000$ and $\delta = \dfrac{5.20x}{\sqrt{R_{E_x}}} = \dfrac{5.20(10)}{\sqrt{253{,}000}} = 0.103 \text{ ft}$

45. A smooth rectangular plate 4 ft wide by 80 ft long moves through 70°F water in the direction of its length. The drag force on the plate (two sides) is 1800 lb. Find (a) the velocity of the plate, (b) the thickness of the boundary layer at the trailing edge and (c) the length x_c of the laminar boundary layer if laminar conditions obtain at the leading edge.

Solution:

(a) For the length of plate and for water the fluid, turbulent flow may be assumed with some degree of confidence. Examining Diagram G, assume $C_D = .002.$

$$\text{Drag force} = 2C_D \rho A V^2/2, \qquad 1800 = C_D(1.94)(4 \times 80)V^2$$

and $\qquad\qquad V^2 = \dfrac{2.90}{C_D} = \dfrac{2.90}{.002},\qquad V = 38.2 \text{ ft/sec}$

Reynolds number $R_E = 38.2(80)/(1.059 \times 10^{-5}) = 289 \times 10^6$. Thus the boundary layer is turbulent, as assumed. Continuing,

$$C_D = \frac{0.455}{(\log_{10} 289 \times 10^6)^{2.58}} = .00186, \qquad V^2 = \frac{2.90}{.00186} = 1560, \qquad V = 39.5 \text{ ft/sec}$$

Recalculation of Reynolds number gives 298×10^6; then

$$C_D = \frac{0.455}{(\log_{10} 298 \times 10^6)^{2.58}} = .00184 \qquad \text{and} \qquad V = 39.7 \text{ ft/sec}$$

This value is within the accuracy anticipated.

(b) The thickness of the boundary layer for turbulent flow is estimated by using equation (15).

$$\frac{\delta}{x} = \frac{0.22}{R_E^{0.167}} \qquad \text{and} \qquad \delta = \frac{0.22(80)}{(298 \times 10^6)^{0.167}} = 0.681 \text{ ft}$$

(c) Assume the critical Reynolds number is about 500,000, i.e., the transition range's lower limit.

$$R_{E_c} = \frac{V x_c}{\nu}, \qquad 500,000 = \frac{39.7 x_c}{1.059 \times 10^{-5}}, \qquad x_c = 0.13 \text{ ft}$$

46. The 10 ft by 4 ft plate of Problem 44 is held in water (50°F) moving at 4 ft/sec parallel to its length. Assuming laminar conditions in the boundary layer at the leading edge of the plate, (a) locate where the boundary layer flow changes from laminar to turbulent, (b) estimate the thickness of the boundary layer at this point, and (c) compute the friction drag of the plate.

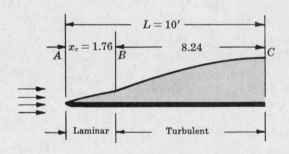

Fig. 11-17

Solution:

(a) Reynolds number $R_E = VL/\nu = 4(10)/(1.410 \times 10^{-5}) = 2,840,000$.

This value of Reynolds number indicates that the boundary layer flow is in the transition range. Assuming the critical Reynolds number is 500,000, the location of the end of the laminar flow conditions may be estimated by using

$$\frac{x_c}{L} = \frac{\text{critical } R_E}{\text{full plate } R_E} \qquad \text{or} \qquad x_c = 10\left(\frac{500,000}{2,840,000}\right) = 1.76 \text{ ft}$$

(b) The thickness of the boundary layer at this point is estimated to be

$$\delta_c = \frac{5.20 x_c}{\sqrt{R_{E_c}}} = \frac{5.20(1.76)}{\sqrt{500,000}} = .0130 \text{ ft}$$

(c) The friction drag may be computed by adding to the drag from the laminar boundary layer up to x_c (in the figure above) the drag from the turbulent boundary layer, B to C. The latter value of drag is obtained by computing the drag for turbulent flow along the entire plate (A to C) less the drag of the fictitious turbulent layer from A to B.

(1) Laminar drag, A to B, on one side.

$$\text{Drag force} = C_D \rho A \frac{V^2}{2} = \frac{1.328}{\sqrt{R_{E_c}}} \rho A \frac{V^2}{2} = \frac{1.328}{\sqrt{500,000}} (1.94)(4 \times 1.76) \frac{4^2}{2} = 0.205 \text{ lb.}$$

(2) Turbulent drag, A to C, if conditions were turbulent for the entire length of the plate.

Drag force $= C_D \rho A \dfrac{V^2}{2}$ (one side)

$$= \frac{.074}{R_E^{0.20}} \rho A \frac{V^2}{2} = \frac{.074}{(2,840,000)^{0.20}}(1.94)(4 \times 10)\frac{4^2}{2} = 2.355 \text{ lb.}$$

(3) Fictitious turbulent drag, A to B.

Drag force $= C_D \rho A \dfrac{V^2}{2}$ (one side)

$$= \frac{.074}{R_{E_c}^{0.20}} \rho A \frac{V^2}{2} = \frac{.074}{(500,000)^{0.20}}(1.94)(4 \times 1.76)\frac{4^2}{2} = 0.585 \text{ lb.}$$

Total drag force (two sides) $= 2[0.205 + (2.355 - 0.585)] = 3.95$ lb.

Had Reynolds number for the entire plate been over 10^7, equation (13) at the beginning of this chapter should be used in part (2) above.

A weighted value of C_D' might be obtained for the entire plate by equating the above total drag force to the drag force expression, as follows.

Total Drag force $= 2C_D' \rho A \dfrac{V^2}{2}$, $3.95 = 2C_D'(1.94)(4 \times 10)\dfrac{4^2}{2}$, $C_D' = .00317$

47. **Measurements on a smooth sphere, 6″ in diameter, in an air stream (68°F) gave a force for equilibrium equal to 0.250 lb. At what velocity was the air moving?**
   Solution:

Total drag $= C_D \rho A V^2/2$,   where C_D = overall drag coefficient.

Since neither Reynolds number nor C_D can be found directly, assume $C_D = 1.00$. Then

$$0.250 = C_D(.00233)\tfrac{1}{4}\pi(\tfrac{1}{2})^2(V^2/2), \qquad V^2 = \frac{1093}{C_D}, \qquad V = 33.1 \text{ ft/sec}$$

Calculate $R_E = \dfrac{Vd}{\nu} = \dfrac{33.1(\tfrac{1}{2})}{16.0 \times 10^{-5}} = 103,000$. From Diagram F, $C_D = 0.59$ (for spheres).

Then $V^2 = \dfrac{1093}{0.59} = 1850$, $V = 43.0$ ft/sec. Anticipating result, use $V = 44.0$ ft/sec.

Recalculate $R_E = \dfrac{Vd}{\nu} = \dfrac{44.0(\tfrac{1}{2})}{16.0 \times 10^{-5}} = 137,500$. From Diagram F, $C_D = 0.56$.

Then $V^2 = 1093/0.56 = 1950$, $V = 44.2$ ft/sec (satisfactory accuracy).

48. **For instantaneous closure of a valve in a pipe line, determine the increase in the pressure produced.**
   Solution:

Let p' equal the change in pressure due to the closure of the valve. The impulse-momentum equation can be employed to evaluate the change in momentum, i.e.,

$$F_x = \frac{wQ}{g}(V_2 - V_1) \qquad \text{in the } x\text{-direction} \tag{A}$$

Neglecting friction, the unbalanced force which causes a change in momentum of the liquid in the pipeline is $p'A$. Equation (A) then becomes

$$-p'A = \frac{w(Ac)}{g}(0 - V_1) \tag{B}$$

where wAc/g represents the mass of liquid having its momentum changed and c represents the celerity of the pressure wave. This pressure wave reduces the velocity to zero as it passes each section. Then

$$p' = \rho c V_1 \tag{C}$$

Equation (C) can be written in terms of pressure head h', i.e.,

$$h' = \frac{cV_1}{g} \tag{D}$$

49. What is the formula for the celerity of the pressure wave due to rapid closing of a valve in a pipeline, considering the pipe to be rigid?

Solution:

By "rapid closure" is meant any time $\leq 2L/c$. In order to establish an expression for celerity c, the principles of work and energy, and of momentum must be utilized.

The kinetic energy of the water will be converted to elastic energy, thereby compressing the water. The kinetic energy is $\frac{1}{2}MV_1^2 = \frac{1}{2}(wAL/g)V_1^2$, where A is the cross-sectional area of the pipe and L is the length of the pipe.

The bulk modulus of elasticity of the water is $E_B = \dfrac{-\Delta p}{(\Delta\text{ volume}) \div (\text{original volume})}$ (psf).

Thus the volume reduction, $\Delta\text{ volume} = \dfrac{(\text{volume})(\Delta p)}{E_B} = \dfrac{(AL)(wh)}{E_B}$.

The work of compression = average pressure intensity times the volume reduction, i.e.,

$$\frac{1}{2}(wAL/g)V_1^2 = \frac{1}{2}wh(ALwh/E_B) \tag{A}$$

or
$$h^2 = V_1^2 E_B/gw \tag{B}$$

The momentum principle yields (neglecting friction),

$$MV_1 - \Sigma(F_x\,dt) = MV_2, \qquad -whA = (wQ/g)(0 - V_1), \qquad whA = (w/g)(Ac)V_1$$

or
$$h = cV_1/g \tag{C}$$

Substituting in (B), we obtain $c^2V_1^2/g^2 = V_1^2 E_B/gw$ from which

$$c = \sqrt{E_B/\rho} \tag{D}$$

50. Develop the formula for the celerity of the pressure wave due to rapid closing of a valve in a pipeline, considering the pipe to be non-rigid.

Solution:

In this solution, the elasticity of the walls of the pipe must be considered, in addition to the factors included in the solution of the preceding problem.

For the pipe, the work done in stretching the pipe walls is equal to the average force exerted in the pipe walls times the strain. From a free body diagram of one-half of the pipe cross section, using $\Sigma Y = 0$, $2T = pdL = whdL$. Also unit strain $\epsilon = \sigma/E$ where $\sigma = pr/t = whr/t$. (See hoop tension in Chapter 2.) In this derivation, head h represents the pressure head above normal, caused by the rapid closing of the valve.

$$\text{Work} = \text{average force} \times \text{strain} = \frac{1}{2}(\frac{1}{2}whdL)(2\pi r\epsilon) \quad \text{in ft lb}$$
$$= \frac{1}{4}whdL(2\pi r)(whr/tE)$$

Adding this value to equation (A) of the preceding problem gives

$$\frac{1}{2}(wAL/g)V_1^2 = \frac{1}{2}wh(ALwh/E_B) + \frac{1}{4}whdL(2\pi whr^2/tE)$$

which, after substituting $h = cV_1/g$ from (C) of Prob. 49, gives

$$\frac{V_1^2}{g} = \frac{c^2V_1^2}{g^2}\left(\frac{w}{E_B} + \frac{wd}{tE}\right)$$

$$\text{Celerity } c = \sqrt{\frac{1}{\rho(1/E_B + d/Et)}} = \sqrt{\frac{E_B}{\rho(1 + E_B d/Et)}}$$

51. Compare the velocities of the pressure waves traveling along a rigid pipe containing (*a*) water at 60°F, (*b*) glycerin at 68°F and (*c*) oil of sp gr 0.800. Use values of bulk modulus for glycerin and oil of 630,000 and 200,000 psi respectively.

Solution:

$$c = \sqrt{\frac{\text{bulk modulus in psf}}{\text{density of fluid}}}$$

(*a*)
$$c = \sqrt{\frac{313,000 \times 144}{1.94}} = 4820 \text{ ft/sec}$$

(*b*)
$$c = \sqrt{\frac{630,000 \times 144}{1.262 \times 62.4/32.2}} = 6080 \text{ ft/sec}$$

(*c*)
$$c = \sqrt{\frac{200,000 \times 144}{0.800 \times 62.4/32.2}} = 4310 \text{ ft/sec}$$

52. In Problem 51, had these liquids been flowing in a 12″ pipe at 4.0 ft/sec and had the flow been stopped suddenly, what increase in pressure could be expected, assuming the pipe to be rigid?

Solution:

The change (increase) in pressure = $\rho c \times$ change in velocity.

(*a*)   Increase in pressure = $1.94(4820)(4 - 0) = 37,400 \text{ lb/ft}^2 = 260$ psi.

(*b*)   Increase in pressure = $2.45(6080)(4) = 59,700 \text{ lb/ft}^2 = 415$ psi.

(*c*)   Increase in pressure = $1.55(4310)(4) = 26,700 \text{ lb/ft}^2 = 185$ psi.

53. A 48″ steel pipe, ⅜″ thick, carries water at 60°F at a velocity of 6 ft/sec. If the pipe line is 10,000 ft long and if a valve at the discharge end is shut in 2.50 sec, what increase in stress in the walls of the pipe could be expected?

Solution:

The pressure wave would travel from valve to inlet end and back again in

$$\text{Time (round trip)} = 2\left(\frac{\text{length of pipe line}}{\text{celerity of pressure wave}}\right)$$

The celerity of the pressure wave for a non-rigid pipe is given by

$$c = \sqrt{\frac{E_B \text{(psf)}}{\rho[1 + (E_B/E)(d/t)]}}$$

where the two ratios are dimensionless when proper units are used.

Taking E for steel = 30×10^6 psi, $c = \sqrt{\dfrac{313,000 \times 144}{1.94\left[1 + \dfrac{313,000}{30 \times 10^6}\left(\dfrac{48}{3/8}\right)\right]}} = 3150$ ft/sec

and   time = $2(10,000/3150) = 6.35$ sec.

But the valve was closed in 2.50 sec. This is equivalent to a *sudden closure*, since the pressure wave upon reaching the closed valve must reverse itself.

Increase in pressure = $\rho c(dV) = 1.94(3150)(6) = 36,700$ psf = 254 psi.

From the hoop tension formula for thin-shelled cylinders,

$$\text{Tensile stress } \sigma = \frac{\text{pressure} \times \text{radius}}{\text{thickness}} = \frac{254 \times 24}{3/8} = 16,300 \text{ psi increase}$$

This increase in stress added to the design value of 16,000 psi approaches the elastic limit of steel. The time of closing of the valve should be increased to at least 6.5 sec, preferably several times 6.35 sec.

For slow closure of valves, where the time is greater than $2L/c$, Norman R. Gibson suggests a method of arithmetical integration. Reference is made to Volume 83 of the Transactions of the American Society of Civil Engineers for 1919.

54. A valve is suddenly closed in a 3″ pipe carrying glycerin at 68°F. The increase in pressure is 100 psi. What is the probable flow in cfs? Use $\rho = 2.45$ and $E_B = 630,000$ psi.

Solution:

The value of the celerity was calculated in problem 51 as 6080 ft/sec.

The change in pressure $= \rho c \times$ change in velocity

$$100 \times 144 = 2.45(6080)V \qquad \text{from which} \qquad V = 0.965 \text{ ft/sec.}$$

Thus $Q = AV = \frac{1}{4}\pi(3/12)^2 \times 0.965 = .0474$ cfs.

55. Air at 80°F flows with a velocity of 20 ft/sec through a 5 ft square ventilation duct. If the control devices are suddenly closed, what force would be exerted on the 5 ft × 5 ft area of closure?

Solution:

For air at 80°F, $\rho = .00228$ slug/ft³ and the celerity

$$c = \sqrt{kgRT} = \sqrt{1.4(32.2)(53.3)(460 + 80)} = 1138 \text{ ft/sec}$$

Then, using $\Delta p = \rho c V$, the force

$$F = \Delta p \times \text{area} = (\rho c V)A = .00228(1138)(20)(5 \times 5) = 1300 \text{ lb}$$

56. A sonar transmitter operates at 2 impulses per second. If the device is held at the surface of fresh water at 40°F and the echo is received midway between impulses, how deep is the water? (The depth is known to be less than 2000 ft.)

Solution:

The celerity of sound in water at 40°F is computed by using

$$c = \sqrt{\frac{\text{bulk modulus}}{\text{density of fluid}}} = \sqrt{\frac{296,000 \times 144}{1.94}} = 4690 \text{ ft/sec}$$

(a) The distance traveled by the sound wave (down and back again) in $\frac{1}{2}$ of $\frac{1}{2}$ sec or $\frac{1}{4}$ sec (one-half impulse time) is

$$2 \times \text{depth} = \text{velocity} \times \text{time}$$
$$= 4690 \times \tfrac{1}{4} \qquad \text{and} \qquad \text{depth} = 586 \text{ ft (least depth)}$$

(b) Had the depth exceeded 586 ft, for the echo to be heard midway between impulses, the sound would have traveled for 3/2 of 1/2 sec or 3/4 sec. Then

$$\text{depth} = \tfrac{1}{2}(4690) \times \tfrac{3}{4} = 1760 \text{ ft}$$

(c) For depths beyond the 2000 ft limit suggested, we obtain

$$\text{depth} = \tfrac{1}{2}(4690) \times \tfrac{5}{4} = 2930 \text{ ft,}$$
$$\text{depth} = \tfrac{1}{2}(4690) \times \tfrac{7}{4} = 4100 \text{ ft, and so on}$$

57. A projectile moves at 2200 ft/sec through still air at 100°F and 14.5 psia. Determine (a) the Mach number, (b) the Mach angle and (c) the drag force for shape B on Diagram H, assuming 8″ in diameter.

Solution:

(a)   Celerity $c = \sqrt{kgRT} = \sqrt{1.4(32.2)(53.3)(460+100)} = 1160$ ft/sec.

$$\text{Mach number } N_M = \frac{V}{c} = \frac{2200}{1160} = 1.90$$

(b)   Mach angle $\alpha = \sin^{-1}\frac{1}{N_M} = \sin^{-1}\frac{1}{1.90} = 31.7°.$

(c)   From Diagram H, shape B, for Mach number of 1.90 the coefficient of drag is 0.60.

The specific weight of the air is $w = \frac{p}{RT} = \frac{14.5 \times 144}{53.3(460+100)} = .0695$ lb/ft³.

Drag force $= C_D\rho A V^2/2 = 0.60(.0695/32.2) \times \frac{1}{4}\pi(8/12)^2 \times (2200)^2/2 = 1090$ lb.

58. From a photograph the Mach angle for a projectile moving through air was 40°. Calculate the speed of the bullet for air conditions in the preceding problem. (Celerity $c = 1160$ ft/sec.)

Solution:

$\text{Sin } \alpha = \frac{c}{V} = \frac{1}{N_M}.$ Then $\sin 40° = \frac{1160}{V}$ and $V = 1800$ ft/sec.

59. What should be the diameter of a sphere, specific gravity 2.50, in order that its freely falling velocity attain the acoustic velocity? Use $\rho = .00237$ for air.

Solution:

For the freely falling body, drag force − weight = 0 and, from Diagram H, $C_D = 0.80$.

For air at 60°F, $c = \sqrt{kgRT} = \sqrt{1.4(32.2)(53.3)(460+60)} = 1120$ ft/sec.

Since   Weight = drag force,

$(2.50 \times 62.4) \times (4\pi/3)(d/2)^3 = 0.80 \times .00237 \times (\pi d^2/4) \times (1120)^2/2,$ $d = 11.4$ ft

Supplementary Problems

60. Show that the momentum correction factor β in Prob. 74 of Chap. 6 is 1.20.

61. Show that the momentum correction factor β in Prob. 72 of Chap. 6 is 1.02.

62. Determine the momentum correction factor β for Prob. 79 in Chap. 6.   *Ans.* $\dfrac{(K+1)^2(K+2)^2}{2(2K+1)(2K+2)}$

63. Show that the momentum correction factor β in Prob. 59 in Chap. 7 is 1.12.

64. A jet of oil, 2″ in diameter, strikes a flat plate held normal to the stream's path. For a velocity in the jet of 80 ft/sec, calculate the force exerted on the plate by the oil, sp gr 0.85.   *Ans.* 230 lb

65. In Problem 64, had the plate been moving at 30 ft/sec in the same direction as the fluid, what force would be exerted on the plate by the oil? At 30 ft/sec in the opposite direction, what force would be exerted?   *Ans.* 89.5 lb, 434 lb

66. A jet of water, 2″ in diameter, exerts a force of 600 lb on a flat plate held normal to the jet's path. What is the rate of discharge?   *Ans.* 2.60 cfs

67. Water flowing at the rate of 1.20 cfs strikes a flat plate held normal to its path. If the force exerted on the plate is 162 lb, calculate the diameter of the stream.   *Ans.* 1.78 in.

68. A jet of water, 2″ in diameter, strikes a curved blade at rest and is deflected 135° from its original direction. Neglecting friction along the blade, find the resultant force exerted on the blade if the jet velocity is 90 ft/sec.   *Ans.* 630 lb, $\theta_x = -22.6°$

69. Had the blade in the preceding problem been moving at 20 ft/sec against the direction of the water, what resultant force is exerted on the blade and what power would be required to maintain the motion? *Ans.* 943 lb, 31.7 hp

70. A stationary blade deflects a 2″ diameter jet moving at 115 ft/sec through 180°. What force does the blade exert on the water?   *Ans.* 1120 lb

71. A horizontal 12″ pipe contracts to a 6″ diameter. If the flow is 4.50 cfs of oil, specific gravity 0.88, and the pressure in the smaller pipe is 38.5 psi, what resultant force is exerted on the contraction, neglecting friction?   *Ans.* 3470 lb

72. A vertical reducing bend (see Fig. 11-18) carries 12.6 cfs of oil, sp gr 0.85, at a pressure of 20.5 psi entering the bend at A. The diameter at A is 16″ and at B is 12″ and the volume between A and B is 3.75 cu ft. Neglecting friction, find the force on the bend.   *Ans.* 5180 lb, $\theta_x = -76.1°$

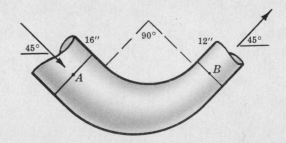

Fig. 11-18

73. A model of a motorboat is driven at 15.0 ft/sec by means of a jet of water 1″ in diameter, ejected directly astern. The velocity of the jet relative to the model is 115 ft/sec. What is the driving force? *Ans.* 122 lb

74. A 2″ diameter nozzle, $c_v = 0.97$, is attached to a tank and discharges a stream of oil, sp gr 0.80, horizontally under a head of 36 ft. What horizontal force is exerted on the tank?   *Ans.* 73.7 lb

75. A toy balloon weighing 0.23 lb is filled with air having a density of .00250 slug/ft³. The small filling tube, $\frac{1}{4}$″ in diameter, is pointed downward and the balloon is released. If the air escapes at the initial rate of 0.30 cfs, what is the instantaneous acceleration (neglecting friction)?   *Ans.* 60.2 ft/sec²

76. A jet-propulsion device is moving upstream in a river with absolute velocity of 28.5 ft/sec. The stream is flowing at 7.5 ft/sec. A jet of water issues from the device at a relative velocity of 60.0 ft/sec. If the flow in the jet is 50 cfs, what thrust is developed on the jet-propulsion device?   *Ans.* 2330 lb

77. What weight will be supported by a wing area of 500 ft^2 at an angle of attack of 4° and an air speed of 100 ft/sec? Use $C_L = 0.65$ and 60°F air.   *Ans.* 3850 lb

78. At what speed should a plane having a wing area of 500 ft^2 and weighing 6000 lb be flown if the angle of attack is 8°. Use $C_L = 0.90$.   *Ans.* 106 ft/sec

79. What wing area should a plane weighing 2000 lb have in order that the plane land at a speed of 35 mph? Use max. $C_L = 1.50$.   *Ans.* 425 ft^2

80. If the drag on a wing of 300 ft^2 area is 680 lb, at what speed is the wing moving for an angle of attack of 7°? Use $C_D = .05$.   *Ans.* 196 ft/sec

81. A 30 mph wind blows at an angle of 8° to the plane of a signboard which is 12 ft long by 2 ft wide. Using values of $C_L = 0.52$ and $C_D = .09$, calculate (a) the force acting on the signboard at right angles to the wind direction and (b) the force acting parallel to the wind direction. Assume standard air at 60°F.   *Ans.* 28.6 lb, 4.95 lb

82. Prove that for a given angle of attack, the drag force on an airfoil is the same for all altitudes. (For a given angle of attack, C_D will not change with altitude.)

83. A wing model of 3 ft span and 4″ chord is tested at a fixed angle of attack in a wind tunnel. The air, at standard pressure and 80°F, has a velocity of 60 mph. The lift and drag are measured at 5.8 lb and 0.50 lb respectively. Determine the lift and drag coefficients.   *Ans.* 0.657, .0566

84. Calculate the Mach number for (a) an airplane moving through standard air at 68°F at 300 mph, (b) a rocket moving through air at 68°F at 2400 mph and (c) a projectile moving through standard air at 68°F at 1200 mph.   *Ans.* 0.391, 3.13, 1.56

85. A turbojet engine takes in air at the rate of 45 lb/sec while moving at 700 ft/sec. If the thrust developed is 2700 lb for an exhaust velocity of 2500 ft/sec, how much fuel is consumed per second?   *Ans.* 2.38 lb/sec

86. Air enters the intake duct of a jet engine at atmospheric pressure and at 500 ft/sec. Fuel is used by the engine at the rate of 1 part of fuel to 50 parts of intake air. The intake duct area is 216 in^2 and the density of the air is .00240 slugs/ft^3. If the velocity of the exhaust gases is 5000 ft/sec and the pressure is atmospheric, what thrust is developed?   *Ans.* 8280 lb

87. An automobile has a projected area of 32.0 ft^2 and moves at 50 mph in still air at 80°F. If $C_D = 0.45$, what horsepower is necessary to overcome air resistance?   *Ans.* 11.8 hp

88. A train travels at 75 mph through standard air at 60°F and is 500 ft long. Consider the surfaces of the train to be 15,000 ft^2 in area and equivalent to a smooth flat plate. For a turbulent leading edge, what is the skin-friction drag?   *Ans.* 375 lb on one side

89. A cylinder, 24″ in diameter and 15 ft long, moves at 30 mph through 60°F water (parallel to its length). What coefficient of drag is indicated if the skin-friction drag is 360 lb.   *Ans.* $C_D = .00204$

90. Calculate the friction drag on a plate 1 ft wide by 3 ft long placed longitudinally (a) in a stream of water at 70°F flowing at 1.0 ft/sec and (b) in a stream of heavy fuel oil at 70°F at 1.0 ft/sec.   *Ans.* .0144 lb, 0.161 lb

91. A balloon 4 ft in diameter weighs 4.0 lb and is acted upon by a buoyant force averaging 5.0 lb. Using $\rho = .00228$ slug/ft^3 and $\nu = 17.0 \times 10^{-5}$ ft^2/sec, estimate the velocity with which it will rise.   *Ans.* 18.7 ft/sec

92. Estimate the terminal velocity of a hailstone $\frac{1}{2}$″ in diameter using air temperature = 40°F and specific gravity of the hailstone = 0.90.   *Ans.* 55.0 ft/sec

93. An object having a projected area of 6.0 ft² moves at 30 mph. If the drag coefficient is 0.30, compute the drag force in 60°F water and in 60°F standard air.   *Ans.* 3380 lb, 4.11 lb

94. A body travels through 60° standard air at 60 mph, and 5.5 horsepower is required to maintain this speed. If the projected area is 13.5 ft², find the drag coefficient.   *Ans.* 0.28

95. A smooth rectangular plate 2 ft wide by 80 ft long moves at 40.0 ft/sec in the direction of its length through oil. Calculate the drag force on the plate and the thickness of the boundary layer at the trailing edge. How long is the laminar boundary layer? Use kinematic viscosity $= 16.0 \times 10^{-5}$ ft²/sec and $w = 53.0$ lb/ft³.   *Ans.* 1120 lb, 1.07 ft, 2.00 ft

96. Assuming a 24″ steel pipe to be rigid, what pressure increase occurs when a flow of 20.0 cfs of oil, specific gravity 0.85 and bulk modulus 250,000 psi, is stopped suddenly?   *Ans.* 341 psi

97. If the pipeline in Problem 96 is 8000 ft long, how much time should be allowed for closing a valve to avoid water hammer?   *Ans.* more than 3.42 sec

98. If a 24″ steel pipe 8000 ft long is designed for a stress of 15,000 psi under a maximum static head of 1085 ft of water, how much will the stress in the walls of the pipe increase when a quick-closing valve stops a flow of 30.0 cfs? ($E_B = 300,000$ psi)   *Ans.* 472 psi

99. Calculate the Mach angle for a bullet moving at 1700 ft/sec through air at 14.3 psi and 60°F.   *Ans.* 41°13′

100. What is the drag force of a projectile (shape *A* in Diagram *H*) 4″ in diameter when it moves at 1900 ft/sec through air at 50°F and 14.3 psi?   *Ans.* 192 lb

Chapter 12

Fluid Machinery

FLUID MACHINERY

Consideration is given here to fundamental principles upon which the design of pumps, blowers, turbines and propellers is based. The essential tools are the principles of impulse-momentum (Chapter 11) and of the forced vortex (Chapter 4), and the laws of similarity (Chapter 5). Modern hydraulic turbines and centrifugal pumps are highly efficient machines with few differences in their characteristics. For each design there is a definite relationship between the speed of rotation N, the discharge or flow Q, the head H, the diameter D of the rotating element and the power P.

FOR ROTATING CHANNELS, the torque and power produced are evaluated by

$$\text{Torque } T \text{ in ft lb} = \frac{wQ}{g}(V_2 r_2 \cos \alpha_2 - V_1 r_1 \cos \alpha_1) \tag{1}$$

and

$$\text{Power } P \text{ in ft lb/sec} = \frac{wQ}{g}(V_2 u_2 \cos \alpha_2 - V_1 u_1 \cos \alpha_1) \tag{2}$$

The development and notation are explained in Problem 1.

WATER WHEELS, TURBINES, PUMPS and BLOWERS

These have certain constants which are commonly evaluated. Details are developed in Problem 5.

1. *The speed factor ϕ is defined as*

$$\phi = \frac{\text{peripheral velocity of rotating element}}{\sqrt{2gH}} = \frac{u}{\sqrt{2gH}} \tag{3}$$

where u = radius of rotating element in ft $\times$ angular velocity in radians/sec = $r\omega$ ft/sec.

This factor is also expressed as

$$\phi = \frac{\text{diameter in inches} \times \text{rpm}}{1840\sqrt{H}} = \frac{D_1 N}{1840\sqrt{H}} \tag{4}$$

2a. *The speed relation* may be expressed as

$$\frac{\text{diameter } D \text{ in ft} \times \text{speed } N \text{ in rpm}}{\sqrt{g \times \text{head } H \text{ in ft}}} = \text{constant } C'_N \tag{5a}$$

Also

$$H = \frac{D^2 N^2}{C_N^2} \tag{5b}$$

in which g is incorporated in the C_N coefficient.

2b. *The unit speed* is defined as the speed of a geometrically similar (homologous) rotating element having a diameter of 1 in., operating under a head of 1 ft. This unit speed (N_u in rpm) is usually expressed in terms of D_1 in inches and N in rpm. Thus

$$N_u = \frac{D \text{ in in.} \times \text{rpm}}{\sqrt{H}} = \frac{D_1 N}{\sqrt{H}} \tag{6a}$$

Also

$$N = N_u \frac{\sqrt{H}}{D_1} \tag{6b}$$

3a. *The discharge relation* may be expressed as

$$\frac{\text{discharge } Q \text{ in cfs}}{(\text{diameter } D \text{ in ft})^2 \sqrt{\text{head } H \text{ in ft}}} = \text{constant } C_Q \tag{7a}$$

Also

$$Q = C_Q D^2 \sqrt{H} = C_Q D^2 \left(\frac{DN}{C_N}\right) = C_Q' D^3 N \tag{7b}$$

The coefficient C_Q may also be expressed in terms of gpm flow units. In taking coefficients from texts or handbooks, units should be checked or mistakes will occur.

If C_Q is made the same for two homologous units, then C_N, C_P and the efficiency will be the same, unless very viscous fluids are involved.

3b. *The unit discharge* is defined as the discharge of an homologous rotating element 1 in. in diameter, operating under a head of 1 ft. The unit flow Q in cfs is written

$$Q_u = \frac{\text{flow } Q \text{ in cfs}}{(\text{diameter } D \text{ in in.})^2 \sqrt{\text{head } H \text{ in ft}}} = \frac{Q}{D_1^2 \sqrt{H}} \tag{8a}$$

Also

$$Q = Q_u D_1^2 \sqrt{H} \tag{8b}$$

4a. *The power relation,* obtained by using values of Q and H in equations (7b) and (5a) is

$$\text{Horsepower } P = \frac{wQH}{550e} = \frac{w(C_Q D^2 \sqrt{H})H}{550e} = C_P D^2 H^{3/2} \tag{9a}$$

Also

$$P = \frac{w(C_Q' D^3 N)}{550e} \times \frac{D^2 N^2}{g(C_N')^2} = C_P' \rho D^5 N^3 \tag{9b}$$

4b. *The unit power* is defined as the power developed by an homologous rotating element 1 in. in diameter, operating under a head of 1 ft. The unit power P_u is

$$P_u = \frac{P}{D_1^2 H^{3/2}} \quad \text{and} \quad P = P_u D_1^2 H^{3/2} \tag{10}$$

SPECIFIC SPEED

The specific speed is defined as the speed of an homologous rotating element of such diameter that it develops 1 horsepower for a head of 1 ft (see Problem 5). The specific speed N_S may be expressed in two forms, as follows:

1. *For turbines:*

$$N_S = \frac{N\sqrt{P}}{\sqrt{\rho}\,(gH)^{5/4}}, \quad \text{which represents the general equation.} \tag{11a}$$

Also $\quad N_S = N_u \sqrt{P_u} = \dfrac{N\sqrt{P}}{H^{5/4}}$ has common application for water turbines. *(11b)*

2. *For pumps and blowers:*

$$N_S = \frac{N\sqrt{Q}}{(gH)^{3/4}} \quad \text{represents the general equation.} \tag{12a}$$

Also $\qquad N_S = N_u\sqrt{Q_u} = \dfrac{N\sqrt{Q}}{H^{3/4}} \quad$ has common application. $\tag{12b}$

EFFICIENCY

Efficiency is expressed as a ratio. It will vary with speed and discharge.

$$\text{\textit{For turbines,} overall efficiency } e \;=\; \frac{\text{power derived from shaft}}{\text{power supplied by water}} \tag{13}$$

$$\text{hydraulic efficiency } e_h \;=\; \frac{\text{power utilized by unit}}{\text{power supplied by water}}$$

$$\text{\textit{For pumps,} efficiency } e \;=\; \frac{\text{power output}}{\text{power input}} \;=\; \frac{wQH}{\text{power input}}. \tag{14}$$

CAVITATION

Cavitation causes the rapid disintegration of metal in pump impellers, turbine runners and blades, Venturi meters and, on occasion, in pipe lines. It occurs when the pressure in the liquid falls below the vapor pressure of that liquid. The reader is referred to such works as "Engineering Hydraulics", Proceedings of the Fourth Hydraulics Conference, for a comprehensive coverage of this specialized subject.

PROPULSION by PROPELLERS

Propulsion by propellers has been the motive power of aircraft and marine craft for some time. Also, propellers are used as fans and as means of producing power from the wind. Propeller design will not be attempted here but expressions for thrust and power are matters of fluid mechanics. Such expressions, developed in Problem 23, are:

$$\text{Thrust } F = \frac{wQ}{g}\,(V_{\text{final}} - V_{\text{initial}}) \quad \text{in lb} \tag{15}$$

$$\text{Power output } P_o = \frac{wQ}{g}(V_{\text{final}} - V_{\text{initial}})\,V_{\text{initial}} \quad \text{in ft lb/sec} \tag{16}$$

$$\text{Power input } P_i = \frac{wQ}{g}\left(\frac{V_{\text{final}}^2 - V_{\text{initial}}^2}{2}\right) \tag{17}$$

$$\text{Efficiency } e = \frac{\text{power output}}{\text{power input}} = \frac{2V_{\text{initial}}}{V_{\text{final}} + V_{\text{initial}}} \tag{18}$$

PROPELLER COEFFICIENTS involve thrust, torque and power. They may be expressed as follows.

$$\text{Thrust coefficient } C_F \;=\; \frac{\text{thrust } F \text{ in lb}}{\rho N^2 D^4} \tag{19}$$

High values of C_F produce good propulsion.

$$\text{Torque coefficient } C_T \;=\; \frac{\text{torque } T \text{ in ft lb}}{\rho N^2 D^5} \tag{20}$$

High values of C_T are common for turbines and windmills.

$$\text{Power coefficient } C_P \;=\; \frac{\text{power } P \text{ in ft lb/sec}}{\rho N^3 D^5} \tag{21}$$

This last coefficient is in the same form as equation $(9b)$ above.

All three coefficients are dimensionless if N is in revolutions per sec.

Solved Problems

1. Determine the torque and power developed by a rotating element (such as pump impeller or turbine runner) under steady flow conditions.

Solution:

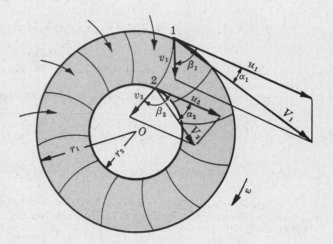

Let Fig. 12-1 represent water entering the curved channels formed by the rotating element at radius r_1 and leaving at radius r_2. The relative velocities of the water with respect to a blade are shown as v_1 entering at (1) and v_2 at exit (2). The linear velocity of the blade is u_1 at (1) and u_2 at (2). The vector diagrams show the absolute velocities of the water (V_1 and V_2).

For the elementary mass of water flowing in dt sec the change in angular momentum is caused by the angular impulse exerted by the runner. That is,

Fig. 12-1

initial angular momentum + angular impulse = final angular momentum

or $\qquad (dM)V_1 \times r_1 \cos\alpha_1 + \text{torque} \times dt = (dM)V_2 \times r_2 \cos\alpha_2$

But $dM = (w/g)Q\,dt$. Substituting and solving for the torque exerted on the water, we obtain

$$\text{torque } T \;=\; \frac{w}{g}Q(V_2 r_2 \cos\alpha_2 - V_1 r_1 \cos\alpha_1)$$

Thus the torque exerted by the fluid on the rotating part is

$$T \;=\; \frac{w}{g}Q(V_1 r_1 \cos\alpha_1 - V_2 r_2 \cos\alpha_2) \quad \text{in ft lb}$$

Power is equal to torque times angular velocity. Then

$$P \;=\; T\omega \;=\; \frac{w}{g}Q(V_1 r_1 \cos\alpha_1 - V_2 r_2 \cos\alpha_2)\omega$$

Since $u_1 = r_1\omega$ and $u_2 = r_2\omega$, the expression becomes

$$P \;=\; \frac{w}{g}Q(V_1 u_1 \cos\alpha_1 - V_2 u_2 \cos\alpha_2) \quad \text{in ft lb/sec} \tag{1}$$

The expressions developed here are applicable to pumps and turbines alike. The important point is that, in the development, point (1) was *upstream* and point (2) was *downstream*.

2. Establish the Bernoulli equation for a rotating turbine runner.

Solution:

Writing the Bernoulli equation, point (1) to point (2) in the figure of Problem 1, yields

$$\left(\frac{V_1^2}{2g} + \frac{p_1}{w} + z_1\right) - H_T - \text{lost head } H_L = \left(\frac{V_2^2}{2g} + \frac{p_2}{w} + z_2\right)$$

From the vector diagram of Problem 1 and the cosine law,

$$V_1^2 = u_1^2 + v_1^2 + 2u_1 v_1 \cos \beta_1$$
$$V_2^2 = u_2^2 + v_2^2 + 2u_2 v_2 \cos \beta_2$$

and

Also, letting $V_1 \cos \alpha_1 = a_1$ and $V_2 \cos \alpha_2 = a_2$, we may evaluate from the vector diagram

$$a_1 = u_1 + v_1 \cos \beta_1 \quad \text{and} \quad a_2 = u_2 + v_2 \cos \beta_2$$

Furthermore,

$$H_T \text{ ft lb/lb} = \frac{wQ}{g}(V_1 u_1 \cos \alpha_1 - V_2 u_2 \cos \alpha_2) \div wQ$$

$$= \frac{1}{g}(u_1 V_1 \cos \alpha_1 - u_2 V_2 \cos \alpha_2) \tag{1}$$

The velocity head and turbine head terms in the above Bernoulli equation would then become

$$\frac{u_1^2 + v_1^2 + 2u_1 v_1 \cos \beta_1}{2g}, \quad \frac{2(u_1 a_1 - u_2 a_2)}{2g}, \quad \frac{u_2^2 + v_2^2 + 2u_2 v_2 \cos \beta_2}{2g}$$

Simplifying and incorporating these terms in the Bernoulli equation produces

$$\left(\frac{v_1^2}{2g} - \frac{u_1^2}{2g} + \frac{p_1}{w} + z_1\right) + \frac{u_2^2}{2g} - H_L = \left(\frac{v_2^2}{2g} + \frac{p_2}{w} + z_2\right)$$

or

$$\left(\frac{v_1^2}{2g} + \frac{p_1}{w} + z_1\right) - \left(\frac{u_1^2 - u_2^2}{2g}\right) - H_L = \left(\frac{v_2^2}{2g} + \frac{p_2}{w} + z_2\right) \tag{2}$$

in which velocities v are relative values and the term in the second parenthesis is designated as the head created by the forced vortex or the centrifugal head.

3. A turbine rotates at 100 rpm and discharges 28.6 cfs. The pressure head at exit is 1.00 ft and the hydraulic efficiency under these conditions is 78.5%. The physical data are: $r_1 = 1.50$ ft, $r_2 = 0.70$ ft, $\alpha_1 = 15°$, $\beta_2 = 135°$, $A_1 = 1.25$ ft², $A_2 = 0.818$ ft², $z_1 = z_2$. Assuming a lost head of 4.00 ft, determine (a) the power delivered to the turbine, (b) the total available head and the head utilized, and (c) the pressure at entrance.

Solution:

(a) Preliminary calculations must be made before substituting in the power equation (equation (1) of Problem 1).

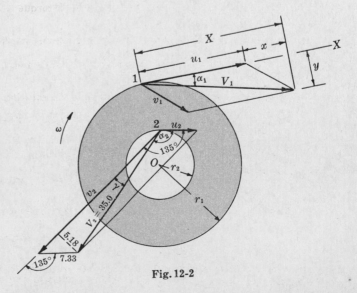

Fig. 12-2

$V_1 = Q/A_1 = 28.6/1.25 = 22.9$ ft/sec, $V_2 = 28.6/0.818 = 35.0$ ft/sec.

$V_1 \cos \alpha_1 = 22.9 \times 0.966 = 22.15$ ft/sec.

$u_1 = 1.50(2\pi)(100/60) = 15.71$ ft/sec, $u_2 = 0.70(2\pi)(100/60) = 7.33$ ft/sec.

From the vector diagram in Fig. 12-2, where $\gamma = \sin^{-1} 5.18/35.0 = 8°31'$, we have

$\alpha_2 = 135° - \gamma = 126°29'$ and $V_2 \cos \alpha_2 = 35.0(-0.595) = -20.8$ ft/sec

Then   power $P = \dfrac{62.4 \times 28.6}{550 \times 32.2} [15.71(22.15) - 7.33(-20.8)] = 50.4$ horsepower.

(b) Efficiency $= \dfrac{\text{output}}{\text{input}} = \dfrac{\text{head utilized}}{\text{head available}}$.

But   head utilized $= \dfrac{\text{horsepower utilized} \times 550}{wQ}$ or $H_T = \dfrac{50.4 \times 550}{62.4 \times 28.6} = 15.55$ ft.

Thus   head available $= 15.55/0.785 = 19.8$ ft

(c) In order to use equation (2) of the preceding problem, we must calculate the two relative velocities. Referring again to the vector diagram above, we obtain

$X = 22.9 \cos 15° = 22.9(0.966) = 22.15$ ft/sec (as above in (a))

$y = 22.9 \sin 15° = 22.9(0.259) = 5.93$ ft/sec

$x = (X - u_1) = 22.15 - 15.71 = 6.44$ ft/sec

$v_1 = \sqrt{(5.93)^2 + (6.44)^2} = \sqrt{76.7} = 8.76$ ft/sec

In similar fashion,

$v_2 = V_2 \cos \gamma + u_2 \cos 45° = 35.0(0.989) + 7.33(0.707) = 39.8$ ft/sec

The Bernoulli equation becomes

$$\left[\frac{(8.76)^2}{2g} + \frac{p_1}{w} + 0 \right] - \left[\frac{(15.71)^2}{2g} - \frac{(7.33)^2}{2g} \right] - 4.00 = \left[\frac{(39.8)^2}{2g} + 1.0 + 0 \right]$$

from which $p_1/w = 31.4$ ft.

4. Determine the value of the head developed by the impeller of a pump.

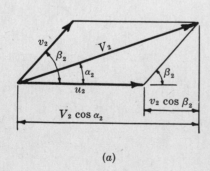

(a)

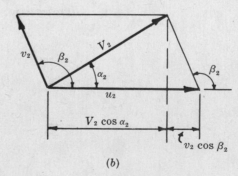

(b)

Fig. 12-3

Solution:

Expression (1) of Problem 1, applied in the direction of flow in a pump (where r_1 is the inner radius, etc.), becomes

$$\text{Power input} = \frac{wQ}{g}(u_2 V_2 \cos \alpha_2 - u_1 V_1 \cos \alpha_1)$$

and the head imparted by the impeller is obtained by dividing by wQ, thus,

$$\text{head } H' = \frac{1}{g}(u_2 V_2 \cos \alpha_2 - u_1 V_1 \cos \alpha_1)$$

In most pumps the flow at point (1) can be assumed to be radial and the value of the $u_1 V_1 \cos \alpha_1$ term is zero. The above equation then becomes

$$\text{head } H' = \frac{1}{g}(u_2 V_2 \cos \alpha_2) \tag{1}$$

It may be seen in Fig. 12-3 (a) and (b) that $V_2 \cos \alpha_2$ can be expressed in terms of u_2 and v_2, thus,

$$V_2 \cos \alpha_2 = u_2 + v_2 \cos \beta_2$$

with due regard to the sign of $\cos \beta_2$. Then

$$\text{head } H' = \frac{u_2}{g}(u_2 + v_2 \cos \beta_2) \tag{2}$$

Furthermore, from the vector triangles,

$$V_2^2 = u_2^2 + v_2^2 - 2u_2 v_2 \cos (180° - \beta_2)$$

from which we may write

$$u_2 v_2 \cos \beta_2 = \tfrac{1}{2}(V_2^2 - u_2^2 - v_2^2)$$

The head equation (2) becomes

$$\text{head } H' = \frac{u_2^2}{2g} + \frac{V_2^2}{2g} - \frac{v_2^2}{2g}$$

The head developed by the pump will be less than this amount by the loss in the impeller and by loss at exit. Then

$$\text{Head developed } H = \left(\frac{u_2^2}{2g} + \frac{V_2^2}{2g} - \frac{v_2^2}{2g}\right) - \text{impeller loss} - \text{exit loss}$$

$$H = \left(\frac{u_2^2}{2g} + \frac{V_2^2}{2g} - \frac{v_2^2}{2g}\right) - K_i \frac{v_2^2}{2g} - K_e \frac{V_2^2}{2g}$$

5. **Evaluate for pumps and turbines (a) the speed factor ϕ, (b) the unit speed N_u, (c) the unit flow Q_u, (d) the unit power P_u and (e) the specific speed.**
   Solution:

(a) By definition, $\phi = \dfrac{u}{\sqrt{2gH}}$. But $u = r\omega = r\dfrac{2\pi N}{60} = \dfrac{\pi D N}{60} = \dfrac{\pi D_1 N}{720}$, where D_1 is the diameter in inches and N is the speed in revolutions per minute. Finally,

$$\phi = \frac{\pi D_1 N}{720} \times \frac{1}{\sqrt{2gH}} = \frac{D_1 N}{1840\sqrt{H}} \tag{1a}$$

(b) If $D_1 = 1$ in. and $H = 1$ ft, we obtain from equation (1a) above the unit speed N_u. Thus

$$N_u = 1840\phi \tag{1b}$$

which is constant for all wheels of like design if ϕ relates to best speed. Also, from (1a) above,

$$N_u = \frac{D_1 N}{\sqrt{H}} \quad \text{in rpm} \tag{2}$$

Thus, for homologous rotating elements, the best speed N varies inversely as the diameter and directly as the square root of H.

(c)　For the tangential turbine, the flow Q through the unit may be expressed as

$$Q = cA\sqrt{2gH} = c\frac{\pi d_1^2}{4 \times 144}\sqrt{2gH} = \frac{c\pi\sqrt{2g}}{576}\Big(\frac{d_1}{D_1}\Big)^2 D_1^2\sqrt{H}$$

$$= (\text{factor})\,D_1^2\sqrt{H} = Q_u D_1^2\sqrt{H} \tag{3}$$

For $D_1 = 1$ in. and $H = 1$ ft, the factor is defined as the unit flow Q_u.

For reaction turbines and pumps, flow Q may be expressed as the product of

$$(c)(A)(\text{velocity component})$$

The velocity component depends upon the square root of H and the sine of angle α_1 (see Fig. 12-1 of Problem 1). Thus flow Q can be written in the form of (3) above.

(d)　Using expression (3) above,

$$\text{power } P = \frac{wQH}{550} = \frac{w(Q_u D_1^2\sqrt{H})H}{550}$$

For $D_1 = 1$ in. and $H = 1$ ft, power $= wQ_u/550 = (\text{factor})$. When the efficiency is included in the power output for turbines and the water horsepower for pumps, the factor becomes the unit power P_u. Then

$$\text{power } P = P_u D_1^2 H^{3/2} \tag{4}$$

(e)　In equation (4) we may substitute for D_1 its value in (2) above, obtaining

$$\text{power } P = P_u\frac{N_u^2 H}{N^2}H^{3/2}$$

Also

$$P_u N_u^2 = \frac{PN^2}{H^{5/2}} \qquad \text{or} \qquad N_u\sqrt{P_u} = \frac{N\sqrt{P}}{H^{5/4}} \tag{5}$$

The term $N_u\sqrt{P_u}$ is called the specific speed N_S. Expression (5) then becomes

$$N_S = \frac{N\sqrt{P}}{H^{5/4}} \quad \text{(for turbines)} \tag{6}$$

If P is replaced by Q by eliminating D in equations (2) and (3), we obtain

$$N_u^2 Q_u = \frac{QN^2}{H^{3/2}}$$

and

$$N_S = \frac{N\sqrt{Q}}{H^{3/4}} \quad \text{(for pumps)} \tag{7}$$

where this specific speed refers to the speed at which 1 cfs would discharge against a 1 ft head.

These are the common expressions for pumps and water wheels. For homologous rotating elements in which different fluids may be utilized, see expressions (9b), (11a) and (12a) at the beginning of this chapter.

6.　A tangential turbine develops 7200 hp at 200 rpm under a head of 790 ft at an efficiency of 82%. (a) If the speed factor is 0.46 compute the wheel diameter, the flow, the unit speed, unit power, unit flow and specific speed. (b) For this turbine, what would be the speed, power and flow under a head of 529 ft? (c) For a turbine having the same design, what size of wheel should be used to develop 3800 horsepower under a 600 ft head and what would be its speed and rate of discharge? Assume no change in efficiency.

Solution:

Referring to Problem 5 for the necessary formulas, we proceed as follows.

(a)　Since $\phi = \dfrac{D_1 N}{1840\sqrt{H}},\quad D_1 = \dfrac{1840\sqrt{790} \times 0.46}{200} = 119$ in.

From　horsepower output $= \dfrac{wQHe}{550},\quad Q = \dfrac{7200 \times 550}{62.4 \times 790 \times 0.82} = 98.0$ cfs.

$$N_u = \frac{ND_1}{\sqrt{H}} = \frac{200 \times 119}{\sqrt{790}} = 847 \text{ rpm}$$

$$P_u = \frac{P}{D_1^2 H^{3/2}} = \frac{7200}{(119)^2 (790)^{3/2}} = .0000229 \text{ hp}$$

$$Q_u = \frac{Q}{D_1^2 \sqrt{H}} = \frac{98.0}{(119)^2 \sqrt{790}} = .000246 \text{ cfs}$$

$$N_s = \frac{N\sqrt{P}}{H^{5/4}} = \frac{200\sqrt{7200}}{(790)^{5/4}} = 4.06 \text{ rpm}$$

(b)   Speed $N = \dfrac{N_u \sqrt{H}}{D_1} = \dfrac{847\sqrt{529}}{119} = 164$ rpm.

Power $P = P_u D_1^2 H^{3/2} = .0000229(119)^2(529)^{3/2} = 3940$ hp.

Flow $Q = Q_u D_1^2 \sqrt{H} = .000246(119)^2 \sqrt{529} = 80.3$ cfs.

The above three quantities might have been obtained by noting that, for the same turbine (D_1 unchanged), the speed varies as $H^{1/2}$, the power varies as $H^{3/2}$ and Q varies as $H^{1/2}$. Thus

$$N = 200\sqrt{\frac{529}{790}} = 164 \text{ rpm}, \qquad P = 7200\left(\frac{529}{790}\right)^{3/2} = 3940 \text{ hp}, \qquad Q = 98.0\sqrt{\frac{529}{790}} = 80.3 \text{ cfs}$$

(c)   From $P = P_u D_1^2 H^{3/2}$ we obtain

$$3800 = .0000229(D_1)^2(600)^{3/2} \quad \text{from which} \quad D_1^2 = 11,300 \quad \text{and} \quad D_1 = 106 \text{ in.}$$

$$N = \frac{N_u \sqrt{H}}{D_1} = \frac{847\sqrt{600}}{106} = 196 \text{ rpm}$$

$$Q = Q_u D_1^2 \sqrt{H} = .000246(11,300)\sqrt{600} = 68.0 \text{ cfs}$$

7.   A turbine develops 144 hp running at 100 rpm under a head of 25 ft.   (a) What power would be developed under a head of 36 ft, assuming the same flow?   (b) At what speed should the turbine run?

Solution:

(a)   Power developed $= wQHe/550$ from which $wQe/550 = \text{hp}/H = 144/25$.

For the same flow (and efficiency), under the 36 ft head we obtain

$$wQe/550 = 144/25 = \text{hp}/36 \quad \text{or} \quad \text{hp} = 208$$

(b) $$N_s = \frac{N\sqrt{P}}{H^{5/4}} = \frac{100\sqrt{144}}{(25)^{5/4}} = 21.4 \text{ rpm}$$

Then $$N = \frac{N_s H^{5/4}}{\sqrt{P}} = \frac{21.4(36)^{5/4}}{\sqrt{208}} = 131 \text{ rpm}$$

8.   An impulse wheel at best speed produces 125 hp under a head of 210 ft.

(a) By what percent should the speed be increased for a 290 ft head?

(b) Assuming equal efficiencies, what power would result?

Solution:

(a)   For the same wheel, the speed is proportional to the square root of the head. Thus

$$N_1/\sqrt{H_1} = N_2/\sqrt{H_2} \quad \text{or} \quad N_2 = N_1\sqrt{H_2/H_1} = N_1\sqrt{290/210} = 1.175 N_1$$

The speed should be increased 17.5%.

(b)   The specific speed relation may be used to obtain the new horsepower produced.

From $N_S = \dfrac{N\sqrt{P}}{H^{5/4}}$ we have $\dfrac{N_1\sqrt{125}}{(210)^{5/4}} = \dfrac{N_2\sqrt{hp_2}}{(290)^{5/4}}$.

Solving for the power produced, $hp_2 = \left[\dfrac{N_1}{1.175 N_1}\sqrt{125}\left(\dfrac{290}{210}\right)^{5/4}\right]^2 = 202$.

The same value of horsepower may be obtained by noting that, for the same wheel, horsepower varies as $H^{3/2}$, giving $hp_2 = 125(290/210)^{3/2} = 202$.

9.   Find the approximate diameter and angular velocity of a Pelton wheel, efficiency 85% and effective head 220 ft, when the flow is 0.95 cfs.   Assume values of $\phi = 0.46$ and $c = 0.975$.

Solution:

For an impulse wheel, the general expression for power is

$$P = \frac{wQHe}{550} = \frac{62.4(cA\sqrt{2gH})He}{550} = \frac{62.4c\pi\sqrt{2g}\,e}{550\times 4\times 144}d^2H^{3/2} = .00411d^2H^{3/2} \qquad (1)$$

where d = diameter of nozzle in inches and the values of c and e are 0.975 and 0.85 respectively. From the data we may also calculate the horsepower from

$$\text{Power} = \frac{wQHe}{550} = \frac{62.4\times 0.95\times 220\times 0.85}{550} = 20.1\ hp$$

Substituting this value in (1) above, we obtain $d = 1.22$ in.   (This same value of diameter d may be calculated by using equation $Q = cA\sqrt{2gH}$ from Chapter 9.)

Now the ratio of the diameter of nozzle to diameter of wheel will be established.   This ratio will result from using the specific speed divided by the unit speed or

$$\frac{N_S}{N_u} = \frac{N\sqrt{P}}{H^{5/4}} \div \frac{ND_1}{\sqrt{H}} = \frac{\sqrt{P}\times\sqrt{H}}{D_1 H^{5/4}}$$

Substituting the value of P from (1) above,

$$\frac{N_S}{N_u} = \frac{\sqrt{.00411d^2H^{3/2}}\sqrt{H}}{D_1 H^{5/4}} = .0642\frac{d}{D_1}$$

But $N_u = 1840\phi$ (see Problem 5 above).   Then

$$N_S = (1840\times 0.46)(.0642\frac{d}{D_1}) = 54.4\frac{d}{D_1} \qquad (2)$$

It will be necessary to assume a value of N_S in (2).   Using $N_S = 2.5$, we have

$$2.5 = \frac{N\sqrt{P}}{H^{5/4}} = \frac{N\sqrt{20.1}}{(220)^{5/4}} \quad\text{or}\quad N = 474\ rpm$$

The speed of an impulse wheel must synchronize with the speed of the generator.   For a 60-cycle generator with 8 pairs of poles, speed $N = 7200/(2\times 8) = 450$ rpm; and with 7 pairs, $N = 7200/(2\times 7) = 515$ rpm.   Using the 7-pair generator for illustration, recalculation produces

$$N_S = \frac{515\sqrt{20.1}}{(220)^{5/4}} = 2.73$$

Then from (2) above, $D_1 = 54.4d/N_S = 54.4(1.22)/2.73 = 24.3$ in.

For the 7-pair generator, $N = 515$ rpm, from above.

10.   The reaction turbines at the Hoover Dam installation have a rated capacity of 115,000 hp at 180 rpm under a head of 487 ft.   The diameter of each turbine is 11 ft and the discharge is 2350 cfs.   Evaluate the speed factor, the unit speed, unit discharge and unit power, and the specific speed.

Solution:

Using equations (4) through (11) at the beginning of the chapter, we obtain the following.

$$\phi = \frac{D_1 N}{1840\sqrt{H}} = \frac{(11 \times 12)180}{1840\sqrt{487}} = 0.585$$

$$N_u = \frac{D_1 N}{\sqrt{H}} = \frac{(11 \times 12)180}{\sqrt{487}} = 1077 \text{ rpm}$$

$$Q_u = \frac{Q}{D_1^2 \sqrt{H}} = \frac{2350}{(132)^2 \sqrt{487}} = .00612 \text{ cfs}$$

$$P_u = \frac{\text{hp}}{D_1^2 H^{3/2}} = \frac{115,000}{(132)^2 (487)^{3/2}} = .000613 \text{ hp}$$

$$N_S = N_u \sqrt{P_u} = 1077\sqrt{.000613} = 26.6$$

11. An impulse wheel rotates at 400 rpm under an effective head of 196 ft and develops 90.0 bhp. For values of $\phi = 0.46$, $c_v = 0.97$, and efficiency $e = 83\%$, determine (a) the diameter of the jet, (b) the flow in cfs, (c) the diameter of the wheel and (d) the pressure head at the 8″ diameter base of the nozzle.

Solution:

(a) The velocity of the jet is $v = c_v \sqrt{2gh} = 0.97\sqrt{64.4 \times 196} = 108.9$ ft/sec.

The flow must be determined before the diameter of the jet can be calculated.

Horsepower developed $= wQHe/550$, $90.0 = 62.4Q(196)(0.83)/550$ and $Q = 4.88$ cfs.

Then, area of jet $= Q/v = 4.88/108.9 = .0448$ ft² and jet diameter $= 0.239$ ft $= 2.87$ in.

(b) Solved under (a).

(c) $\phi = \dfrac{D_1 N}{1840\sqrt{H}}$, $0.46 = \dfrac{D_1(400)}{1840\sqrt{196}}$ and $D_1 = 29.6$ in.

(d) Effective head $h = (p/w + V^2/2g)$, where p and V are average values of pressure and velocity measured at the base of the nozzle. The value of $V_8 = Q/A_8 = 4.88/0.349 = 13.98$ ft/sec.

Then $\dfrac{p}{w} = h - \dfrac{V_8^2}{2g} = 196 - \dfrac{(13.98)^2}{2g} = 193$ ft.

12. A Pelton wheel develops 6000 brake horsepower under a net head of 400 ft at a speed of 200 rpm. Assuming $c_v = 0.98$, $\phi = 0.46$, efficiency $= 88\%$ and the jet diameter-wheel diameter ratio of 1/9, determine (a) the flow required, (b) the diameter of the wheel, (c) the diameter and the number of jets required and (d) the specific speed.

Solution:

(a) Water horsepower $= wQH/550$, $6000/0.88 = 62.4Q(400)/550$ and $Q = 150.5$ cfs.

(b) Jet velocity $v = c_v\sqrt{2gh} = 0.98\sqrt{64.4(400)} = 157.5$ ft/sec.

Peripheral velocity $u = \phi\sqrt{2gh} = 0.46\sqrt{64.4(400)} = 73.9$ ft/sec.

Then $u = r\omega = \pi DN/60$, $73.9 = \pi D(200/60)$ and $D = 7.05$ ft.

(c) Since $d/D = 1/9$, $d = 7.05/9 = 0.783$ ft diameter.

Number of jets $= \dfrac{\text{flow } Q}{\text{flow per jet}} = \dfrac{Q}{A_{\text{jet}} v_{\text{jet}}} = \dfrac{150.5}{\frac{1}{4}\pi(0.783)^2(157.5)} = 1.98$. Use 2 jets.

(d) The specific speed for the 2 nozzles is $N_S = \dfrac{N\sqrt{P}}{H^{5/4}} = \dfrac{200\sqrt{6000}}{(400)^{5/4}} = 8.6$.

13. At the Pickwick plant of TVA the propeller-type turbines are rated at 48,000 hp at 81.8 rpm under a 43 ft head. The discharge diameter is 292.3 in. For a geometrically similar turbine to develop 36,000 hp under a 36 ft head, what speed and diameter should be used? What percentage change in flow is probable?

Solution:

The specific speed of geometrically similar turbines can be expressed as

$$N_s = \frac{N\sqrt{P}}{H^{5/4}}. \quad \text{Then} \quad \frac{81.8\sqrt{48,000}}{(43)^{5/4}} = \frac{N\sqrt{36,000}}{(36)^{5/4}} \quad \text{and} \quad N = 75.6 \text{ rpm}$$

The same result may be obtained by calculating N_u, then P_u and N_s. Apply these values to the turbine to be designed. Thus

$$N_u = \frac{D_1 N}{\sqrt{H}} = \frac{292.3(81.8)}{\sqrt{43}} = 3640$$

$$P_u = \frac{P}{D_1^2 H^{3/2}} = \frac{48,000}{(292.3)^2 (43)^{3/2}} = .0020$$

$$N_s = N_u \sqrt{P_u} = 3640\sqrt{.0020} = 162.8$$

and

$$N = \frac{N_s H^{5/4}}{\sqrt{P}} = \frac{162.8(36)^{5/4}}{\sqrt{36,000}} = 75.6 \text{ rpm, as above}$$

For the diameter of the new turbine, using $N_u = \dfrac{D_1 N}{\sqrt{H}}$, $D_1 = \dfrac{N_u \sqrt{H}}{N} = \dfrac{3640\sqrt{36}}{75.6} = 289$ in.

For the percentage change in flow Q, the flow relation for Pickwick and new turbines is

$$\text{new} \ \frac{Q}{D_1^2 H^{1/2}} = \text{Pickwick} \ \frac{Q}{D_1^2 H^{1/2}}, \quad \frac{Q_{\text{Pick}}}{(292.3)^2 (43)^{1/2}} = \frac{Q_{\text{new}}}{(289)^2 (36)^{1/2}}$$

and new $Q = 0.893 Q_{\text{Pick}}$ or about 11% decrease in Q.

14. A model turbine, diameter 15 in., develops 12 hp at a speed of 1500 rpm under a head of 25 ft. A geometrically similar turbine 75 in. in diameter will operate at the same efficiency under a 49 ft head. What speed and power would be expected?

Solution:

From expression (5a) at the beginning of this chapter,

$$C_N' = \frac{ND}{\sqrt{gH}} = \text{constant for homologous turbines}$$

Thus $\text{model} \ \dfrac{ND}{\sqrt{gH}} = \text{prototype} \ \dfrac{ND}{\sqrt{gH}}$, $\quad \dfrac{1500 \times 15}{\sqrt{g \times 25}} = \dfrac{N \times 75}{\sqrt{g \times 49}} \quad$ and $\quad N = 420$ rpm.

From expression (9a), $C_P = \dfrac{P}{D^2 H^{3/2}} = \text{constant.}$ Thus

$$\text{model} \ \frac{P}{D^2 H^{3/2}} = \text{prototype} \ \frac{P}{D^2 H^{3/2}}, \quad \frac{12}{(15)^2 (25)^{3/2}} = \frac{P}{(75)^2 (49)^{3/2}}, \quad P = 822 \text{ hp}$$

15. A reaction turbine, 20 in. in diameter, when running at 600 rpm developed 261 bhp when the flow was 26.2 cfs. The pressure head at entrance to the turbine was 91.50 ft and the elevation of the turbine casing above tailwater level was 6.26 ft. The water enters the turbine with a velocity of 12.0 ft/sec. Calculate (a) the effective head, (b) the efficiency, (c) the speed expected under a head of 225 ft, and (d) the brake horsepower and the discharge under the 225 ft head.

Solution:

(a) Effective head $H = \dfrac{p}{w} + \dfrac{V^2}{2g} + z = 91.50 + \dfrac{(12.0)^2}{2g} + 6.26 = 100.0$ ft.

(b)   Power supplied by water $= wQH/550 = 62.4(26.2)(100.0)/550 = 297$ horsepower.

$$\text{Efficiency} = \frac{\text{power from shaft}}{\text{power supplied}} = \frac{261}{297} = 87.8\%.$$

(c)   For the same turbine, the $\dfrac{ND_1}{\sqrt{H}}$ ratio is constant. Then $\dfrac{N \times 20}{\sqrt{225}} = \dfrac{600 \times 20}{\sqrt{100}}$ or $N = 900$ rpm.

(d)   For the same turbine, the $\dfrac{P}{D_1^2 H^{3/2}}$ and $\dfrac{Q}{D_1^2 \sqrt{H}}$ ratios are also constant. Then

$$\frac{P}{(20)^2 (225)^{3/2}} = \frac{261}{(20)^2 (100)^{3/2}}, \quad P = 883 \text{ hp} \quad \text{and} \quad \frac{Q}{(20)^2 \sqrt{225}} = \frac{26.2}{(20)^2 \sqrt{100}}, \quad Q = 39.3 \text{ cfs.}$$

16. A pump impeller, 12 in. in diameter, discharges 5.25 cfs when running at 1200 rpm. The blade angle β_2 is 160° and the exit area A_2 is 0.25 ft². Assuming losses of $2.8(v_2^2/2g)$ and $0.38(V_2^2/2g)$, compute the efficiency of the pump (exit area A_2 is measured normal to v_2).

Solution:

The absolute and relative velocities at exit must be calculated first. Velocities u_2 and v_2 are

$$u_2 = r_2\omega = (6/12)(2\pi \times 1200/60) = 62.8 \text{ ft/sec}, \quad v_2 = Q/A_2 = 5.25/0.25 = 21.0 \text{ ft/sec}$$

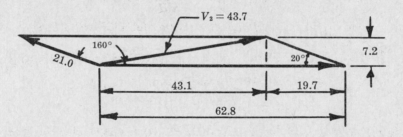

Fig. 12-4

From the vector diagram shown in Fig. 12-4 above, the value of the absolute velocity at exit is $V_2 = 43.7$ ft/sec. From Problem 4,

$$\text{head furnished by impeller,} \quad H' = \frac{u_2^2}{2g} - \frac{v_2^2}{2g} + \frac{V_2^2}{2g} = \frac{(62.8)^2}{2g} - \frac{(21.0)^2}{2g} + \frac{(43.7)^2}{2g} = 84.1 \text{ ft.}$$

Head delivered to water, $H = H' - \text{losses} = 84.1 - \left(2.8\dfrac{(21.0)^2}{2g} + 0.38\dfrac{(43.7)^2}{2g}\right) = 53.6$ ft.

Efficiency $e = H/H' = 53.6/84.1 = 63.8\%$.

The value of H' might have been calculated by means of the commonly-used expression

$$H' = \frac{u_2}{g}(u_2 + v_2 \cos \beta_2) = \frac{62.8}{g}[62.8 + 21.0(-0.940)] = 84.1 \text{ ft}$$

17. A centrifugal pump discharged 300 gpm against a head of 55 ft when the speed was 1500 rpm. The diameter of the impeller was 12.5 in. and the brake horsepower was 6.0. A geometrically similar pump 15.0 in. in diameter is to run at 1750 rpm. Assuming equal efficiencies, (a) what head will be developed, (b) how much water will be pumped and (c) what brake horsepower will be developed?

Solution:

(a)   From the speed relation, the $\dfrac{DN}{\sqrt{H}}$ ratios for model and prototype are equal. Then

$$\frac{12.5 \times 1500}{\sqrt{55}} = \frac{15.0 \times 1750}{\sqrt{H}} \qquad \text{and} \qquad H = 108 \text{ ft}$$

(b) From the discharge relation, the $\dfrac{Q}{D^2\sqrt{H}}$ ratios are equal. Then

$$\frac{300}{(12.5)^2\sqrt{55}} = \frac{Q}{(15.0)^2\sqrt{108}} \quad \text{and} \quad Q = 605 \text{ cfs}$$

Another useful discharge relation is $\dfrac{Q}{D^3N}$ = constant, from which

$$\frac{Q}{(15.0)^3(1750)} = \frac{300}{(12.5)^3(1500)} \quad \text{and} \quad Q = 605 \text{ cfs}$$

(c) The speed relation, $\dfrac{P}{D^5N^3}$ = constant, may be used for model and prototype. Then

$$\frac{P}{(15.0)^5(1750)^3} = \frac{6.0}{(12.5)^5(1500)^3} \quad \text{and} \quad P = 23.6 \text{ hp}$$

18. A 6 in. pump delivers 1300 gpm against a head of 90 ft when rotating at 1750 rpm.
The head-discharge and efficiency curves are shown in Fig. 12-5 below. For a geometrically similar 8 in. pump running at 1450 rpm and delivering 1800 gpm, determine
(a) the probable head developed by the 8 in. pump. (b) Assuming a similar efficiency
curve for the 8 in. pump, what power would be required to drive the pump at the
1800 gpm rate?

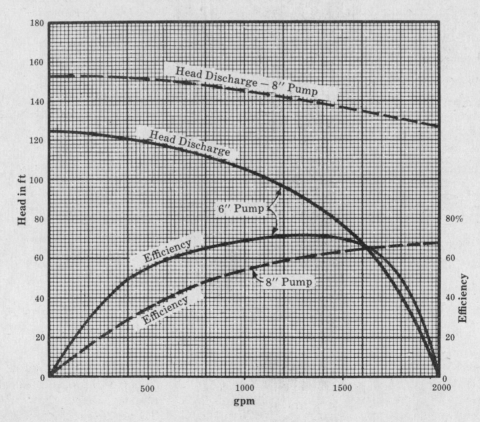

Fig. 12-5

Solution:
(a) The homologous pumps will have identical characteristics at corresponding flows. Choose several
rates of flow for the 6 in. pump and read off the corresponding heads. Calculate the values of
Q and H so that a curve for the 8 in. pump can be plotted. One such calculation is detailed below
and a table of values established by similar determinations.

Using the given 1300 gpm and the 90 ft head, we obtain from the speed relation

$$H_8 = (D_8/D_6)^2(N_8/N_6)^2 H_6 = (8/6)^2(1450/1750)^2 H_6 = 1.222 H_6 = 1.222(90) = 110.0 \text{ ft}$$

From the flow relation, $\dfrac{Q}{D^3 N} = $ constant, we obtain

$$Q_8 = (D_8/D_6)^3(N_8/N_6)Q_6 = (8/6)^3(1450/1750)Q_6 = 1.963 Q_6 = 1.963(1300) = 2550 \text{ gpm}$$

Additional values, which have been plotted as the dotted line in Fig. 12-5 above, are as follows.

For 6″ Pump at 1750 rpm			For 8″ Pump at 1450 rpm		
Q in gpm	H in ft	Efficiency	Q in gpm	H in ft	Efficiency
0	124	0	0	151.6	0
500	119	54%	980	145.5	54%
800	112	64%	1570	134.5	64%
1000	104	68%	1960	127.0	68%
1300	90	70%	2550	110.0	70%
1600	66	67%	3140	80.6	67%

From the head-discharge curve, for $Q = 1800$ gpm the head is 130 ft.

(b) The efficiency of the 8″ pump would probably be somewhat higher than that of the 6″ pump at comparable rates of flow. For this case, the assumption is that the efficiency curves are the same at comparable rates of flow. The table above lists the values for the flows indicated. The figure above gives the efficiency curve for the 8″ pump and, for the 1800 gpm flow, the value is 67%. Then

$$P = \frac{wQH}{550e} = \frac{62.4[1800/(60 \times 7.48)](130)}{550(0.67)} = 88.5 \text{ horsepower required}$$

19. It is required to deliver 324 gpm against a head of 420 ft at 3600 rpm. Assuming acceptable efficiency of pump at specific speeds of the impeller between 1200 and 4000 rpm when flow Q is in gpm units, how many pumping stages should be used?

Solution:

For 1 stage, $N_S = \dfrac{N\sqrt{Q}}{H^{3/4}} = \dfrac{3600\sqrt{324}}{(420)^{3/4}} = 697$. This value is too low.

Try 3 stages. Then the head/stage $= 420/3 = 140$ ft and $N_S = \dfrac{3600\sqrt{324}}{(140)^{3/4}} = 1590$.

Compare this value with the value for 4 stages, for which $H = 420/4 = 105$ ft, i.e., with

$$N_S = \frac{3600\sqrt{324}}{(105)^{3/4}} = 1975$$

The latter specific speed seems attractive. However, in practice, the additional cost of the 4 stage pump might outweigh the increased efficiency of the unit. An economic cost study should be made.

20. In order to predict the behavior of a small oil pump, tests are to be made on a model using air. The oil pump is to be driven by a 1/20 hp motor at 1800 rpm and a 1/4 hp motor is available to drive the air pump at 600 rpm. Using the sp gr of oil at 0.912 and the density of air constant at .00238 slug/ft³, what size model should be built?

Solution:

Using the power relation, we obtain   prototype $\dfrac{P}{\rho D^5 N^3} = $ model $\dfrac{P}{\rho D^5 N^3}$. Then

$$\frac{1/20}{0.912(62.4/32.2)D_p^5(1800)^3} = \frac{1/4}{.00238 D_m^5(600)^3} \quad \text{and} \quad \frac{D_m}{D_p} = \frac{10}{1}$$

The model should be 10 times as large as the oil pump.

21. A pump, running at 1750 rpm, has a head-discharge curve as shown in Fig. 12-6 below. The pump is to pump water through 1500 ft of 6 in. pipe, $f = .025$. The static lift is 40.0 ft and minor losses may be neglected. Calculate the flow and head under these conditions.

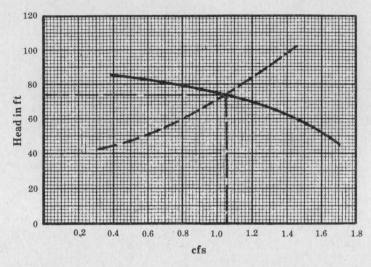

Fig. 12-6

Solution:

The loss of head through the pipe line will increase with increased flow. A curve can be drawn indicating the total head pumped against as a function of the flow (shown dotted). But

$$\text{head pumped against} = \text{lift} + \text{pipe loss}$$

$$= 40.0 + .025\left(\frac{1500}{6/12}\right)\frac{V^2}{2g} = 40.0 + 75.0\frac{V^2}{2g}$$

We may evaluate this head as follows:

$Q =$	0.40	0.60	0.80	1.00	1.20	cfs
$V = Q/A =$	2.04	3.06	4.07	5.10	6.11	ft/sec
$75\, V^2/2g =$	4.8	10.9	19.2	30.3	43.5	ft lost
Total head $=$	44.8	50.9	59.2	70.3	83.5	ft

Fig. 12-6 indicates that when the flow is 1.05 cfs the head developed by the pump will equal the total head being pumped against, i.e., 74 ft.

For the economic size of pipe calculation, see Chapter 8, Problem 18.

22. What is the power ratio of a pump and its 1/5 scale model if the ratio of the heads is 4 to 1?

Solution:

For geometrically similar pumps, $\dfrac{P}{D^2 H^{3/2}}$ for pump $= \dfrac{P}{D^2 H^{3/2}}$ for model. Then

$$\frac{P_p}{(5D)^2(4H)^{3/2}} = \frac{P_m}{D^2 H^{3/2}} \quad \text{and} \quad P_p = 25(4)^{3/2}P_m = 200P_m$$

23. Develop the expressions for thrust and power output of a propeller, the velocity through the propeller, and the efficiency of the propeller.

Solution:

(a) Using the principle of impulse-momentum, the thrust F of the propeller changes the momentum of mass M of air in Fig. 12-7. The propeller may be stationary in a fluid moving with velocity of approach V_1 or it may be moving to the left at velocity V_1 through quiescent fluid. Thus, neglecting vortices and friction,

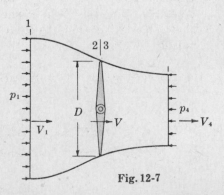

Fig. 12-7

$$\text{Thrust } F = \frac{wQ}{g}(\Delta V) = \frac{wQ}{g}(V_4 - V_1) \tag{1a}$$

$$= \frac{w}{g}(\tfrac{1}{4}\pi D^2 V)(V_4 - V_1) \tag{1b}$$

(b) The power output is simply $P = $ thrust force $\times$ velocity

$$= \frac{wQ}{g}(V_4 - V_1)V_1 \tag{2}$$

(c) The thrust F is also equal to $(p_3 - p_2)(\tfrac{1}{4}\pi D^2)$. Therefore from (1b),

$$p_3 - p_2 = \frac{w}{g}V(V_4 - V_1) \tag{3}$$

Principles of work and kinetic energy, using a unit of 1 ft³ and assuming no lost head, produce the following:

$$\text{Initial KE/ft}^3 + \text{work done/ft}^3 = \text{final KE/ft}^3$$

$$\tfrac{1}{2}(w/g)V_1^2 + (p_3 - p_2) = \tfrac{1}{2}(w/g)V_4^2$$

from which
$$p_3 - p_2 = \frac{w}{g}\left(\frac{V_4^2 - V_1^2}{2}\right) \tag{4}$$

The same result may be obtained by applying the Bernoulli equation between 1 and 2, and 3 and 4 and solving for $(p_3 - p_2)$. Note that $(p_3 - p_2)$ represents lb/ft² $\times$ ft/ft or ft lb/ft³.

Equating (3) and (4) yields

$$V = \frac{V_1 + V_4}{2} = \frac{V_1 + (V_1 + \Delta V)}{2} = V_1 + \frac{\Delta V}{2} \tag{5}$$

indicating that the velocity through the propeller is the average of the velocities ahead and behind the propeller.

Flow of fluid Q may be in terms of this velocity V, as follows:

$$Q = AV = \tfrac{1}{4}\pi D^2 V = \tfrac{1}{4}\pi D^2\left(\frac{V_1 + V_4}{2}\right) \tag{6a}$$

or
$$Q = \tfrac{1}{4}\pi D^2(V_1 + \tfrac{1}{2}\Delta V) \tag{6b}$$

(d) The efficiency of the propeller is

$$e = \frac{\text{power output}}{\text{power input}} = \frac{(wQ/g)(V_4 - V_1)V_1}{\tfrac{1}{2}(wQ/g)(V_4^2 - V_1^2)} = \frac{2V_1}{V_4 + V_1} = \frac{V_1}{V} \tag{7}$$

the denominator representing the change in kinetic energy created by the power input.

24. A propeller model, 15 in. in diameter, developed a thrust of 50 lb at a speed of 10 ft/sec in water. (a) What thrust should a similar 75 in. propeller develop at the same velocity through the water? (b) At a velocity of 20 ft/sec? (c) What would be the slipstream velocity in (b)?

Solution:

(a) Linear velocity $V = r\omega$ or varies as DN. Then we may write

$$V_m \propto 15N_m \qquad \text{and} \qquad V_p \propto 75N_p$$

Since the velocities are equal, $15N_m = 75N_p$.

Using the thrust coefficient relation, equation (19), we obtain

$$\frac{F}{\rho N^2 D^4} \text{ model} = \frac{F}{\rho N^2 D^4} \text{ prototype}, \qquad \frac{50}{\rho\left(\frac{75}{15} N_p\right)^2 (15)^4} = \frac{F}{\rho (N_p)^2 (75)^4}, \qquad F = 1250 \text{ lb}$$

In equation (*19*), diameter D is in ft and N is in revolutions per sec. However, when the ratios are equated to each other, as long as the same units are used for corresponding items (ft/ft, in./in., rpm/rpm), a correct solution results.

(*b*) Here $V_m \propto 15 N_m$ and $(2V_m = V_p) \propto 75 N_p$. These values give $30 N_m = 75 N_p$. Then

$$\frac{50}{\rho\left(\frac{75}{30} N_p\right)^2 (15)^4} = \frac{F}{\rho N_p^2 (75)^4} \qquad \text{and} \qquad F = 5000 \text{ lb}$$

Note: The above linear velocity – angular velocity – diameter relationship may be written

$$\frac{V}{ND} \text{ for model} = \frac{V}{ND} \text{ for prototype} \tag{1}$$

This relation is called the advance – diameter ratio inasmuch as V/N is the distance the propeller advances in one revolution.

(*c*) The slipstream velocity (or velocity change) can be obtained by solving expression (*6b*) in the preceding problem for ΔV after substituting $\frac{F}{\rho \Delta V}$ for Q (from equation (*1a*)). Then

$$\frac{F}{\rho \Delta V} = (\tfrac{1}{4}\pi D^2) V_1 + (\tfrac{1}{4}\pi D^2)(\tfrac{1}{2}\Delta V) \qquad \text{and} \qquad (\Delta V)^2 + 2V_1 \Delta V - \frac{8F}{\rho \pi D^2} = 0$$

Solving for ΔV gives, as the real root,

$$\Delta V = -V_1 + \sqrt{V_1^2 + \frac{8F}{\rho \pi D^2}} \tag{2}$$

From the values above, using D in ft units,

$$\Delta V = -20.0 + \sqrt{(20.0)^2 + \frac{8 \times 5000}{1.94\pi (75/12)^2}} = 3.85 \text{ ft/sec} \quad \text{or} \quad V_4 = 23.85 \text{ ft/sec}$$

25. Determine the thrust coefficient of a propeller which is 4 in. in diameter, revolves at 1800 rpm and develops a thrust of 2.50 lb in fresh water.

Solution:

$$\text{Thrust coefficient} = \frac{F}{\rho N^2 D^4} = \frac{2.50}{1.94(1800/60)^2 (4/12)^4} = 0.116.$$

The coefficient is dimensionless when F is in lb, N in revolutions/sec and D in ft.

26. The power and thrust coefficients of an 8 ft diameter propeller moving forward at 100 ft/sec at a rotational speed of 2400 rpm are .068 and .095 respectively. (*a*) Determine the power requirement and thrust in air ($\rho = .00237$ slug/ft³). (*b*) If the advance-diameter ratio for maximum efficiency is 0.70, what is the air speed for the maximum efficiency?

Solution:

(*a*) Power $P = C_{P}\rho N^3 D^5$ in ft lb/sec $= \dfrac{.068(.00237)(2400/60)^3 (8)^5}{550} = 615$ horsepower.

Thrust $F = C_F \rho N^2 D^4$ in lb $= .095(.00237)(2400/60)^2 (8)^4 = 1475$ lb.

(*b*) Since $V/ND = 0.70$, $V = 0.70(2400/60)(8) = 224$ ft/sec.

27. An airplane flies at 180 mph in still air, $w = .0750$ lb/ft³. The propeller is 5.5 ft in diameter and the velocity of the air through the propeller is 320 ft/sec. Determine (a) the slipstream velocity, (b) the thrust, (c) the horsepower input, (d) the horsepower output, (e) the efficiency and (f) the pressure difference across the propeller.

Solution:

Using the expressions developed in Problem 23 above, we obtain, from (5),

(a) $V = \frac{1}{2}(V_1 + V_4)$, $320 = \frac{1}{2}[180(5280/3600) + V_4]$, $V_4 = 376$ ft/sec (relative to fuselage).

(b)   Thrust $F = \dfrac{w}{g} Q(V_4 - V_1) = \dfrac{.0750}{32.2}[\frac{1}{4}\pi(5.5)^2(320)](376 - 264) = 1982$ lb.

(c)   Power input $P_i = FV/550 = 1982(320)/550 = 1153$ horsepower.

(d)   Power output $P_o = FV_1/550 = 1982(264)/550 = 952$ horsepower.

(e)   Efficiency $e = 952/1153 = 82.6\%$

   or, from equation (7) of Problem 23, $e = \dfrac{2V_1}{V_4 + V_1} = \dfrac{2(264)}{376 + 264} = 82.6\%$.

(f)   Pressure difference $= \dfrac{\text{thrust } F}{\text{area } (\frac{1}{4}\pi D^2)} = \dfrac{1982}{\frac{1}{4}\pi(5.5)^2} = 83.5$ psf

   or, from equation (4) of Problem 23,   pressure difference $= \dfrac{.0750}{32.2}\left(\dfrac{(376)^2 - (264)^2}{2}\right) = 83.5$ psf.

Supplementary Problems

28. An impulse wheel operates under an effective head of 625 ft. The jet diameter is 4 in. For values of $\phi = 0.45$, $c_v = 0.98$, $\beta = 160°$ and $v_2 = 0.85(V_1 - u)$, compute the power input to the shaft.
Ans. 1040 hp

29. An impulse wheel is to develop 2500 hp under an effective head of 900 ft. The nozzle diameter is 5 in., $c_v = 0.98$, $\phi = 0.46$ and $D/d = 10$. Calculate the efficiency and the speed of rotation.
Ans. 76.3%, 508 rpm

30. A turbine model, built to a scale of 1:5, was found to develop 4.25 bhp at a speed of 400 rpm under a head of 6 ft. Assuming equivalent efficiencies, what speed and power of the full-size turbine can be expected under a head of 30 ft?   Ans. 179 rpm, 1190 hp

31. From the following data, determine the diameter of the impulse wheel and its speed of rotation: $\phi = 0.46$, $e = 82\%$, $c_v = 0.98$, $D/d = 12$, head = 1296 ft, and power delivered = 4800 hp.
Ans. 61 in., 500 rpm

32. A reaction turbine running at best speed produced 34.0 bhp at 620 rpm under a 100 ft head. If the efficiency proved to be 70.0% and the speed ratio $\phi = 0.75$, determine (a) the diameter of the wheel, (b) the flow in cfs, (c) the characteristic speed N_s and (d) for a head of 196 ft, the brake horsepower and the discharge.   Ans. 22.3 in., 4.28 cfs, 11.42 rpm, 93.3 hp and 6.00 cfs

33. Under conditions of maximum efficiency a 50 in. turbine developed 300 hp under a head of 15 ft and at 95 rpm. At what speed should a homologous turbine 25 in. in diameter operate under a head of 25 ft? What power should be developed?   Ans. 245 rpm, 161 hp

34. A 60″ impulse wheel develops 625 brake horsepower when operating at 360 rpm under a head of 400 ft. (a) Under what head should a similar wheel operate at the same speed in order to develop 2500 brake horsepower? (b) For the head just calculated, what diameter should be used?
    Ans. 697 ft, 79.2 in.

35. The speed ratio ϕ of a turbine is 0.70 and the specific speed is 20.0. Determine the diameter of the runner in order that 2500 hp will be developed under a head of 324 ft.   *Ans.* 42.2 in.

36. A turbine test produced the following data: brake horsepower = 22.5, head = 16.0 ft, N = 140 rpm, runner diameter 36 in. and Q = 14.0 cfs. Calculate the power input, the efficiency, the speed ratio and the specific speed.   *Ans.* 25.4 hp, 88.5%, 0.685, 20.8

37. A centrifugal pump rotates at 600 rpm. The following data are taken: $r_1 = 2″$, $r_2 = 8″$, radial $A_1 = 12\pi$ in², radial $A_2 = 30\pi$ in², $\beta_1 = 135°$, $\beta_2 = 120°$, radial flow at entrance to blades. Neglecting friction, calculate the relative velocities at entrance and exit and the power transmitted to the water.
    Ans. 14.8 ft/sec, 4.83 ft/sec, 16.0 hp

38. What size centrifugal pump, running at 730 rpm, will pump 9 cfs against a head of 36 ft, using a value of $C_N = 1450$?   *Ans.* 12″

39. A centrifugal pump delivers 2.50 cfs against a 25 ft head at 1450 rpm and requires 9.0 hp. If the speed is reduced to 1200 rpm, calculate the flow, head and power, assuming the same efficiency.
    Ans. 2.08 cfs, 17.1 ft, 5.14 hp

40. An 80 in. diameter propeller rotates at 1200 rpm in an air stream moving at 132 ft/sec. Tests indicate a thrust of 720 lb and power absorbed as 220 hp. Calculate, for density of air of .00237 slug/ft³, the thrust and power coefficients.   *Ans.* 0.383, 0.483

41. A 5 ft diameter propeller moves through water at 30 ft/sec and develops a thrust of 3500 lb. What is the increase in the velocity of the slipstream?   *Ans.* 2.9 ft/sec

42. An 8 in. propeller developed a thrust of 16 lb at 140 rpm and a water velocity of 12 ft/sec. For a similar propeller in a ship moving at 24 ft/sec, what size propeller would be required for a thrust of 40,000 lb? At what speed should it rotate?   *Ans.* 200 in., 11.2 rpm

43. In a wind tunnel a fan produces an air velocity of 75 ft/sec when the fan rotates at 1200 rpm. (a) What velocity will be produced if the fan rotates at 1750 rpm? (b) If a 3.25 hp motor drives the fan at 1200 rpm, what size motor is required to drive the fan at 1750 rpm?   *Ans.* 109.4 ft/sec, 10.1 hp

44. What size motor is required to supply 90,000 ft³/min of air to a wind tunnel, if the losses in the tunnel are 5.67 in. of water and if the fan efficiency is 68%? Use $w_{air} = .0750$ lb/ft³.   *Ans.* 118 hp

45. A 9 ft diameter propeller moves through air, $w = .0763$ lb/ft³, at 300 ft/sec. If 1200 hp is delivered to the propeller, what thrust is developed and what is the efficiency of the propeller?
    Ans. 2030 lb, 92.2%

APPENDIX

Tables and Diagrams

TABLE 1

(A) APPROXIMATE PROPERTIES OF SOME GASES

Gas	Specific Weight w at 68°F, 1 atm. lb/ft³	Gas Constant R ft/°Rankine	Adiabatic Exponent k	Kinematic Viscosity ν at 68°F, 1 atm. ft²/sec
Air	0.0752	53.3	1.40	16.0×10^{-5}
Ammonia	.0448	89.5	1.32	16.5
Carbon dioxide	.1146	34.9	1.30	9.1
Methane	.0416	96.3	1.32	19.3
Nitrogen	.0726	55.1	1.40	17.1
Oxygen	.0830	48.3	1.40	17.1
Sulfur dioxide	.1695	23.6	1.26	5.6×10^{-5}

(B) SOME PROPERTIES OF AIR AT ATMOSPHERIC PRESSURE

Temperature °F	Density ρ slug/ft³	Specific Weight w lb/ft³	Kinematic Viscosity ν ft²/sec	Dynamic Viscosity μ lb sec/ft²
0	0.00268	0.0862	12.6×10^{-5}	3.28×10^{-7}
20	.00257	.0827	13.6	3.50
40	.00247	.0794	14.6	3.62
60	.00237	.0763	15.8	3.74
68	.00233	.0752	16.0	3.75
80	.00228	.0735	16.9	3.85
100	.00220	.0709	18.0	3.96
120	.00215	.0684	18.9×10^{-5}	4.07×10^{-7}

(C) MECHANICAL PROPERTIES OF WATER AT ATMOSPHERIC PRESSURE

Temp. °F	Density slug/ft³	Specific Wt. lb/ft³	Dynamic Viscosity lb sec/ft²	Surface Tension lb/ft	Vapor Pressure lb/in² abs.	Elastic Modulus lb/in²
32	1.94	62.4	3.75×10^{-5}	0.00518	0.08	287000
40	1.94	62.4	3.24	.00514	0.12	296000
50	1.94	62.4	2.74	.00508	0.17	305000
60	1.94	62.4	2.36	.00504	0.26	313000
70	1.94	62.3	2.04	.00497	0.36	319000
80	1.93	62.2	1.80	.00492	0.51	325000
90	1.93	62.1	1.59	.00486	0.70	329000
100	1.93	62.0	1.42	.00479	0.96	331000
120	1.92	61.7	1.17×10^{-5}	.00466	1.70	332000

TABLE 2

SPECIFIC GRAVITY AND KINEMATIC VISCOSITY OF CERTAIN LIQUIDS

(Kinematic Viscosity = tabular value × 10^{-5})

Temp. °F	Water** Spec. Grav.	Water** Kin. Visc. ft²/sec	Commercial Solvent Spec. Grav.	Commercial Solvent Kin. Visc. ft²/sec	Carbon Tetrachloride Spec. Grav.	Carbon Tetrachloride Kin. Visc. ft²/sec	Medium Lubricating Oil Spec. Grav.	Medium Lubricating Oil Kin. Visc. ft²/sec
40	1.000	1.664	0.728	1.61	1.621	0.810	0.905	477
50	1.000	1.410	.725	1.48	1.608	.750	.900	280
60	0.999	1.217	.721	1.37	1.595	.700	.896	188
70	.998	1.059	.717	1.26	1.582	.650	.891	125
80	.997	0.930	.713	1.17	1.569	.607	.888	94
90	.995	.826	.709	1.10	1.555	.560	.885	69
100	.993	.739	.705	1.03	1.542	.530	.882	49.2
110	.991	.667	.702	0.96	1.520	.500	.874	37.5
120	.990	.610					.866	29.3
150	.980	.475					.865	16.1

Temp. °F	Dust-Proofing Oil* Spec. Grav.	Dust-Proofing Oil* Kin. Visc. ft²/sec	Medium Fuel Oil* Spec. Grav.	Medium Fuel Oil* Kin. Visc. ft²/sec	Heavy Fuel Oil* Spec. Grav.	Heavy Fuel Oil* Kin. Visc. ft²/sec	Regular Gasoline* Spec. Grav.	Regular Gasoline* Kin. Visc. ft²/sec
40	0.917	80.9	0.865	6.55	0.918	444	0.738	0.810
50	.913	56.5	.861	5.55	.915	312	.733	.765
60	.909	40.8	.858	4.75	.912	221	.728	.730
70	.905	30.6	.854	4.12	.908	157	.724	.690
80	.902	23.4	.851	3.65	.905	114	.719	.660
90	.898	18.5	.847	3.19	.902	83.6	.715	.630
100	.894	14.9	.843	2.78	.899	62.7	.710	.600
110	.890	12.2	.840	2.27	.895	48.0	.706	.570

Some Other Liquids

Liquid and Temperature	Spec. Grav.	Kin. Visc. ft²/sec
Turpentine at 68°F	0.862	1.86
Linseed oil at 86°F	0.925	38.6
Ethyl alcohol at 68°F	0.789	1.65
Benzene at 68°F	0.879	0.802
Glycerin at 68°F	1.262	711
Castor oil at 68°F	0.960	1110
Light machinery oil at 62°F	0.907	147

*Kessler & Lenz, University of Wisconsin, Madison.
**ASCE Manual 25.

TABLE 3

FRICTIONAL FACTORS f FOR WATER ONLY

(Temperature range about 50°F to 70°F)

For old pipe   — approximate range ϵ from .004 ft to .020 ft
For average pipe — approximate range ϵ from .002 ft to .003 ft
For new pipe   — approximate range ϵ from .0005 ft to .0010 ft

(f = tabular value $\times 10^{-4}$)

Diameter and Type of Pipe		VELOCITY (ft/sec)										
		1	2	3	4	5	6	8	10	15	20	30
4″	Old, comm.	435	415	410	405	400	395	395	390	385	375	370
	Average, comm.	355	320	310	300	290	285	280	270	260	250	250
	New pipe	300	265	250	240	230	225	220	210	200	190	185
	Very smooth	240	205	190	180	170	165	155	150	140	130	120
6″	Old, comm.	425	410	405	400	395	395	390	385	380	375	365
	Average, comm.	335	310	300	285	280	275	265	260	250	240	235
	New pipe	275	250	240	225	220	210	205	200	190	180	175
	Very smooth	220	190	175	165	160	150	145	140	130	120	115
8″	Old, comm.	420	405	400	395	390	385	380	375	370	365	360
	Average, comm.	320	300	285	280	270	265	260	250	240	235	225
	New pipe	265	240	225	220	210	205	200	190	185	175	170
	Very smooth	205	180	165	155	150	140	135	130	120	115	110
10″	Old, comm.	415	405	400	395	390	385	380	375	370	365	360
	Average, comm.	315	295	280	270	265	260	255	245	240	230	225
	New pipe	260	230	220	210	205	200	190	185	180	170	165
	Very smooth	200	170	160	150	145	135	130	125	115	110	105
12″	Old, comm.	415	400	395	395	390	385	380	375	365	360	355
	Average, comm.	310	285	275	265	260	255	250	240	235	225	220
	New pipe	250	225	210	205	200	195	190	180	175	165	160
	Very smooth	190	165	150	140	140	135	125	120	115	110	105
16″	Old, comm.	405	395	390	385	380	375	370	365	360	350	350
	Average, comm.	300	280	265	260	255	250	240	235	225	215	210
	New pipe	240	220	205	200	195	190	180	175	170	160	155
	Very smooth	180	155	140	135	130	125	120	115	110	105	100
20″	Old, comm.	400	395	390	385	380	375	370	365	360	350	350
	Average, comm.	290	275	265	255	250	245	235	230	220	215	205
	New pipe	230	210	200	195	190	180	175	170	165	160	150
	Very smooth	170	150	135	130	125	120	115	110	105	100	95
24″	Old, comm.	400	395	385	380	375	370	365	360	355	350	345
	Average, comm.	285	265	255	250	245	240	230	225	220	210	200
	New pipe	225	200	195	190	185	180	175	170	165	155	150
	Very smooth	165	140	135	125	120	120	115	110	105	100	95
30″	Old, comm.	400	385	380	375	370	365	360	355	350	350	345
	Average, comm.	280	255	250	245	240	230	225	220	210	205	200
	New pipe	220	195	190	185	180	175	170	165	160	155	150
	Very smooth	160	135	130	120	115	115	110	110	105	100	95
36″	Old, comm.	395	385	375	370	365	360	355	355	350	345	340
	Average, comm.	275	255	245	240	235	230	225	220	210	200	195
	New pipe	215	195	185	180	175	170	165	160	155	150	145
	Very smooth	150	135	125	120	115	110	110	105	100	95	90
48″	Old, comm.	395	385	370	365	360	355	350	350	345	340	335
	Average, comm.	265	250	240	230	225	220	215	210	200	195	190
	New pipe	205	190	180	175	170	165	160	155	150	145	140
	Very smooth	140	125	120	115	110	110	105	100	95	90	90

TABLE 4

TYPICAL LOSS OF HEAD ITEMS

(Subscript 1 = Upstream and Subscript 2 = Downstream)

Item	Average Lost Head
1. From Tank to Pipe — flush connection (entrance loss)	$0.50\dfrac{V_2^2}{2g}$
— projecting connection	$1.00\dfrac{V_2^2}{2g}$
— rounded connection	$0.05\dfrac{V_2^2}{2g}$
2. From Pipe to Tank (exit loss)	$1.00\dfrac{V_1^2}{2g}$
3. Sudden Enlargement	$\dfrac{(V_1-V_2)^2}{2g}$
4. Gradual Enlargement (see Table 5)	$K\dfrac{(V_1-V_2)^2}{2g}$
5. Venturi Meters, Nozzles and Orifices	$\left(\dfrac{1}{c_v^2}-1\right)\dfrac{V_2^2}{2g}$
6. Sudden Contraction (see Table 5)	$K_c\dfrac{V_2^2}{2g}$
7. Elbows, Fittings, Valves* Some typical values of K are: 45° Bend .... 0.35 to 0.45 90° Bend .... 0.50 to 0.75 Tees .... 1.50 to 2.00 Gate Valves (open) .... about 0.25 Check Valves (open) .... about 3.0	$K\dfrac{V^2}{2g}$

*See Hydraulics Handbooks for details.

TABLE 5

VALUES OF K*

Contractions and Enlargements

Sudden Contraction		Gradual Enlargement for Total Angles of Cone						
d_1/d_2	K_c	4°	10°	15°	20°	30°	50°	60°
1.2	0.08	0.02	0.04	0.09	0.16	0.25	0.35	0.37
1.4	.17	.03	.06	.12	.23	.36	.50	.53
1.6	.26	.03	.07	.14	.26	.42	.57	.61
1.8	.34	.04	.07	.15	.28	.44	.61	.65
2.0	.37	.04	.07	.16	.29	.46	.63	.68
2.5	.41	.04	.08	.16	.30	.48	.65	.70
3.0	.43	.04	.08	.16	.31	.48	.66	.71
4.0	.45	.04	.08	.16	.31	.49	.67	.72
5.0	.46	.04	.08	.16	.31	.50	.67	.72

*Values from "King's Handbook of Hydraulics" — McGraw-Hill Book Company.

TABLE 6

SOME VALUES OF HAZEN-WILLIAMS COEFFICIENT C_1

Extremely smooth and straight pipes 140

New, smooth cast iron pipes 130

Average cast iron, new riveted steel pipes 110

Vitrified sewer pipes 110

Cast iron pipes, some years in service 100

Cast iron pipes, in bad condition 80

TABLE 7

DISCHARGE COEFFICIENTS FOR VERTICAL SHARP-EDGED CIRCULAR ORIFICES

For Water at 60°F Discharging into Air at Same Temperature

Head in Feet	Orifice Diameter in Inches					
	0.25	0.50	0.75	1.00	2.00	4.00
0.8	0.647	0.627	0.616	0.609	0.603	0.601
1.4	.635	.619	.610	.605	.601	.600
2.0	.629	.615	.607	.603	.600	.599
4.0	.621	.609	.603	.600	.598	.597
6.0	.617	.607	.601	.599	.597	.596
8.0	.614	.605	.600	.598	.596	.595
10.0	.613	.604	.600	.597	.596	.595
12.0	.612	.603	.599	.597	.595	.595
14.0	.611	.603	.598	.596	.595	.594
16.0	.610	.602	.598	.596	.595	.594
20.0	.609	.602	.598	.596	.595	.594
25.0	.608	.601	.597	.596	.594	.594
30.0	.607	.600	.597	.595	.594	.594
40.0	.606	.600	.596	.595	.594	.593
50.0	.605	.599	.596	.595	.594	.593
60.0	.605	.599	.596	.594	.593	.593

Source: F. W. Medaugh and G. D. Johnson, Civil Engr., July 1940, p. 424.

TABLE 8

SOME EXPANSION FACTORS Y FOR COMPRESSIBLE FLOW THROUGH FLOW-NOZZLES AND VENTURI METERS

p_2/p_1	k	Ratio of Diameters (d_2/d_1)				
		0.30	0.40	0.50	0.60	0.70
0.95	1.40	0.973	0.972	0.971	0.968	0.962
	1.30	.970	.970	.968	.965	.959
	1.20	.968	.967	.966	.963	.956
0.90	1.40	0.944	0.943	0.941	0.935	0.925
	1.30	.940	.939	.936	.931	.918
	1.20	.935	.933	.931	.925	.912
0.85	1.40	0.915	0.914	0.910	0.902	0.887
	1.30	.910	.907	.904	.896	.880
	1.20	.902	.900	.896	.887	.870
0.80	1.40	0.886	0.884	0.880	0.868	0.850
	1.30	.876	.873	.869	.857	.839
	1.20	.866	.864	.859	.848	.829
0.75	1.40	0.856	0.853	0.846	0.836	0.814
	1.30	.844	.841	.836	.823	.802
	1.20	.820	.818	.812	.798	.776
0.70	1.40	0.824	0.820	0.815	0.800	0.778
	1.30	.812	.808	.802	.788	.763
	1.20	.794	.791	.784	.770	.745

For $p_2/p_1 = 1.00$, $Y = 1.00$.

TABLE 9

A FEW AVERAGE VALUES OF n FOR USE IN THE KUTTER'S AND MANNING'S FORMULAS AND m IN THE BAZIN FORMULA

Type of Open Channel	n	m
Smooth cement lining, best planed timber	0.010	0.11
Planed timber, new wood-stave flumes, lined cast iron	0.012	0.20
Good vitrified sewer pipe, good brickwork, average concrete pipe, unplaned timber, smooth metal flumes	0.013	0.29
Average clay sewer pipe and cast iron pipe, average cement lining	0.015	0.40
Earth canals, straight and well maintained	0.023	1.54
Dredged earth canals, average condition	0.027	2.36
Canals cut in rock	0.040	3.50
Rivers in good condition	0.030	3.00

TABLE 10

VALUES OF C FROM THE KUTTER FORMULA

Slope S	n	Hydraulic Radius R in Feet														
		0.2	0.3	0.4	0.6	0.8	1.0	1.5	2.0	2.5	3.0	4.0	6.0	8.0	10.0	15.0
.00005	.010	87	98	109	123	133	140	154	164	172	177	187	199	207	213	220
	.012	68	78	88	98	107	113	126	135	142	148	157	168	176	182	189
	.015	52	58	66	76	83	89	99	107	113	118	126	138	145	150	159
	.017	43	50	57	65	72	77	86	93	98	103	112	122	129	134	142
	.020	35	41	45	53	59	64	72	80	84	88	95	105	111	116	125
	.025	26	30	35	41	45	49	57	62	66	70	78	85	92	96	104
	.030	22	25	28	33	37	40	47	51	55	58	65	74	78	84	90
.0001	.010	98	108	118	131	140	147	158	167	173	178	186	196	202	206	212
	.012	76	86	95	105	113	119	130	138	144	148	155	165	170	174	180
	.015	57	64	72	81	88	92	103	109	114	118	125	134	140	143	150
	.017	48	55	62	70	75	80	88	95	99	104	111	118	125	128	135
	.020	38	45	50	57	63	67	75	81	85	88	95	102	107	111	118
	.025	28	34	38	43	48	51	59	64	67	70	77	84	89	93	98
	.030	23	27	30	35	39	42	48	52	55	59	64	72	75	80	85
.0002	.010	105	115	125	137	145	150	162	169	174	178	185	193	198	202	206
	.012	83	92	100	110	117	123	133	139	144	148	154	162	167	170	175
	.015	61	69	76	84	91	96	105	110	114	118	124	132	137	140	145
	.017	52	59	65	73	78	83	90	97	100	104	110	117	122	125	130
	.020	42	48	53	60	65	68	76	82	85	88	94	100	105	108	113
	.025	30	35	40	45	50	54	60	65	68	70	76	83	86	90	95
	.030	25	28	32	37	40	43	49	53	56	59	63	69	74	77	82
.0004	.010	110	121	128	140	148	153	164	171	174	178	184	192	197	193	203
	.012	87	95	103	113	120	125	134	141	145	149	153	161	165	168	172
	.015	64	73	78	87	93	98	106	112	115	118	123	130	134	137	142
	.017	54	62	68	75	80	84	92	98	101	104	110	116	120	123	128
	.020	43	50	55	61	67	70	77	83	86	88	94	99	104	106	110
	.025	32	37	42	47	51	55	60	65	68	70	75	82	85	88	92
	.030	26	30	33	38	41	44	50	54	57	59	63	68	73	75	80
.001	.010	113	124	132	143	150	155	165	172	175	178	184	190	195	197	201
	.012	88	97	105	115	121	127	135	142	145	149	154	160	164	167	171
	.015	66	75	80	88	94	98	107	112	116	119	123	130	133	135	141
	.017	55	63	68	76	81	85	92	98	102	105	110	115	119	122	127
	.020	45	51	56	62	68	71	78	84	87	89	93	98	103	105	109
	.025	33	38	43	48	52	55	61	65	68	70	75	81	84	87	91
	.030	27	30	34	38	42	45	50	54	57	59	63	68	72	74	78
.01	.010	114	125	133	143	151	156	165	172	175	178	184	190	194	196	200
	.012	89	99	106	116	122	128	136	142	145	149	154	159	163	166	170
	.015	67	76	81	89	95	99	107	113	116	119	123	129	133	135	140
	.017	56	64	69	77	82	86	93	99	103	105	109	115	118	121	126
	.020	46	52	57	63	68	72	78	84	87	89	93	98	102	105	108
	.025	34	39	44	49	52	56	62	65	68	70	75	80	83	86	90
	.030	27	31	35	39	43	45	51	55	58	59	63	67	71	73	77

TABLE 11*

VALUES of DISCHARGE FACTOR K in $Q = (K/n)y^{8/3}S^{1/2}$ for TRAPEZOIDAL CHANNELS

(y = depth of flow, b = bottom width of channel)

Side Slopes of Channel Section (horizontal to vertical)

y/b	Vertical	$\frac{1}{4}:1$	$\frac{1}{2}:1$	$\frac{3}{4}:1$	$1:1$	$1\frac{1}{2}:1$	$2:1$	$2\frac{1}{2}:1$	$3:1$	$4:1$
.01	146.7	147.2	147.6	148.0	148.3	148.8	149.2	149.5	149.9	150.5
.02	72.4	72.9	73.4	73.7	74.0	74.5	74.9	75.3	75.6	76.3
.03	47.6	48.2	48.6	49.0	49.3	49.8	50.2	50.6	50.9	51.6
.04	35.3	35.8	36.3	36.6	36.9	37.4	37.8	38.2	38.6	39.3
.05	27.9	28.4	28.9	29.2	29.5	30.0	30.5	30.9	31.2	32.0
.06	23.0	23.5	23.9	24.3	24.6	25.1	25.5	26.0	26.3	27.1
.07	19.5	20.0	20.4	20.8	21.1	21.6	22.0	22.4	22.8	23.6
.08	16.8	17.3	17.8	18.1	18.4	18.9	19.4	19.8	20.2	21.0
.09	14.8	15.3	15.7	16.1	16.4	16.9	17.4	17.8	18.2	19.0
.10	13.2	13.7	14.1	14.4	14.8	15.3	15.7	16.2	16.6	17.4
.11	11.83	12.33	12.76	13.11	13.42	13.9	14.4	14.9	15.3	16.1
.12	10.73	11.23	11.65	12.00	12.31	12.8	13.3	13.8	14.2	15.0
.13	9.80	10.29	10.71	11.06	11.37	11.9	12.4	12.8	13.3	14.1
.14	9.00	9.49	9.91	10.26	10.57	11.1	11.6	12.0	12.5	13.4
.15	8.32	8.80	9.22	9.57	9.88	10.4	10.9	11.4	11.8	12.7
.16	7.72	8.20	8.61	8.96	9.27	9.81	10.29	10.75	11.20	12.1
.17	7.19	7.67	8.08	8.43	8.74	9.28	9.77	10.23	10.68	11.6
.18	6.73	7.20	7.61	7.96	8.27	8.81	9.30	9.76	10.21	11.1
.19	6.31	6.78	7.19	7.54	7.85	8.39	8.88	9.34	9.80	10.7
.20	5.94	6.40	6.81	7.16	7.47	8.01	8.50	8.97	9.43	10.3
.22	5.30	5.76	6.16	6.51	6.82	7.36	7.86	8.33	8.79	9.70
.24	4.77	5.22	5.62	5.96	6.27	6.82	7.32	7.79	8.26	9.18
.26	4.32	4.77	5.16	5.51	5.82	6.37	6.87	7.35	7.81	8.74
.28	3.95	4.38	4.77	5.12	5.43	5.98	6.48	6.96	7.43	8.36
.30	3.62	4.05	4.44	4.78	5.09	5.64	6.15	6.63	7.10	8.04
.32	3.34	3.77	4.15	4.49	4.80	5.35	5.86	6.34	6.82	7.75
.34	3.09	3.51	3.89	4.23	4.54	5.10	5.60	6.09	6.56	7.50
.36	2.88	3.29	3.67	4.01	4.31	4.87	5.38	5.86	6.34	7.28
.38	2.68	3.09	3.47	3.81	4.11	4.67	5.17	5.66	6.14	7.09
.40	2.51	2.92	3.29	3.62	3.93	4.48	4.99	5.48	5.96	6.91
.42	2.36	2.76	3.13	3.46	3.77	4.32	4.83	5.32	5.80	6.75
.44	2.22	2.61	2.98	3.31	3.62	4.17	4.68	5.17	5.66	6.60
.46	2.09	2.48	2.85	3.18	3.48	4.04	4.55	5.04	5.52	6.47
.48	1.98	2.36	2.72	3.06	3.36	3.91	4.43	4.92	5.40	6.35
.50	1.87	2.26	2.61	2.94	3.25	3.80	4.31	4.81	5.29	6.24
.55	1.65	2.02	2.37	2.70	3.00	3.55	4.07	4.56	5.05	6.00
.60	1.46	1.83	2.17	2.50	2.80	3.35	3.86	4.36	4.84	5.80
.70	1.18	1.53	1.87	2.19	2.48	3.03	3.55	4.04	4.53	5.49
.80	.982	1.31	1.64	1.95	2.25	2.80	3.31	3.81	4.30	5.26
.90	.831	1.15	1.47	1.78	2.07	2.62	3.13	3.63	4.12	5.08
1.00	.714	1.02	1.33	1.64	1.93	2.47	2.99	3.48	3.97	4.93
1.20	.548	.836	1.14	1.43	1.72	2.26	2.77	3.27	3.76	4.72
1.40	.436	.708	.998	1.29	1.57	2.11	2.62	3.12	3.60	4.57
1.60	.357	.616	.897	1.18	1.46	2.00	2.51	3.00	3.49	4.45
1.80	.298	.546	.820	1.10	1.38	1.91	2.42	2.91	3.40	4.36
2.00	.254	.491	.760	1.04	1.31	1.84	2.35	2.84	3.33	4.29
2.25	.212	.439	.700	.973	1.24	1.77	2.28	2.77	3.26	4.22

*Values from King's "Handbook of Hydraulics", 4th edition, McGraw-Hill Co.

TABLE 12*

VALUES of DISCHARGE FACTOR K' in $Q = (K'/n)b^{8/3}S^{1/2}$ for TRAPEZOIDAL CHANNELS

(y = depth of flow, b = bottom width of channel)

Side Slopes of Channel Section (horizontal to vertical)

y/b	Vertical	$\frac{1}{4}:1$	$\frac{1}{2}:1$	$\frac{3}{4}:1$	$1:1$	$1\frac{1}{2}:1$	$2:1$	$2\frac{1}{2}:1$	$3:1$	$4:1$
.01	.00068	.00068	.00069	.00069	.00069	.00069	.00069	.00069	.00070	.00070
.02	.00213	.00215	.00216	.00217	.00218	.00220	.00221	.00222	.00223	.00225
.03	.00414	.00419	.00423	.00426	.00428	.00433	.00436	.00439	.00443	.00449
.04	.00660	.00670	.00679	.00685	.00691	.00700	.00708	.00716	.00723	.00736
.05	.00946	.00964	.00979	.00991	.01002	.01019	.01033	.01047	.01060	.01086
.06	.0127	.0130	.0132	.0134	.0136	.0138	.0141	.0143	.0145	.0150
.07	.0162	.0166	.0170	.0173	.0175	.0180	.0183	.0187	.0190	.0197
.08	.0200	.0206	.0211	.0215	.0219	.0225	.0231	.0236	.0240	.0250
.09	.0241	.0249	.0256	.0262	.0267	.0275	.0282	.0289	.0296	.0310
.10	.0284	.0294	.0304	.0311	.0318	.0329	.0339	.0348	.0358	.0376
.11	.0329	.0343	.0354	.0364	.0373	.0387	.0400	.0413	.0424	.0448
.12	.0376	.0393	.0408	.0420	.0431	.0450	.0466	.0482	.0497	.0527
.13	.0425	.0446	.0464	.0480	.0493	.0516	.0537	.0556	.0575	.0613
.14	.0476	.0502	.0524	.0542	.0559	.0587	.0612	.0636	.0659	.0706
.15	.0528	.0559	.0585	.0608	.0627	.0662	.0692	.0721	.0749	.0805
.16	.0582	.0619	.0650	.0676	.0700	.0740	.0777	.0811	.0845	.0912
.17	.0638	.0680	.0716	.0748	.0775	.0823	.0866	.0907	.0947	.1026
.18	.0695	.0744	.0786	.0822	.0854	.0910	.0960	.1008	.1055	.1148
.19	.0753	.0809	.0857	.0899	.0936	.1001	.1059	.1115	.1169	.1277
.20	.0812	.0876	.0931	.0979	.1021	.1096	.1163	.1227	.1290	.1414
.22	.0934	.1015	.109	.115	.120	.130	.139	.147	.155	.171
.24	.1061	.1161	.125	.133	.140	.152	.163	.173	.184	.204
.26	.119	.131	.142	.152	.160	.175	.189	.202	.215	.241
.28	.132	.147	.160	.172	.182	.201	.217	.234	.249	.281
.30	.146	.163	.179	.193	.205	.228	.248	.267	.287	.324
.32	.160	.180	.199	.215	.230	.256	.281	.304	.327	.371
.34	.174	.198	.219	.238	.256	.287	.316	.343	.370	.423
.36	.189	.216	.241	.263	.283	.319	.353	.385	.416	.478
.38	.203	.234	.263	.288	.312	.353	.392	.429	.465	.537
.40	.218	.253	.286	.315	.341	.389	.434	.476	.518	.600
.42	.233	.273	.309	.342	.373	.427	.478	.526	.574	.668
.44	.248	.293	.334	.371	.405	.467	.525	.580	.633	.740
.46	.264	.313	.359	.401	.439	.509	.574	.636	.696	.816
.48	.279	.334	.385	.432	.474	.553	.625	.695	.763	.897
.50	.295	.355	.412	.463	.511	.598	.679	.757	.833	.983
.55	.335	.410	.482	.548	.609	.722	.826	.926	1.025	1.22
.60	.375	.468	.557	.640	.717	.858	.990	1.117	1.24	1.49
.70	.457	.592	.722	.844	.959	1.17	1.37	1.56	1.75	2.12
.80	.542	.725	.906	1.078	1.24	1.54	1.83	2.10	2.37	2.90
.90	.628	.869	1.11	1.34	1.56	1.98	2.36	2.74	3.11	3.83
1.00	.714	1.022	1.33	1.64	1.93	2.47	2.99	3.48	3.97	4.93
1.20	.891	1.36	1.85	2.33	2.79	3.67	4.51	5.32	6.11	7.67
1.40	1.07	1.74	2.45	3.16	3.85	5.17	6.42	7.64	8.84	11.2
1.60	1.25	2.16	3.14	4.14	5.12	6.99	8.78	10.52	12.2	15.6
1.80	1.43	2.62	3.93	5.28	6.60	9.15	11.6	14.0	16.3	20.9
2.00	1.61	3.12	4.82	6.58	8.32	11.7	14.9	18.1	21.2	27.3
2.25	1.84	3.81	6.09	8.46	10.8	15.4	19.8	24.1	28.4	36.7

*Values from King's "Handbook of Hydraulics", 4th edition, McGraw-Hill Co.

TABLE 13

AREAS of CIRCLES

Internal Diameter		Area		Internal Diameter		Area	
(in.)	(ft)	(in²)	(ft²)	(in.)	(ft)	(in²)	(ft²)
1.00	0.0833	0.785	0.00545	10.00	0.833	78.54	0.545
1.25	.1042	1.227	.00852	12.00	1.000	113.1	.785
1.50	.1250	1.767	.01227	14.00	1.167	153.9	1.069
1.75	.1458	2.405	.01670	16.00	1.333	201.1	1.396
2.00	.1667	3.142	.0218	18.00	1.500	254.5	1.767
2.25	.1875	3.976	.0276	20.00	1.667	314.2	2.182
2.50	.2083	4.909	.0341	24.00	2.00	452.4	3.142
2.75	.229	5.940	.0412	30.00	2.50	706.9	4.909
3.00	.250	7.069	.0491	36.00	3.00	1,018	7.069
3.50	.292	9.621	.0668	42.00	3.50	1,385	9.621
4.00	.333	12.57	.0873	48.00	4.00	1,810	12.57
4.50	.375	15.90	.1105	54.00	4.50	2,290	15.90
5.00	.417	19.64	.1364	60.00	5.00	2,827	19.64
6.00	.500	28.27	.1964	72.00	6.00	4,072	28.27
8.00	.667	50.27	.3491	84.00	7.00	5,542	38.49

TABLE 14

WEIGHTS and DIMENSIONS of CAST IRON PIPE

Nom. Diam. of Pipe (in.)	Class A Pipe (100 ft Head)			Class B Pipe (200 ft Head)			Class C Pipe (300 ft Head)		
	Wall thickness (in.)	Inside Diam. (in.)	Weight (lb/ft)	Wall thickness (in.)	Inside Diam. (in.)	Weight (lb/ft)	Wall thickness (in.)	Inside Diam. (in.)	Weight (lb/ft)
4	0.42	3.96	20.0	0.45	4.10	21.7	0.48	4.04	23.3
6	.44	6.02	30.8	.48	6.14	33.3	.51	6.08	35.8
8	.46	8.13	42.9	.51	8.03	47.5	.56	8.18	52.1
10	.50	10.10	57.1	.57	9.96	63.8	.62	10.16	70.8
12	.54	12.12	72.5	.62	11.96	82.1	.68	12.14	91.7
14	.57	14.16	89.6	.66	13.98	102.5	.74	14.17	116.7
16	.60	16.20	108.3	.70	16.00	125.0	.80	16.20	143.8
18	.64	18.22	129.2	.75	18.00	150.0	.87	18.18	175.0
20	.67	20.26	150.0	.80	20.00	175.0	.92	20.22	208.3
24	.76	24.28	204.2	.89	24.02	233.3	1.04	24.24	279.2
30	.88	29.98	291.7	1.03	29.94	333.3	1.20	30.00	400.0
36	.99	35.98	391.7	1.15	36.00	454.2	1.36	35.98	545.8
42	1.10	42.00	512.5	1.28	41.94	591.7	1.54	42.02	716.7
48	1.26	47.98	666.7	1.42	47.96	750.0	1.71	47.98	908.3
54	1.35	53.96	800.0	1.55	54.00	933.3	1.90	54.00	1,142
60	1.39	60.02	916.7	1.67	60.06	1,104	2.00	60.20	1,342
72	1.62	72.10	1,282	1.95	72.10	1,547	2.39	72.10	1,904
84	1.72	84.10	1,636	2.22	84.10	2,104			

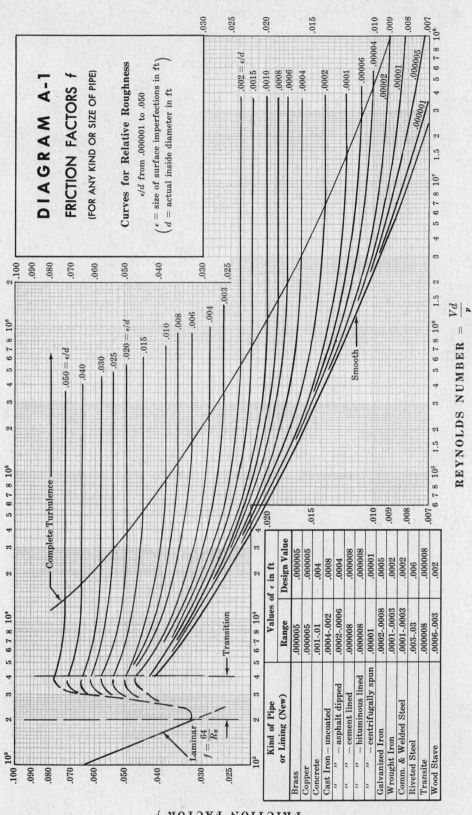

DIAGRAM A-1

FRICTION FACTORS f

(FOR ANY KIND OR SIZE OF PIPE)

Curves for Relative Roughness

ϵ/d from .000001 to .050

$\left(\begin{array}{l}\epsilon = \text{size of surface imperfections in ft} \\ d = \text{actual inside diameter in ft}\end{array}\right)$

Kind of Pipe	Values of ϵ in ft	
or Lining (New)	Range	Design Value
Brass	.000005	.000005
Copper	.000005	.000005
Concrete	.001-.01	.004
Cast Iron — uncoated	.0004-.002	.0008
" " — asphalt dipped	.0002-.0006	.0004
" " — cement lined	.000008	.000008
" " — bituminous lined	.000008	.000008
" " — centrifugally spun	.00001	.00001
Galvanized Iron	.0002-.0008	.0005
Wrought Iron	.0001-.0003	.0002
Comm. & Welded Steel	.0001-.0003	.0002
Riveted Steel	.03-.03	.006
Transite	.000008	.000008
Wood Stave	.0006-.003	.002

REYNOLDS NUMBER $= \dfrac{Vd}{\nu}$

FRICTION FACTOR f

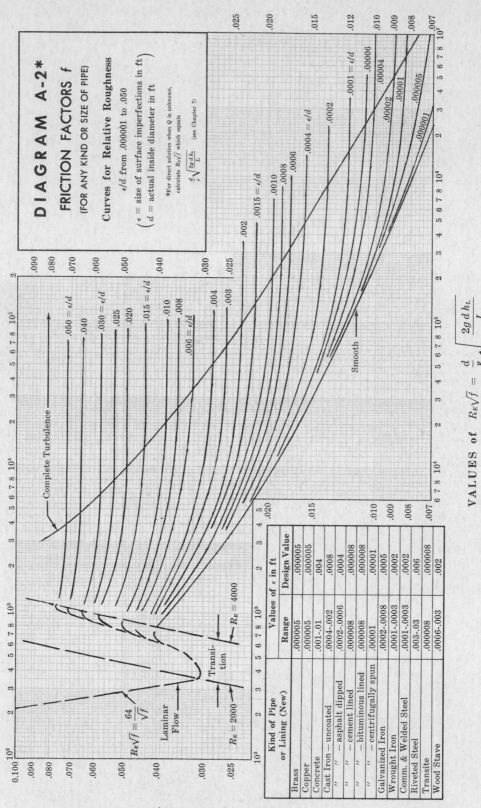

DIAGRAM A-2*
FRICTION FACTORS f
(FOR ANY KIND OR SIZE OF PIPE)

Curves for Relative Roughness

ϵ/d from .000001 to .050

$\left(\begin{array}{l}\epsilon = \text{size of surface imperfections in ft}\\ d = \text{actual inside diameter in ft}\end{array}\right)$

*For direct solution when Q is unknown, calculate $R_E\sqrt{f}$ which equals $\frac{d}{\nu}\sqrt{\frac{2g\,d\,h_L}{L}}$ (see Chapter 7)

FRICTION FACTOR f

VALUES of $R_E\sqrt{f} = \frac{d}{\nu}\sqrt{\frac{2g\,d\,h_L}{L}}$

Kind of Pipe or Lining (New)	Values of ϵ in ft	
	Range	Design Value
Brass	.000005	.000005
Copper	.000005	.000005
Concrete	.001-.01	.004
Cast Iron — uncoated	.0004-.002	.0008
" " — asphalt dipped	.0002-.0006	.0004
" " — cement lined	.000008	.000008
" " — bituminous lined	.000008	.000008
" " — centrifugally spun	.00001	.00001
Galvanized Iron	.0002-.0008	.0005
Wrought Iron	.0001-.0003	.0002
Comm. & Welded Steel	.0001-.0003	.0002
Riveted Steel	.003-.03	.006
Transite	.000008	.000008
Wood Stave	.0006-.003	.002

DIAGRAM B

FLOW CHART

HAZEN-WILLIAMS FORMULA, $C_1 = 100$

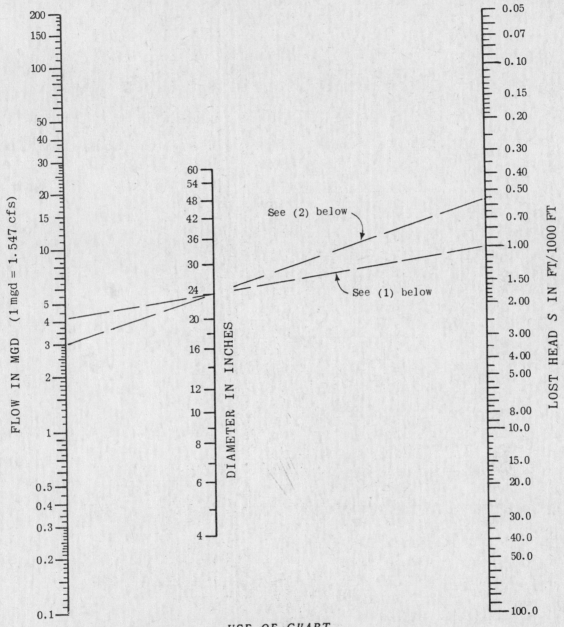

USE OF CHART

(1) Given $D = 24''$, $S = 1.0$ ft/1000 ft, $C_1 = 120$; find flow Q.

Chart gives $Q_{100} = 4.2$ mgd.

For $C_1 = 120$, $Q = (120/100)(4.2) = 5.0$ mgd.

(2) Given $Q = 3.6$ mgd, $D = 24''$, $C_1 = 120$; find Lost Head.

Change Q_{120} to Q_{100}: $Q_{100} = (100/120)(3.6) = 3.0$ mgd.

Chart gives $S = 0.55$ ft/1000 ft.

260

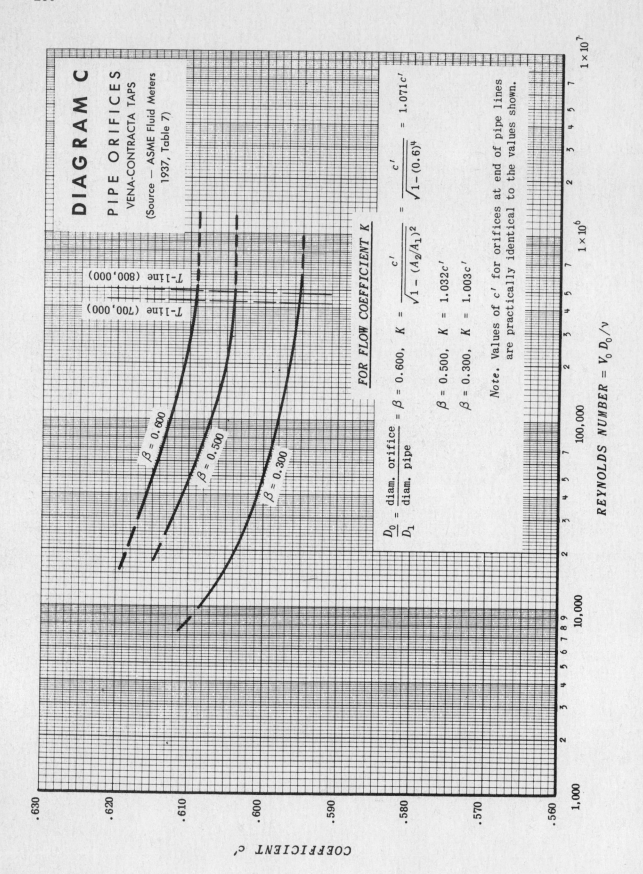

DIAGRAM C

PIPE ORIFICES
VENA-CONTRACTA TAPS

(Source — ASME Fluid Meters
1937, Table 7)

$\beta = 0.600$

$\beta = 0.500$

$\beta = 0.300$

T-line (800,000)

T-line (700,000)

FOR FLOW COEFFICIENT K

$\dfrac{D_0}{D_1} = \dfrac{\text{diam. orifice}}{\text{diam. pipe}} = \beta = 0.600, \quad K = \dfrac{c'}{\sqrt{1-(A_2/A_1)^2}} = \dfrac{c'}{\sqrt{1-(0.6)^4}} = 1.071c'$

$\beta = 0.500, \quad K = 1.032c'$

$\beta = 0.300, \quad K = 1.003c'$

Note. Values of c' for orifices at end of pipe lines
are practically identical to the values shown.

REYNOLDS NUMBER $= V_0 D_0/\nu$

COEFFICIENT c'

.630   .620   .610   .600   .590   .580   .570   .560

1,000

10,000

100,000

1×10^6

1×10^7

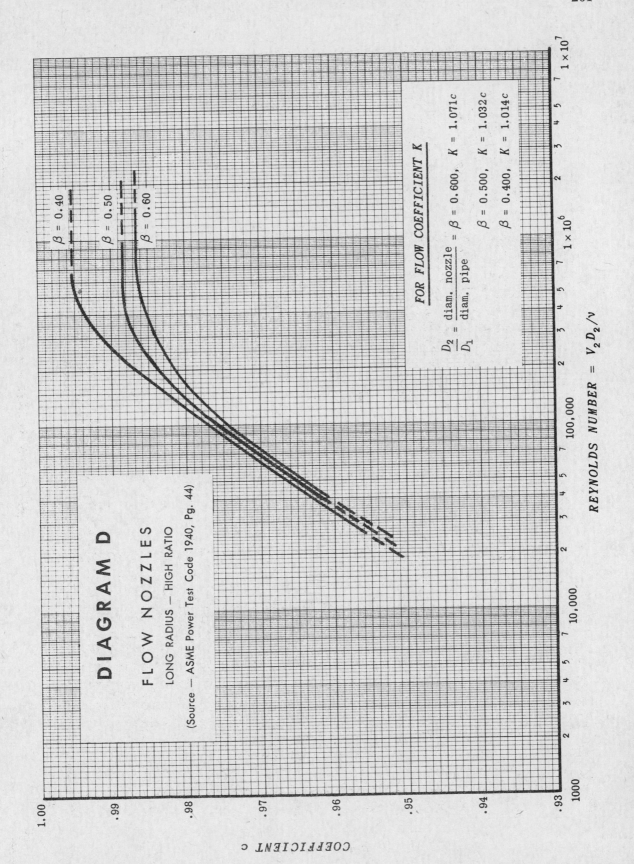

DIAGRAM D

FLOW NOZZLES

LONG RADIUS — HIGH RATIO

(Source — ASME Power Test Code 1940, Pg. 44)

$\beta = 0.40$

$\beta = 0.50$

$\beta = 0.60$

FOR FLOW COEFFICIENT K

$\dfrac{D_2}{D_1} = \dfrac{\text{diam. nozzle}}{\text{diam. pipe}} = \beta = 0.600, \quad K = 1.071c$

$\beta = 0.500, \quad K = 1.032c$

$\beta = 0.400, \quad K = 1.014c$

REYNOLDS NUMBER $= V_2 D_2 / \nu$

COEFFICIENT c

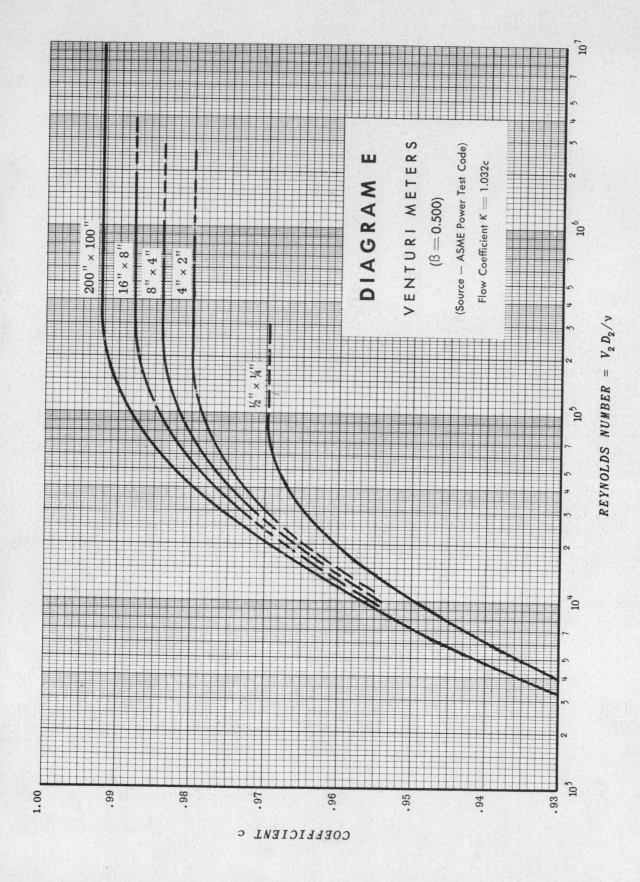

DIAGRAM E

VENTURI METERS

($\beta = 0.500$)

(Source — ASME Power Test Code)

Flow Coefficient $K = 1.032c$

200" × 100"

16" × 8"

8" × 4"

4" × 2"

½" × ¼"

REYNOLDS NUMBER = $V_2 D_2 / \nu$

COEFFICIENT c

263

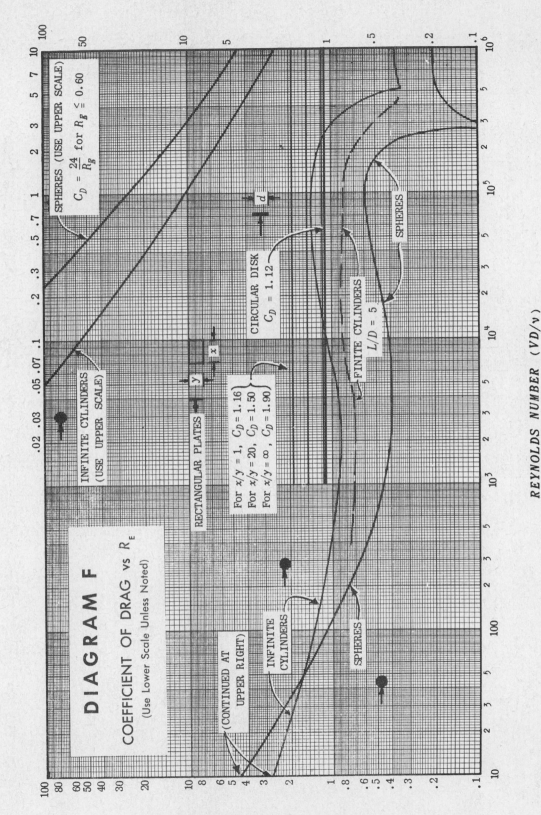

DIAGRAM F

COEFFICIENT OF DRAG vs R_E

(Use Lower Scale Unless Noted)

REYNOLDS NUMBER (VD/ν)

COEFFICIENT OF DRAG C_D

DIAGRAM G

DRAG COEFFICIENTS FOR SMOOTH, FLAT PLATES

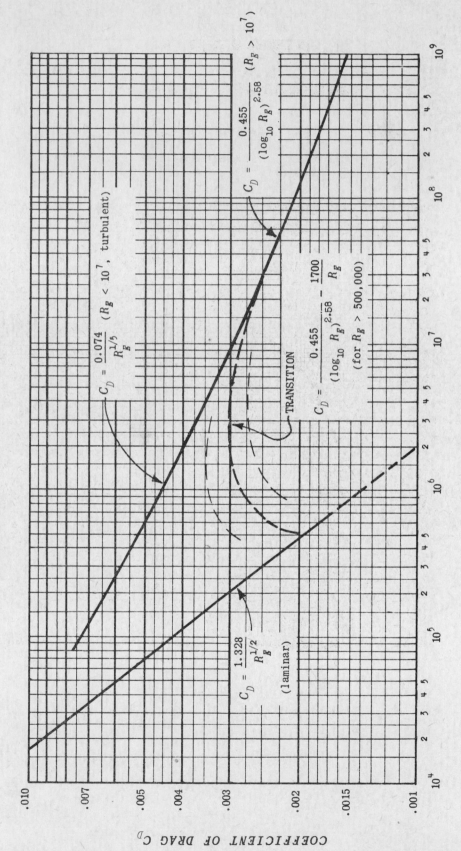

$C_D = \dfrac{0.074}{R_E^{1/5}}$ $(R_E < 10^7,\ \text{turbulent})$

$C_D = \dfrac{0.455}{(\log_{10} R_E)^{2.58}}$ $(R_E > 10^7)$

$C_D = \dfrac{0.455}{(\log_{10} R_E)^{2.58}} - \dfrac{1700}{R_E}$

(for $R_E > 500{,}000$)

TRANSITION

$C_D = \dfrac{1.328}{R_E^{1/2}}$

(laminar)

REYNOLDS NUMBER (Vx/ν)

COEFFICIENT OF DRAG C_D

DIAGRAM H

DRAG COEFFICIENTS AT SUPERSONIC VELOCITIES

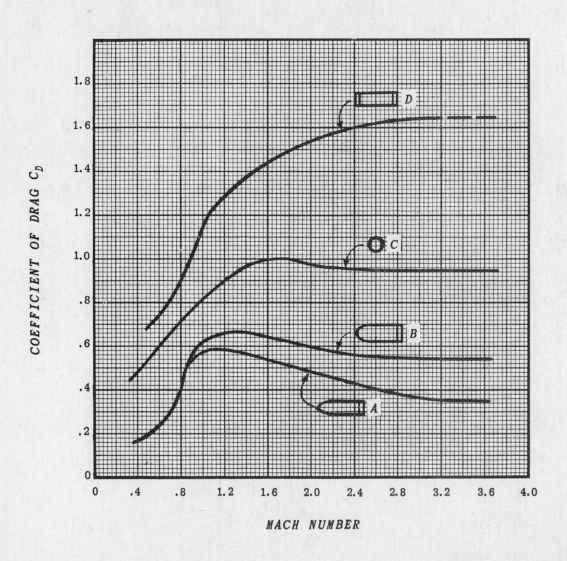

INDEX

Catalog

If you are interested in a list of SCHAUM'S
OUTLINE SERIES in Science, Mathematics,
Engineering and other subjects, send your name
and address, requesting your free catalog, to:

SCHAUM'S OUTLINE SERIES, Dept. C
McGRAW-HILL BOOK COMPANY
1221 Avenue of Americas
New York, N.Y. 10020